JOHN VON NEUMANN
AND NORBERT WIENER

From Mathematics to the
Technologies of Life and Death

Steve J. Heims

The MIT Press
Cambridge, Massachusetts,
and London, England

© 1980 by
The Massachusetts Institute of
Technology

This book was set in VIP Hel-
vetica by Achorn Graphic Services, Inc., and printed and
bound by Halliday Lithograph in
the United States of America.

**Library of Congress Cataloging
in Publication Data**

Heims, Steve J
  John von Neumann and
Norbert Wiener.

  Bibliography: p.
  Includes index.
  1. von Neumann, John,
1903–1957. 2. Wiener,
Norbert, 1894–1964.
3. Mathematicians—United
States—Biography. I. Title.
QA28.H44   510′.92′2 [B]
80-16185
ISBN 0-262-08105-9

To Suzanne
and to my daughter Leila

CONTENTS

PREFACE

The background and context of the story told in this book is a historical transformation that gained momentum abruptly in August 1945, when Hiroshima and Nagasaki were devastated by atomic bombs. To many American men and women, scientists among them, these events seemed to change the world radically. The discontinuity, whether it was in material and political realities or in people's perception of them, led some scientists to alter sharply the direction of their work and thought. It also led to conflicts, arguments, moralizing, ridicule, and the emergence of a new, military-minded political group that had its way in the end.

The historical reality was of course more subtle and complex than such a summary can suggest; one of its important features was the awareness, especially among scientists, that the changed circumstances were of our own making. Consequently, an admission that the change had been for the worse would lead to our seeking scapegoats or it would oblige us to confront our own motives and examine our own actions. That awareness would serve to remind us of our collective power to radically alter conditions on earth and would confront us with new responsibilities.

The modern transformation of society had been long in preparation, though hardly noticed. The production and military use of atomic bombs in 1945 only accelerated it. The stage had been set by the interaction of two distinct kinds of revolutions, separated by several centuries of European history. The first was the scientific revolution associated with the seventeenth century and the names of Galileo, Descartes,

Leibniz, and especially Isaac Newton. Its essential innovation was the successful description of major aspects of the material world through numerical mathematical relationships, apparently universal and valid for all times just as the theorems of Euclid's geometry were thought to be. This was the decisive century for modern physics, as well as for other sciences that took physics as a model. It gave credibility to a mathematical or mechanical view of the world. The second event was the technological and industrial revolution, which received its major impetus with the rise of modern capitalism in late eighteenth-century England and set a precedent for subsequent technological-industrial developments. The legacy from these two great revolutions has passed through many hands; it has changed, has grown, has adapted, while the rest of the world has changed and has adapted to it.

The first half of the twentieth century saw refinement and revision of the Newtonian world picture (relativity theory, quantum theory, statistical physics), as well as continued pressure for technological innovation and the spread of industrialization. In the weapons field, for example, the submarine, the tank, military aircraft, poison gas, biological warfare, rockets, and nuclear weapons came to the fore. In the second half of the twentieth century, it is the pressures arising in part from the very success and demonstrated power of mathematization, scientific knowledge, technical innovation, and industrialization that have shaped the new cultural transformation, which is different from either the scientific or the industrial revolution. The new transformation derives from the conscious awareness and diverse evaluations of the effects of the two earlier revolutions. It consists of a transformation of the patterns of relationship between technology and people. The reverberations from this transformation include a transvaluation of scientific activity, of mathematization and technique, of technological innovation, and of industrialization. Nor does it leave prevailing political and economic patterns uncriticized. Insofar as scientific activity is concerned, Jerome Ravetz suitably identifies the changed situation: "As the world of science has grown in size and in power, its deepest problems have changed from the epistemological to the social. . . . The

social consequences of the Industrial Revolution were very
deep, and those of the present change in science, while not
comparable in detail, will be equally so."[1] In the earlier period
scientific and technological development proceeded dynam-
ically, whereas considerations of the social and political
impact of science and technology largely reflected the inertia of traditional social patterns. But by the mid-twentieth
century the situation was reversed: reflections on social and
political implications were slowly gaining ground, whereas
the continued focus on scientific and technological "progress" was now in large part an expression of the mindless,
though powerful, inertia of tradition.[2]

For many scientists (when I use the word "scientist" I include mathematicians) the successful construction of atom
bombs and their dramatic military-political use constituted a
watershed in this radical cultural transformation. How did they
respond?

John von Neumann and Norbert Wiener were mathematicians whose professional lives encompassed both sides
of the watershed. Both of them responded quickly and actively to the events of World War II. Before that war each had
become accustomed to being a "mathematical genius"; after
the war, more than any other mathematicians of their generation in America, they played active and prominent roles outside of mathematics proper. This adds a special interest to
the story of their lives, and through that story we can examine
concretely something of the incipient transformation, as manifested in two involved, highly intelligent human beings, as
well as in the tensions among their differing modes of response.

Each man is sufficiently interesting to deserve a full-length
biography, and sooner or later someone will be inspired to
undertake that task. The present work is quasi-biographical;
the genre to which it belongs is that of the double biographical essay, in that I have selected some elements of Wiener's
and von Neumann's lives and work for discussion in roughly
chronological order. Separate chapters on Wiener's and von
Neumann's childhoods are followed by a discussion of some
of their mathematical work in the 1920s, placing it within a

broad history of ideas. One cannot hope to understand a mathematician without at least some appreciation of his mathematics. Though their origins and early education are described, in the belief that these had a strong influence on their adult attitudes and activities, the influence of other mathematicians on their work is discussed only very briefly. Nor is any effort made to give a comprehensive description of their mathematical works. Had my primary viewpoint been that of the history of mathematics, these topics would have been put at the center.[3] The discussion of the mathematics is confined to chapters 3 to 5 and can be omitted by the reader without loss of continuity. However, these chapters are written for the wide audience that appreciates concepts and ideas. Consequently, it is hoped that they will not be forbidding even to the reader untrained in mathematics. Chapters 6 and 7 compare working habits, style, and philosophy, especially as related to the 1920s, and are based on the heuristic hypothesis that these elements provide the links between personal lives and impersonal mathematics or science.

The early chapters of the book examine separately various ingredients—origins, mathematical ideas, style—that are part of the makeup of each man, as do some of the later chapters, which emphasize their respective impact on society, their social function, and their relation to technology. Most of the middle chapters treat von Neumann and Wiener jointly. The breadth of the activities in which they engaged is illustrated by the fact that chapter 11 is essentially political history. In the late 1940s technology, politics, and biology came to the fore. I have conjectured that to some extent political attitudes and attitudes toward technology are spun out of the same cloth as attitudes toward certain issues within science, as expressions of the same personality. Yet only to some extent: whereas attitudes within science also express a relation to the prevailing interests and fashions within the scientific community, attitudes pertaining to politics and technology reflect for each mathematician both the experience of his youth and the objective conditions governing technology and politics in his mature years.

The difficulty with emphasizing certain themes in Wiener's and von Neumann's lives is in doing justice to the complexities, the contradictions, and all the other subtleties that a single person and his situation normally contain. Thus I make no claim to definitiveness or completeness, but in describing a person or a situation can any historian ever make such a claim? Personal impressions of each of the two men, related by people who knew them, are interspersed throughout the book. These impressions, as well as the two men's life stories, serve as a partial antidote against abstraction, for they tend to convey something of the whole, living, talking, acting human beings that Wiener and von Neumann were.

In spite of continuities and persistent themes, it would have been awkward to take such diverse topics as mathematics, philosophy, technology, political function, and personal and creative life and insist on serving them up as a homogenized blend throughout the book. Instead the model for presentation chosen resembles more nearly that of the Cubist painter who paints on one canvas many views of a face whose various facets may differ in shape, content, texture, and color, but nevertheless complement each other in elaborating a central, complex reality.

The kinds of themes that occupy center stage at various points in the narrative are distinct from Benjamin Franklin's characterization of science and technical innovation two centuries earlier as "philosophical Experiments that let light into the Nature of Things, tend to increase the Power of Man over Matter, and multiply the Conveniencies or Pleasures of Life."[4] The conveniences and pleasures of life are certainly a boon. They improve and embellish living, but they do not touch the core of life itself. In the twentieth century, however, we find that science and technology have come to affect people, and in particular the protagonists, so much more deeply that the appropriate metaphors are existential ones, with "birth," "creation," "death," and "power over death" prominent among them. However, the reference to "technologies of life and death" in the title is intended as literal, not metaphoric.

This brings me to a difficult but unavoidable aspect of this essay. As author I consciously or by default choose an attitude toward a number of historiographic issues, at least for purposes of exposition. Even though the historian must depict and where possible comprehend events in terms of prior and related events, "causes," and explanations at the same time that he seeks to understand the protagonists with the help of empathic identification, such a description of the historian's task seems to me incomplete. I am also concerned with the meanings of actions and events, and consequently with values. When a mathematician formulates and proves a deep theorem, I credit him with the creation of a particular value in his own time. Although historians might be able to give some explanation for why he was led to the theorem, I view the mathematician not only as a product of history but also and simultaneously as a responsible person who chose to do what he did, and did it. At certain junctures in their lives when Norbert Wiener and John von Neumann are confronted with personal but consequential choices in the face of several suitable options available to them, the act of choosing serves each of them consciously to shape his own future and define his own character. Thus, notwithstanding the limitations on their freedom imposed by circumstances, historical forces, and their own natures, I take the view that they acted as responsible men.

In the latter portion of the book, the impetus the nuclear weapons race received in the early 1950s has a prominent place. It would be frivolous and disingenuous on my part to pretend to be neutral concerning the arms race that began with the first atomic bomb. All of us who pay taxes are at least in some little way participants, and no one alive today is immune from its psychological, political, and physical effects. Because I have seen some of the survivors in the city of Hiroshima, and worked as a theoretical physicist, the issue is more vivid for me than it is for most people.[5] Although hindsight might tend to vindicate the early actions and decisions to escalate the nuclear arms race, in that the most feared disaster, nuclear holocaust, has not occurred so far, these actions and decisions are nevertheless likely to lead to unimaginable human suffering and destruction within a few generations. In this regard, reasonable expectations in 1950 did not differ markedly from those of today. Although the major governmental decisions involved only a handful of men,[6] it is possible to adopt a Tolstoyan viewpoint and attribute the events entirely to the grand sweep of inexorable historical forces, but this would ignore the other side of the paradox of personal freedom. However, after all the understanding and explaining is done, even when one reckons with the political, social, and psychological pressures on people in the decade following World War II, there is one conclusion: the series of actions and decisions of those within the US and Soviet governments acting in concert to promote the maximal acceleration of the nuclear arms race is dreadful and, in the light of foreseeable probable consequences affecting future generations, grossly inhumane. To say any less is to succumb to what G. B. Shaw has called the "Devil's sentimentality."[7] Again, I view the human actors as responsible and concern myself with the question of how it happens that honorable and sophisticated men knowingly engage in actions likely eventually to give rise to destruction and misery of a severity and on a scale beyond what we can imagine or describe.

With this as my viewpoint, I will attempt to be factual and impartial, and to remember that not only are the decisionmakers and their advisers people like ourselves but that the past, present, and likely future victims of nuclear weapons are people like ourselves as well.

Only in the epilogue is the story of von Neumann and Wiener brought directly to bear on the present. Without entering into the philosophical conundrum of the freedom of the will, I again find it convenient to adopt a language there according to which scientists, technologists, all of us are viewed as having some individual freedom of choice.

ACKNOWLEDGMENTS

I am grateful to the circumstances and to the diverse people that in one way or another contributed to the evolution of this book. Lewis Coser showed me in a few conversations in the mid-1960s how one might use intellectual tools to understand the social function of scientists. Toward the end of the 1960s H. Millard Clements suggested that a series of conferences held in 1946–53 on the interdisciplinary topics of cybernetics and teleological mechanisms might serve as a starting point for examining the nature of recent scientific activity and of scientists. Since I was then a physics professor and it was difficult to bootleg time for pursuing the study of science rather than engaging in science itself, I engaged a work-study student, Janice O'Hare (later Janice Weiss, now unfortunately deceased), who began exploring biographical material on the conference participants. Her enthusiasm for the project and her independent spirit helped strengthen my resolve to pursue the study in earnest. It was only much later that I decided to write a book on just two of the conferees and neglect the others.

Throughout the 1970s my contact with universities was sporadic. I particularly appreciated the openness and friendliness of the people who made up the Technology Studies Group in the Humanities Department at MIT; they made an outsider welcome. In addition, Silvan Schweber of Brandeis University has been helpful through conversation and in other ways. I also thank John Weiss for a critical reading of chapter 3, Leonard Richardson for reading an early, more technical version of chapter 4, Morris Schwartz for reading chapter 12,

William Woodward for reading chapter 13, and Lorenz Finison for reading both chapters 12 and 13.

Those who encouraged the project in a more personal
way, such as by listening to my monologues, include Sherwin
Lehrer, Walter Grant, Jonathan Bayliss, and many others. Suzanne Nichelson helped me to sustain my commitment to
the research and writing over the years. Patricia Gordon
helped me in the research one summer, and around 1971
Michael Donnelly assisted the research, and still later Perdita
Connolly gave me some editorial help. I was fortunate to be
able to take the time for the preliminary study and the research and writing. A one-year fellowship from the National
Science Foundation was important.

I think it is appropriate to mention that although I held a
number of one-semester teaching positions during the past
few years, I also received a modest income during some of
the in-between periods from unemployment compensation.
The value of periods of unemployment deserves recognition
in a society in which the concept of full employment is an
anachronism.

In gathering material for this book, I spoke to a lot of people,
and I thank them all for their kindness in taking time to talk
with me. I especially appreciate those close friends and relatives of Wiener and von Neumann who have generously allowed me to use their personal recollections for my biographical purposes.

The staffs of the MIT archives, the Niels Bohr Library, and
the Library of Congress archives have been helpful in making
source materials available to me. The staff of the ERDA archives was also helpful, although the procedure and the
classification of material made my access limited.

JOHN VON NEUMANN
AND NORBERT WIENER

1 WIENER'S YOUTH: HIS FATHER'S IMAGE

To appreciate Norbert Wiener, it is necessary to understand the extraordinary intellectual environment of his upbringing. Norbert described his father, Leo Wiener, as an energetic, individualistic, robust, adventurous, and romantic figure. This description is consonant with Leo Wiener's own self-description in his recollections which were serialized in a Boston newspaper in 1910.[1] Leo Wiener was born in 1862 in Bialystok, Russia (now part of Poland), the son of a Jewish schoolteacher who had wanted to replace the Yiddish commonly spoken in his community with literary German. Among Leo Wiener's "respectable" relatives were rabbis on the one hand and successful businessmen and bankers on the other. He displayed a general intellectual precocity and an exceptional gift for languages; before he was ten he had learned to speak (besides Yiddish) German, Russian, and French. After receiving a solid gymnasium education which emphasized Latin and Greek, he went to the University of Warsaw to study medicine, although he soon dropped out of that curriculum, apparently not finding it to his taste. Next he tried studying mechanical engineering at the Berlin Polytechnicum, but again he found himself discontented and drifting. In these student years he took up vegetarianism and swore off drink, biases he would retain the rest of his life and later impose on his son Norbert. He was nineteen when he decided to head for America with the intention of founding a vegetarian-humanitarian-socialist community, in the expectation that some of his Berlin vegetarian friends would eventually come to join him. By the time he arrived on the strange new continent his capital had dwindled down to twenty-five cents, and his desire for travel and adventure was conditioned by the need to earn money.[2] In spite of his considerable education—he had picked up a knowledge of Serbian, Greek, Danish, Dutch, Polish, Italian, and various German dialects before leaving Europe and had studied English on the boat to America—during his first few years in America he worked as a laborer, a farm-hand, a janitor, a deliverer of bakery goods, and a peddler.

Leo Wiener was to abandon the plan for a utopian community. He eventually became a teacher of Greek, Latin, and mathematics at a high school in Kansas City. By his own account, he was an enthusiastic teacher, successfully imposing his own views on his students:

The tense striving of the parents after material betterment was mellowing in their children into spiritual alertness, and they were as wax in the hands of the educational sculptor. . . . I taught not according to rules learned by rote . . . but by instilling in them the sentiments which actuated my own life, by imparting, not knowledge, but the desire for knowledge. . . . They, too, were my friends, and they entered into my own enthusiasm and fell in line with many of my views of life and culture. They shared with me the joy of outdoor life, and crowds of boys, and sometimes bevies of girls, too, followed me to my favorite haunts among the charming hills and into the deep woods in which the neighborhood of Kansas City abounds. . . .[3]

He became a leader in the small intellectual community of Kansas City and thereby also acquired social mobility:

My somewhat dramatic appearance on the Kansas City stage and my usefulness as a private teacher of languages and literatures opened for me the doors of those to whom usually only an ancestry of successful pork packers or liquor dealers, or at least a genealogic tree with its roots in the Mayflower served as an admission card into the company of the select.[4]

He incidentally continued to study languages: Gaelic, American Indian languages, and Bantu. His effort to learn some Chinese was abortive.

From Kansas City he moved to Boston, and in 1896 was appointed instructor in Slavic languages and literature at Harvard University. Within the next few years Leo Wiener devoted
considerable effort to promoting the appreciation of Yiddish literature among English-speaking readers. Yiddish, or
Judeo-German, he explained, was a group of dialects spoken
by the Jews of German origin in Russia, Austria, and Ruma-
nia.[5] He extolled the high quality of much Yiddish literature
but at the same time confidently predicted that the Yiddish language is "in America . . . certainly doomed to extinction."[6] He did not lament this prospect, although his *History
of Yiddish Literature in the Nineteenth Century* and his translations of Yiddish literature into English seem intended as
monuments to a dying language. In particular, he introduced
to American readers the Yiddish poems of Morris Rosenfeld.
The first stanza of the first poem in the collection "In the
Sweatshop" anticipates a theme—the human use of human
beings—that was to interest Leo's son Norbert a half-century
later:

The machines in the shop roar so wildly / that often I forget in
the roar that I am; / I am lost in the terrible tumult, / my ego
disappears, I am a machine. / I work, and work, and work
without end; / I am busy, and busy, and busy at all time. / For
what? and for whom? I know not, I ask not! / How should a
machine ever come to think?[7]

Leo Wiener discontinued working on Jewish literature around
1902. As a believer in the assimilation of Jews into the American melting pot, he took no part in the activities of the Boston
Jewish community and in fact angrily objected to those whose
first loyalty was to the Jewish community or to "being a Jew."[8]
His work on Yiddish literature was followed by extensive
translations of Russian literature into English, including the
complete works of Tolstoy,[9] whose ideals of human compassion and respect for the oppressed and the undervalued
had inspired Wiener's own humanitarianism. Leo Wiener's
knowledge of a multitude of languages was not so much
an end in itself as a means for the interpretation of cul-

tural history, the subject of most of his writings. He was one of the first scholars in the field of Slavic studies in America and in 1911 became a full professor at Harvard. His is a success story with a Horatio Alger flavor—the young, albeit educated, Russian Jew who arrives in America as a penniless immigrant and achieves the heights of eminence in the academic world. Although even in literary Boston Wiener's accomplishments as a scholar did not make him a member of the social elite, he did attain a level of success that would have been unreachable if meritocracy in the American social structure had not prevailed over xenophobia and over tendencies to inherited aristocracy.

"That then," Norbert Wiener says in *Ex-Prodigy,* his autobiography, "was the strange young man who became my father and teacher."[10] It was after he became a teacher of languages in Kansas City that he met Norbert's mother, Bertha Kahn, and married her in 1893. Wiener's autobiography, so full of his father, says so little about his mother that the omission becomes noteworthy. She was the daughter of Henry Kahn, "a German-Jewish immigrant from the Rhineland and a department store owner in St. Joseph, Missouri."[11] A small, pretty woman, she is remembered as a practical, sociable, and "folksy" housewife by some of the people in the small country town where she and her husband later settled. Norbert Wiener says of her:

In the family of divided roots and Southern gentility into which she was born, etiquette played a perhaps disproportionately large part, and trespassed on much of the ground which might be claimed by principle. It is small wonder, then, that my mother had, and conceived that she had, a very heavy task in reducing my brilliant and absent-minded father, with his enthusiasms and his hot temper, to an acceptable measure of social conformity.[12]

The picture of herself as a Southern belle always stayed with Bertha Wiener, at least in the opinion of some of her neighbors and of her son Norbert, who in spite of his father's "defection from the Jewish community" was later to blame only his mother for denying her own (and thus her family's)

Jewishness and for holding outspoken anti-Semitic prejudices:

My mother's attitude toward the Jews and all unpopular
groups was different. Scarcely a day went by in which we did
not hear some remark about the gluttony of the Jews or the
bigotry of the Irish or the laziness of the Negroes. . . . If the
maintenance of my identity as a Jew had not been forced on
me as an act of integrity, and if the fact that I was of Jewish
origin had been known to me, but surrounded by no family-
imposed aura of emotional conflict, I could and would have
accepted it as a normal fact of my existence, of no exceptional importance either to myself or to anyone else.[13]

He never quite forgave his mother: hatred of Jewishness
was hatred of him. She, on whose love he had depended, had
betrayed him. When he belatedly discovered that he was
Jewish, as we will see later, it became another element in his
sense of himself as an outsider. But his most *intense* experience of being an outsider derived without doubt from his role
as a child prodigy.

Leo Wiener's own accomplishments had been prodigious,
but in spite of his having risen from peddler to Harvard professor, his linguistic theories never attained for him the recognition among scholars that he sought. Leo Wiener made
no bones about his intentional molding of Norbert, the eldest (b. November 25, 1894), and of his sisters to make them
geniuses; he aired his educational ideas publicly in the *Boston Evening Record* and in the *American Journal of Pediatrics*
and his ideas were reported as follows in the *American
Magazine* of July 1911:

Professor Leo Wiener, of Harvard University . . . believes that
the secret of precocious mental development lies in early
training. . . . He is the father of four children, ranging in age
from four to sixteen; and he has had the courage of his convictions in making them the subjects of an educational experiment. The results have . . . been astounding, more especially in the case of his oldest son, Norbert.

This lad, at eleven, entered Tufts College, from which he
graduated in 1909, when only fourteen years old. He then entered the Harvard Graduate School. . . .

The article goes on to quote Wiener, in speaking about Norbert:

His sisters, Constance and Bertha, promise to make almost as remarkable a record. . . . It is all nonsense to say, as some people do, that Norbert and Constance and Bertha are unusually gifted children. They are nothing of the sort. If they know more than other children of their age, it is because they have been trained differently. . . .

Just what method have I used? . . . Instead of leaving them [children] to their own devices—or, worse still, repressing them, as is generally done—they should be encouraged to use their minds to think for themselves, to come as close as they can to the intellectual level of their parents. . . . It requires, though, on the part of the parents, a constant watchfulness over their words and actions. When in the presence of their children they should use only the best English, must discuss subjects of real moment and in a coherent, logical way; must make the children feel that they consider them capable of appreciating all that is said. . . . What is no less important, every child should be carefully studied to determine aptitudes. One child will have a natural bent for mathematics, another for reading, another for drawing, and so forth. Whatever it is, it can be utilized by the parent as affording a line of least resistance along which to begin the educational process. Take the case of my boy Norbert. When he was 18 months old, his nurse-girl one day amused herself by making letters in the sand of the seashore. She noticed that he was watching her attentively, and in fun she began to teach him the alphabet. Two days afterward she told me, in great surprise, that he knew it perfectly. Thinking that this was an indication that it would not be hard to interest him in reading, I started teaching him how to spell at the age of three. In a very few weeks he was reading quite fluently, and by six was acquainted with a number of excellent books, including works by Darwin, Ribot, and other scientists, which I had put in his hands in order to instill in him something of the scientific spirit. I did not expect him to understand everything he read, but I encouraged him to question me about what he did not understand, and while endeavoring to make things clear to him, I tried to make him feel that he could, if he would, work out his difficulties unaided. The older he grew the more I insisted on this, on the one hand keeping up his interest by letting him see that I was interested in everything he was doing, and on the other encouraging him constantly to think for himself.[14]

Under his father's guidance, Norbert not only read voraciously and extensively but learned his first mathematics,
Latin and German, and some biology, among many other
things. Norbert had very mixed feelings about this means of
education:

So you can make your child a genius, can you? Yes, as you
can make a blank canvas into a painting by Leonardo or a
ream of clean paper into a play by Shakespeare. My father
could give me only what my father had: his sincerity, his brilliance, his learning, and his passion. These qualities are not
to be picked up on every street corner.
 Galatea needs a Pygmalion. What does the sculptor do except remove the surplus marble from the block, and make the
figure come to life with his own brain and out of his own love?
And yet, if the stone be spaulded and flawed, the statue will
crumble under the mallet and chisel of the artist. Let those
who choose to carve a human soul to their own measure be
sure that they have a worthy image after which to carve it, and
let them know that the power of molding an emerging intellect
is a power of death as well as a power of life. A strong drug is
a strong poison. The physician who ventures to use it must
first be sure he knows the dosage.[15]

These quotations show something of the intense and complex
relationship between father and son, doubtless the crucial
one in the son's intellectual development. The debate once
fashionable among social scientists about nature versus nurture could hardly take a more personal form than it did in this
case. The metaphor of the sculptor who makes the figure
come to life "with his own brain and out of his own love" is certainly an extraordinary one for a father-son relationship. It has
overtones of one of Norbert's childhood ideas about the process of birth after his sister Bertha was born. "I was full of fan-
cies about what birth might mean, and had a weird idea that if
one could put a doll, say a doll made out of a medicine bottle,
through the proper course of incantation, one could make a
baby out of it."[16]
 The image of being created by one's father also implies a
denial of the mother's contribution. (In fact, however, Norbert's
dependence on his mother continued well beyond the years

of childhood.) The power of this image is reflected in Norbert Wiener's frequent reference to Galatea and Pygmalion and his later fascination with the golem in Jewish legend.[17] In particular, as we shall see, his probing, original analysis of the relation of a scientific and technological creator to his mechanical creations reinvokes the mythological theme of Pygmalion or the golem. He does not state it quite so baldly, but the feeling that emerges unmistakably from his autobiography is that Norbert, although of normal birth, felt himself, or at least his mind, to be somehow directly and especially created, as God created Adam. It is probable that this sense of special creation and his father's dominating love and care, and his awareness that he became similar to his own father and his equal, were sources of Norbert's confidence and of his consciousness that he had the capacity for exceptional scholarly accomplishment. He was chosen to be a genius; if he did not live up to this expectation, he would be an abysmal failure.

A child is not a piece of marble to be sculpted at will. Norbert also actively modeled himself after his image of his father, even as he was being pressured to do so. The image was that of a scholar who sought intellectual achievement and recognition and who at the same time held to high ideals of human conduct and integrity. This also meant that Norbert rejected the temptation to fall in with what he perceived to be the values of his mother's family, wealth and social success. In his later years he would be much concerned with temptations to achieve worldly fame and fortune, which he saw as threatening to seduce scientists and engineers from their proper function.

Leo Wiener's standards were always very high. When Norbert made mistakes in his studies, his father was often extremely critical and harsh, as he remembered later:

Algebra was never hard for me, although my father's way of teaching it was scarcely conducive to peace of mind. Every mistake had to be corrected as it was made. He would begin the discussion in an easy, conversational tone. This lasted exactly until I made the first mathematical mistake. Then the gentle and loving father was replaced by the avenger of blood. The first warning he gave me of my unconscious delinquency was a very sharp and aspirated "What!" and if I did not follow this by coming to heel at once, he would admonish me, "Now do this again!" By this time I was weeping and terrified. Almost inevitably I persisted in sin, or what was worse, corrected an admissible statement into a blunder. Then the last shreds of my father's temper were torn. . . .

The very tone of my father's voice was calculated to bring me to a high pitch of emotion, and when this was combined with irony and sarcasm, it became a knout with many lashes. My lessons often ended in a family scene. Father was raging, I was weeping and my mother did her best to defend me, although hers was a losing battle. She suggested at times that the noise was disturbing the neighbors and that they had come to the door to complain, and this may have put a measure of restraint on my father without comforting me in the least. There were times for many years when I was afraid that the unity of the family might not be able to stand these stresses, and it is just in this unity that all of a child's security lies.[18]

Yet along with these harsh educational methods the father encouraged and identified with the son's strong, eager curiosity and spontaneous interest in everything he saw. Norbert equally remembered his enjoyment of the nature walks, the mushroom-hunting trips, the visits to foundries, factories, and museums, and the travels to Europe accompanied by his father. It is here one recognizes something of the budding interdisciplinary scientist, who, unlike most people, never lost his childish curiosity and his zest for knowledge. As a small boy Norbert "longed to be a naturalist."[19] He also read voraciously: "I had full liberty to roam in what was the very catholic and miscellaneous library of my father. At one period or other the scientific interests of my father had covered most of the imaginable subjects of study."[20]

Contributing to the intellectual atmosphere of the Wiener home were Leo's friends, including Harvard colleagues such as mathematicians Bôcher and Osgood, physiologist Cannon, and philosopher William James; lively conversations on scholarly topics were a natural part of everyday discourse.

Norbert's education under his father, with all its obsession and intensity, love and loneliness, disregard of convention, insistence on knowledge and learning, tyranny, and expectation made for a Faustian man, a man with a Nobel Prize complex, a truth seeker, a grandiose person, a lonely man passionately using intellect to deal with ultimate problems. But at age nine, as all through his life, his intellectual sophistication and maturity far outstripped his social and emotional development.

It was at this time, when Norbert was nine years old, that his family moved away from Cambridge to a farm about an hour's train ride from Boston. His parents decided to have him attend the public high school in Ayer, a town clustered around a railroad junction not far from the Wiener farm. Because of his intense earlier education at home, Norbert was in many respects intellectually more advanced than his classmates, although they were seven years older than he. Yet he was still a child among adolescents, and the older boys and girls indulged this precocious child with good-natured tolerance. One of the teachers, Miss Leavitt, took him under her wing,[21] and while all the other students sat in desks arranged in rows, she permitted Norbert to bring his chair right up to her desk, and throughout the class he sat next to her. On one occasion she even held him on her lap. When he had to walk from one class to another, he would sometimes take Miss Leavitt's hand, and she would take him to his next class. The vegetarian lunch that he brought from home he would usually consume at her house or her married sister's house. She was a kindly, maternal schoolteacher, who responded with affection to the bright young boy and with sympathy to his unique situation.

The consideration that teachers and students showed him made Norbert recall this period as a happy one. He experienced the prerogative of special freedom, both intellectually and emotionally. He could show off his knowledge without stimulating envy or resentment, and he could and did cry or throw a tantrum without much disapprobation. One of his most pleasurable activities was when with some playmates his own age he explored the forests and streams in the neighborhood or experimented with mechanical contraptions and devices, occasionally getting into minor mischief. In the evening his father would oversee his studies, even while he was attending

Ayer High School, so that there was continuity in Leo Wiener's direction of Norbert's education. Though at school he was the baby, at home he was the oldest child, with two younger sisters (b. 1898, 1902) and then also a baby brother (b. 1906).

However benign his life at school, he was socially an outsider. This was highlighted by the occasion of his graduation; he was eleven years old, while the other students were seventeen or eighteen. The class consisted of fifteen boys and five girls. At the graduation ceremony two girls were placed next to Norbert and given the task of gently poking him should he start falling asleep, for it was past his accustomed bedtime. His graduation party was reported in the local paper:

Professor and Mrs. Wiener entertained the teachers and senior class of the High School at their home . . . Friday afternoon, in honor of their son Norbert, who is a member of that class, although he is but eleven years old. Fishing, boating, music and dancing furnished abundant entertainment. Refreshments were served on the piazza. All agreed they had a very enjoyable afternoon.[22]

Notwithstanding the last sentence, Norbert did not enjoy his party and recalled that "I was rather an outsider at the feast. I sat on one side of the room in the kneehole of the desk and watched the dancing as a ritual in which I had no part."[23] Others recall that he became tired at some point during the party and was sent to bed long before it ended.

Certain kinds of experiences with his peers upset Norbert. He was awkward and clumsy, he lacked motor control and manual dexterity, his eyesight was bad, and he was short and fat. He could not catch a ball properly, nor could he throw snowballs. To compound the difficulties, having spent most of his life within the hothouse atmosphere of his family, he was somewhat ignorant of the customs and ways of children his own age. It was almost inevitable that he would be teased and would inspire the resourcefulness of practical jokesters. In those circumstances he would be at their mercy, quite helpless and deeply hurt. He was very slow to learn the ways of the world, and even when he came to understand such things intellectually, he retained a kind of emotional ignorance and childish vulnerability.

Though many found the adult Norbert Wiener lovable, as a mature man he was a moody person. At times delightful and generous, with a high good humor, he at other times reflected personal tensions in ways that made him a difficult man, hypersensitive to slights and alternating between conceit and self-deprecation. As his autobiography shows, he had an inner core of pain. When something touched it—especially if he, with his sense of vulnerability, suspected someone had betrayed him or taken advantage of him—he could be extremely harsh, even to old friends. He left a string of abruptly broken friendships.

Norbert was eleven when his father enrolled him at Tufts College in Medford, Massachusetts. His parents moved from the farm to Medford so he could live at home. In college he became fascinated with biology. He submitted his first article to a magazine; it described his own detailed observations of anthills.[24] One incident is of particular interest: while a student at Tufts, Norbert instigated among several of his classmates a secret scientific experiment, requiring the vivisection of a guinea pig. The boys tied with a thread or string a certain artery and sought to observe how circulation was restored by the creation of new pathways between blood vessels. "The surgery was botched," recalls Wiener, "as we had not properly separated the artery from the accompanying vein and nerve, and the animal died." This unsuccessful illegal experiment to observe a particular case of homeostasis had a special poignancy for him, because it flagrantly violated the message of the antivivisectionist tracts to which his father subscribed. Wiener describes the event in his autobiography with strong overtones of shame for his "confusion," and explicitly mentions his sense of guilt at the time. The vivisectionist episode has some of the flavor of Gandhi's secret and guilty youthful experiments at violating his family's vegetarian practices by eating meat to make himself grow bigger and stronger, but in its genuinely scientific motivation it is in the tradition of the Renaissance anatomist Vesalius's theft of human skeletons.

Aside from the interest that biology would have for an intellectual boy entering adolescence, it was also a field of study where Norbert could expect to go beyond his father's relatively limited knowledge.[25] When after graduating from Tufts in mathematics he started graduate school at Harvard, he chose to major in biology, despite his father's skepticism. But his lack of manual dexterity so interfered with his laboratory work that after only one semester his father decided, against Norbert's wishes, that a career in biology was not for him.

This semester at Harvard was also important in that it opened his eyes to the social class structure at Harvard University—his father's employer—which reflected sharply the circumstances of the transformation New England was undergoing at the time. He was exposed to the snobbishness and seemingly shallow values of the Harvard "gentlemen" from the "good families" of Boston's Back Bay and felt himself a miserable misfit for reasons that went beyond the fact of his age and his prodigy status. He was learning who he was not, as part of the slower process of finding out who he was or was to become. These social forces had of course been part of the context in which the Wiener family had lived all along, even if on the farm they had been pushed into the background by the local rural environment. But as a sensitive teenager and Harvard graduate student, Norbert was forced to confront them directly.

The early-nineteenth-century population of New England, and Boston in particular, had been dominated by the descendants of Puritans from England, although by that time the harsh Calvinism of the early settlers had largely given way to the more liberal Unitarianism. Out of this community had come a great flowering of American literature, as well as the pre-Civil War abolitionist movement, one expression of its high democratic ideals. There also developed a group who not only considered Boston "the hub of the universe," but also regarded themselves as a natural political, economic, social, and cultural elite. Political and economic life in New England was largely controlled by a small number of families, all of them of Anglo-Saxon Protestant ancestry.

The latter half of the nineteenth century saw large waves of immigration of Irish Catholics to New England, as well as many other groups, including Russian Jews. It was also a time of rapid industrialization. Under the double impact of industrialization and immigration, the New England aristocracy felt their style of life and their hegemony threatened. Indeed, the new Irish gained political power in Boston, and commercial and industrial development was promoted by recent immigrants. Faced with the loss of status, the decline of power, and the changing quality of New England life, many leading New Englanders adopted a rearguard tactic consisting of an arrogant belief in the superiority of the early arrivals in America compared to the newcomers, and of the Protestant English and Teutonic stock relative to Irish, Russians, Italians, Catholics, and Jews. The upwardly mobile Jew in particular became, for many among the old upper class, the symbol of the *nouveau riche* American and of the new materialist and industrial emphasis in New England, a symbol deeply resented and loathed. The writings of Henry Adams and Brooks Adams, brothers from an aristocratic New England Yankee family that had already included two presidents of the United States, express the keen experience of this displacement. They had been nurtured in the liberal literary world of James Russell Lowell and Charles Eliot Norton in Boston, but they responded to the crumbling of a familiar world in a manner acceptable to their time, by making the Jew a scapegoat and adopting an intense anti-Semitism.[26]

Harvard College during the late nineteenth century was governed by Boston Brahmins and drew a large portion of its faculty and student body from the upper-class Back Bay families. When Leo Wiener joined the Harvard faculty, the college was expanding rapidly. Enrollment had doubled between 1892 and 1906, so that by 1906 there were more than 2,000 students. Eliot was president of the college and was enthusiastic about its growth—"Why should not Harvard College expand indefinitely like the United States?" he asked.[27] Under Eliot the college shifted from its earlier emphasis on literature and "developing the moral character of students" to an emphasis on empirical knowledge, science, and graduate research with the introduction of the elective system of courses, an innovation that expressed Eliot's support for open exploration and individuality.[28] Although himself from an upper-class

Boston family, Eliot favored diversity at the college and in the
New England population generally and actively opposed ethnocentric or racist measures. These were live and controversial issues at Harvard. Thus it was not unusual, when in a large undergraduate history course various "races" (today one would use a different word) were compared, that the "uniquely Anglo-Saxon trait of 'love of liberty' " was shown to be needed for self-government.[29] It was three Harvard graduates in the 1890s who initiated a political organization to inhibit the immigration of foreigners from objectionable races, the Immigration Restriction League of Boston, which became very powerful in New England. Whereas Eliot was an articulate foe of the league, his successor in 1909 (the year Norbert entered Harvard), A. Lawrence Lowell, became an active member and officer.[30]

It was after one term at Harvard, when Norbert went to Cornell to study philosophy (his first venture away from home), that he discovered that both his parents were Jews:

When I became aware of my Jewish origin, I was shocked. Later, when I looked up my mother's maiden name and found that Kahn is merely a variant of Cohen, I was doubly shocked. . . . As I reasoned it out to myself, I was a Jew and if the Jews were marked by those characteristics which my mother found so hateful, why I must have those characteristics myself, and share them with all those dear to me. . . . I looked in the mirror and there was no mistake: the bulging myopic eyes, the slightly everted nostrils, the dark, wavy hair, the thick lips. They were all there.[31]

The recognition of his own Jewishness advanced Norbert's growing independence and a break with his parents' attitudes, especially with his mother's. As an atheist, cut off from any overt Jewish tradition (even though he later developed a truculent pride in his Jewish ancestors, particularly his purported ancestor the physician and philosopher Moses Maimonides), he, like his father, always remained outside the Jewish community. It seems the closest connection he found was through reading Heine: "There are passages from Heine, especially in his *Disputation* and his *Prinzessin Sabbath,* that relate and express the religious exaltation of the Jew, which I cannot recite without tears."[32] But even without up-holding any specifically Jewish tradition, he participated in some of the general elements of that tradition and lived one variety of the Jewish experience. His sense of being an outsider, and even at times of being persecuted, were part of that. So was his commitment to science and learning, and especially to the universal truths of mathematics, which no politics or prejudice can destroy. So was his generous, liberal, humanitarian outlook and spontaneous feeling of human fellowship, which overrode all narrow loyalties, and a certain independence from and even anger at the values of the prevailing elite. So were his intense, volatile temperament and his passionate involvement with whatever concerned him.

At Cornell Norbert, an adolescent sixteen, lived at the house of a professor friend of his father's. Notwithstanding his discovery that he was a Jew and the attendant conflict with his parents, he was desperately lonely away from home. His mother had always seen to his bathing regularly and dressing properly, and he was quite at a loss by himself in these matters. Socially, too, he felt doubly unsure without his parents' guidance. As to his studies, he missed the discipline his father had imposed, something he had not learned to do himself. But through the mails Norbert kept up intellectual contact with his father the best he could. He inquired about the details of his father's latest linguistic discoveries and raised such topics as the "origin of the Aryan verb and noun inflection from the Sumerian." He loyally reported in detail to his father that he had come across an encyclopedia article by a man who misrepresented his (Leo Wiener's) position concerning some word origins. Occasionally Norbert wrote on philosophical topics to his father in Latin.[33]

After a year at Cornell, thoroughly shaken by his discoveries about himself, Norbert returned home and enrolled in the doctoral program in philosophy at Harvard. He wrote his dissertation in mathematical logic, under the direction of Karl Schmidt, who was standing in for the ailing Josiah Royce. With considerable effort and anxiety, Norbert passed all the examinations and completed an acceptable dissertation while still eighteen years old.

He spent the following two years as a postdoctoral student and scholar, studying especially with Bertrand Russell and the mathematician G. H. Hardy in Cambridge, England, but also with Hilbert in Göttingen, Germany, and others. He was far away from home now, in Europe, and had to cope largely on his own, even though his overbearing parents continued to advise him and arrange for him insofar as it was in their power.

In the early part of the twentieth century the leading centers of philosophy, mathematics, and science were in Europe, not in America; a few postdoctoral years of learning in Europe were customary to complete the training of a serious young American scholar. Norbert described these as years of emancipation. He was being taken seriously by some of the leading philosophers and mathematicians in the world. He was on his own in England and Germany, albeit under the tutelage of Russell; he made friendships among other traveling scholars and, as he says, learned to meet people both like himself and different "and to get on with them," to "comport" himself "as a social being."[34] Under the stimulus of Russell, Wiener did productive work in symbolic logic during these years, work in which he always used the notation and formulations of Russell and Whitehead's *Principia Mathematica* as his starting point. I have not attempted to trace all the ways in which Russell may have influenced Wiener—this would require a lengthy technical discussion of the philosophy of mathematics—but even such specific items as Russell's calling to Wiener's attention the Einstein theory of Brownian motion[35] (a theory important to Wiener's later work; see chapter 3) or Wiener's heuristic use of Russellian paradoxes later on are indications that Wiener was affected by Russell's ideas.[36] As far as Russell himself is concerned, the impression Wiener made on him is recorded in a letter he wrote to a friend:

At the end of Sept. an infant prodigy named Wiener, Ph.D. (Harvard), aged 18, turned up with his father who teaches Slavonic languages there, having first come to America to found a vegetarian communist colony, and having abandoned that intention for farming, and farming for the teaching of various subjects, (say) mathematics, Roman Law, and minerology, in various universities. The youth has been flattered, and thinks himself God Almighty—there is a perpetual contest between him and me as to which is to do the teaching.[37]

But among all his teachers, at Cambridge, Harvard, and Göttingen, Wiener singled out G. H. Hardy as having had the greatest effect on his subsequent mathematical work:

Hardy's course . . . was a revelation to me . . . [in his] attention to rigor. . . . In all my years of listening to lectures in mathematics, I have never heard the equal of Hardy for clarity, for interest, or for intellectual power. If I am to claim any man as my master in my mathematical training, it must be G. H. Hardy.[38]

More specifically, Hardy "introduced me to the Lebesgue integral, which was to lead directly to the main achievement of my early career."

The anxious issue for Wiener at Cambridge was, What does Russell (or Moore, or Hardy) think of my work?, and the mood he expressed in his weekly letters home seemed to depend largely on his perceptions of Russell's, Hardy's, and Moore's opinions of him at that moment. His father tried to assure him by mail: "Hardy and Russell, who seem indifferent to you, I am sure think well of you." At the same time Leo Wiener continued in his role as disciplinarian by correspondence: "Above all else work!" he advised his son,[39] who still depended on his father's exhortations for discipline. To his mother Norbert reported having his pants pressed, taking regular sponge baths, his daily diet, and getting a haircut. He was writing papers on symbolic logic and dealing adequately with the practical matters of everyday life, even though he at times felt disheartened, as when he wrote home that "I feel that I am regarded by the men here as something of a fool and a bore."[40]

Norbert's view at that time of Russell's work in symbolic logic is of considerable interest, for it presages a fundamental difference in outlook between the two men:

His type of mathematical analysis he [Russell] applies as a sort of Procrustean bed to the facts, and those that contain more than his system provides for, he lops short, and those that contain less, he draws out. He is, nevertheless, within his limitations, a wonderfully accurate thinker.[41]

In a few years Norbert, once out from under the direct influence of Russell, discontinued work in symbolic logic and moved into other branches of mathematics. He remained consistently skeptical of the analytic philosophy that flourished at Cambridge, instead evolving a point of view closer to existential philosophy.

During the years 1915–19, Norbert held a variety of jobs, which for one reason or another were temporary: an assistantship in philosophy at Harvard University, which ended because he was not promoted to instructor; an instructorship in mathematics at the University of Maine; a job with the *Encyclopedia Americana* helping to compile short articles; a wartime job as computer mathematician calculating ballistic firing tables at the Aberdeen Proving Ground in Maryland; a hitch as a private in the army; and a job as feature writer for a Boston newspaper. But in spite of all this worldly experience, he remained vulnerable and sensitive. When away from Boston he wrote to his sister Constance with humor and affection; Constance's fiancé was a mathematician, and Norbert used her as a go-between to communicate his mathematical ideas to the Harvard mathematicians. At the University of Maine the problem of maintaining discipline in the classroom overwhelmed him; he wrote of the students there that "their sole interests were in football and in nagging the lives out of their professors. As I was young, nervous, and responsive I was their chosen victim."[42] In the army, too, he was the butt of practical jokes. When one of the soldiers with whom he apparently had a running feud persuaded Norbert that army regulations required he shave off his precious mustache, Norbert naively did so; later he discovered that he had been tricked and naturally was furious.[43]

In 1919, after the end of the War, Norbert Wiener's lifelong association with the mathematics department of the Massachusetts Institute of Technology began with his appointment

as instructor. While his teaching was always uneven, the research on functional integration that he began in 1919 and
pursued for a number of years, and his early work in potential
theory, were major mathematical achievements. He had em-
barked on the process—a process with which he was impatient—of becoming established as one of the leading American mathematicians.

MIT was a training school for civil and mechanical engineers, and Wiener was enjoined to relate his mathematics to
engineering problems. He did this with great enthusiasm and
success. His work was so much guided by aesthetic considerations that Wiener regarded the experience of doing mathematical work as identical to that of creating works of art.[44] In
this he was in complete agreement with his teacher and friend
Hardy, who examined this motivation in mathematics most
fully.[45] But unlike Hardy, Wiener was stimulated by problems
in engineering and physics; when in his first years at MIT he
collaborated with the inventive electrical engineer Vannevar
Bush, the latter found that until he worked with Wiener he had
not known "that a mathematician and an engineer could have
such good times together."[46] Wiener's intellectual interests
had been extremely broad from his early years and never lost
their wide scope. But he had now committed himself (more
precisely, his father had committed him) to a career in mathematics, and it was in this field that he now had to make good
the promise and expectation implicit in his status of ex-
prodigy.

It is highly probable that Wiener first encountered John von
Neumann, who was nine years younger than he, in Göttingen
during the years 1924–26. Wiener had stayed in Göttingen,
the Mecca for mathematicians, during the summers of 1924
and 1925. A Guggenheim Fellowship had allowed him to take
a year off from MIT, and he also spent the summer of 1926
there. John von Neumann was at that time a twenty-two-year-
old mathematical genius, until spring 1926 still formally enrolled as a doctoral student in Budapest, who not only spent
the latter half of 1926 in Göttingen but had also visited there for
extended periods in the two previous years. Thus both men

were part of the small group of young mathematicians and
theoretical physicists who, though not regular members of the
Göttingen faculty, were nevertheless part of the mathematical
community during their extended visits. The members of this
select community participated in the same seminars and in
frequent discussions in the town's coffeehouses and beer
gardens, and went for walks in the surrounding countryside.

It was a time of strong nationalism in Germany, the prelude
to Hitler's rise. Although German professors, with their exalted
social position, generally tended to be hostile toward the
democratic government of Weimar Germany, at least among
physicists the universities of Göttingen and Berlin stood out
as centers of relative liberalism. Theory was particularly emphasized there, and—worse yet, according to the reactionary
physicists—often carried out by Jews.[47] This was in fact the
typical situation of those engaged in sophisticated cultural
pursuits in Weimar Germany: such pursuits were carried out
with a sense of precariousness and uneasiness, lest vulgar
right-wing nationalism gain ascendancy and destroy all.[48]
Norbert Wiener was quite aware of these currents,[49] but in his
concern for achieving recognition he perhaps did not see the
objective significance of the harshness that characterized the
Göttingen mathematicians at that time.[50] The dominant mood
was perceived by one other American visitor as "a miserable
German mood, bitter, sullen . . . discontented and angry."[51]

His letters to his sisters and brother-in-law again reveal
Norbert as affectionate, charming, and full of playful humor. In
the summer of 1925 he lectured to the assembled mathematicians in Göttingen about his own work and wrote home
exultantly that the great Hilbert himself had described it as
"sehr schön" (beautiful) and others too had liked it. These
were not empty words, for the administrative head of the
Göttingen mathematics department, Richard Courant, told
Wiener that if he came back in 1926 he would be given the
position of visiting professor and provided with a research assistant and all the amenities. To Wiener this confirmation from
some of the leading mathematicians in the world was what
he had been looking for. A happy and exuberant tone runs

through his letters home during this interval, as well as praise of the likable Courant.

In the period between Wiener's departure from Göttingen in 1925 and his return in 1926, Courant had become animated by the vision of obtaining Rockefeller money for a new building, a "mathematical institute," at Göttingen. He also came to anticipate that the American visitor Norbert Wiener might be helpful in obtaining the Rockefeller money. Meanwhile Leo Wiener was seeking to exploit his son's invitation to Göttingen as a route to obtaining a hearing for his own theories from the German linguists. In the end Norbert, with his hypersensitivity to being taking advantage of, found himself in the midst of personal and political circumstances beyond his ken and beyond his mastery. As a result his stay in Göttingen became very unhappy, and out of this experience came an abiding hostility toward the enterpreneurial Courant. He later accused Courant of plagiarizing certain mathematical ideas. (Not only Wiener, but also his MIT colleague Jesse Douglas at one point felt plagiarized by Courant.) As late as 1942, sixteen years after his unhappy Göttingen visit, when the American Mathematical Society was about to bestow an honor on Courant (clear indication that most American mathematicians did not share Wiener's view), Wiener came close to resigning in protest.[52] He also gave vent to his disappointment with Courant by writing a novel (never published) about a professor who appropriates young men's ideas as his own, recognizably based on Courant.

The Courant incident is typical of Wiener. From time to time throughout his life Wiener was aggrieved by a colleague's actions, and his impulsive anger and later sense of enmity would be total. It happened in the 1950s in a complicated situation involving Walter Pitts and Warren McCulloch. Wiener's great friendship of many decades with Vannevar Bush also ended with a quarrel, and then Bush too found himself the protagonist of a never-to-be-published novel. The Harvard mathematician W. F. Osgood is reputed also to have been the subject of one of Wiener's unpublished novels, in which social exclusion and Harvard snobbery played a prominent role.

If Wiener made one lifelong enemy in Göttingen, he also made one lifelong friend: Dirk Struik, a Dutch mathematician of Wiener's age. The two liked each other from the start and spent many hours talking and hiking in the nearby mountains. Wiener was influential in later bringing Struik to MIT, where their friendship continued and where they also collaborated on mathematical research.[53] When a quarter-century later Struik, an intellectual Marxist, was indicted in the courts in the spirit of the anticommunist witch-hunts of that time, Wiener happened to be out of the country, but when he heard the news he immediately wrote to the president of MIT to urge that the institute stand by Struik; if they failed to do so, he, Wiener—by then MIT's pet and pride—would resign.[54] This loyalty too is characteristic of Wiener. Wiener's relation to von Neumann, on the other hand, which emerged out of their work at Göttingen, did not take the form of either close friendship or enmity but rather that of competition and a mutual respect for each other's mathematical work.

Perhaps the most challenging problem discussed in Göttingen in 1925 was that of a comprehensive mathematical description of atomic physics. As Wiener put it, "By 1925 . . . the world was clamoring for a theory of quantum effects which would be a unified whole and not a patchwork."[55] Max Born, professor of theoretical physics at Göttingen, had been occupied with this problem, and he and Wiener began to work together on it. Born came to MIT for a year after the summer of 1925 as a visiting professor, and a joint article he wrote with Wiener took a crucial step toward a "unified description of quantum effects."[56] Quantum mechanics was attracting all of the best and the brightest among mathematically inclined physicists and physically inclined mathematicians at that time, when Wiener wrote that "young men like Heisenberg himself, Dirac, Wolfgang Pauli and John von Neumann were making new discoveries almost every day. This feverish atmosphere is not one in which I function well."[57]

Although later on Wiener did further research in quantum theory, it was out of the mainstream. He generally tended to shy away from highly competitive areas of work; as others made rapid progress in formulating quantum theory, Born and Wiener's joint article was soon superseded by more complete treatments. Göttingen was the center of this "feverish atmosphere," and because of the many new discoveries by young physicists the years 1925–27 came to be known as the "miracle years" there.[58] The prevailing attitude among this group of young men, to which von Neumann belonged, was one of optimism, even cockiness, that they would master the mysteries of the quantum by formal mathematics.

Hilbert had begun thinking about the mathematical formulation of quantum theory, and in 1926 worked with von Neumann and the physicist Nordheim on the problem. In their joint article they made use of the new Wiener-Born idea, giving credit to them, and developed the mathematical formulation further. Since Wiener and von Neumann were both now in Göttingen, it is very likely that von Neumann looked up Wiener and they discussed the topic. It would be left to von Neumann to give a definitive and completely satisfactory formulation of quantum theory a few years later. Von Neumann, unlike Wiener, enjoyed being in the midst of an active, highly competitive mathematical field.

While Wiener was in the emotionally strenuous process of establishing himself in the world of mathematicians, he also sought to establish himself in another way by moving out of his parents' house, taking a wife, and starting a family life of his own. In Cambridge his parents had frequently given teas on Sundays, which offered Norbert an opportunity to meet young women. The match his mother had hoped to arrange was between Norbert and a young German-born woman, Margaret Engelmann. Margaret, a capable, practical woman, would be good for Norbert, for he could transfer some of his emotional needs and dependency onto her, or so his parents felt. In Europe Norbert had encountered a young woman astrophysicist, a person more like himself in temperament and interests, and he pursued this romantic interest with hopes of matrimony. But ultimately he accepted for himself his parents' choice, and in 1926 Norbert Wiener and Margaret Engelmann set out for a honeymoon in Europe, visiting Göttingen and other places. It would be a stable marriage that lasted as long as Norbert lived, and it would produce two daughters.

Although one is struck by the weighty role of Norbert's parents in arranging the pattern of his life, in choosing his profession and even his wife, it must be remembered that he emerged as a man with vitality, zest, and enormous originality. The external amenities, an orderly pattern of life, and discipline were always difficult for him to impose on himself.

2 VON NEUMANN'S YOUTH: MATHEMATICAL REASONING POWER AS AMULET

Unlike Wiener, the adult John von Neumann was a man with an easy mastery of practical matters—a man of the world. His extraordinariness lay in his mental abilities. These were so dazzling that some of his admiring colleagues were at a loss to describe them in ordinary human terms. Thus, for example, Eugene Wigner, himself a gifted theoretical physicist and Nobel laureate, said that, confronted with von Neumann's mind, "One had the impression of a perfect instrument whose gears were machined to mesh accurately to a thousandth of an inch."[1] A number of his colleagues viewed him as different from ordinary mortals. Jokingly they spoke of him as a demigod, but when they spoke of him as personifying a higher stage in human development, further evolved than the rest of us, they meant it.[2]

This picture of John von Neumann as worldly and endowed with extraordinary mental powers is probably one-sided. Another aspect was apparently revealed in his intimate personal life, and it is not the purpose of this book to describe that in detail. Some of his personal friends, especially women, even while conscious of his charm, speak of his childishness, his insensitivity to feelings, and his arrested emotional development.[3] But, they add, a true genius in some areas of life must be allowed to be limited in other respects.

To begin to understand John von Neumann, who later became an important historical figure (and not just in mathematics), we heed the comment of economist William Fellner, also a Hungarian and a longtime friend of von Neumann, to the effect that many traits of von Neumann noted by Americans are just those of a good Budapester of his time and social class; that he was very much a Budapest type.[4] His genius is a biological anomaly, but the direction and manner of its development reflect the setting of his youth, early-twentieth-century Budapest. What was this setting like? And what were the social, political, and economic conditions that shaped John von Neumann?

Imagine the city of Budapest at the time when the Hapsburg Franz Joseph I was not only the emperor of Austria but the ruling King of Hungary. The city is divided in two by the picturesque Danube River. On the right bank is Buda, the beautiful old city, built on hills. Some of its houses date from medieval times and are Gothic in style, other buildings along the narrow, winding, cobblestone streets are of the Renaissance, while still others date from the seventeenth century, the Hungarian baroque. There are churches, a splendid palace, and remnants of castles in this old center of European culture.

The left bank of the Danube is flat, in sharp contrast to the hills of the opposite bank. Originally known as the town of Pest, this was, in the early years of the twentieth century, the rapidly growing modern part of the city, where its government buildings, apartment houses, elegant shops, and its largest banks and businesses were located. Though one found there the intense pace of modern competitive commercial activity, there were yet many indications of cultured and leisurely pursuits. The small downtown section of Pest is bordered on one side by the Danube, and on the other by the outer Ring (*Körút* in Hungarian), a semicircular boulevard that begins and ends at the Danube. On the embankment along the river are pleasant cafés, where people still found time to sit and talk with a view of the hills of Buda and of steamships going by on the river. The parliament building, some of the major ministries, art museums, and opera house lie between the Ring and the river. The opera house and the museum were assiduously attended. Well-kept parks with many statues of Hungary's heroes and wide, tree-lined streets are also a part of this section of the city. The financial center and the select residential areas were contained within the Ring. Transportation at that time was either by horse-drawn cab or electric streetcar, although walking was still a popular means of getting from one part of the city to another. In fact walking, whether along the Danube or on the forested Margaret Island in the river, was a favorite pastime.

The social structure of Hungary in the early part of the twentieth century was still largely feudal. Most of the population was rural, consisting of poor peasant farmers, many of whom did not speak Hungarian but were by language and ethnic identity Croatians, Rumanians, Serbs, Slovaks, or some other nationality.[5] Hungary, at this period, was the lesser partner in the Austro-Hungarian Empire and was politically dependent on Vienna. This lack of political autonomy served a useful purpose for Hungary within the framework of European power politics by providing an effective alliance against a feared and aggressive Russia. However, the non-Hungarian-speaking minorities, who were rather badly treated, were often not in the first instance loyal to the Hungarian government and were a source of political instability despite the government's attempts to inculcate Hungarian language and loyalty. The generally cosmopolitan aristocracy tended to identify with the court in Vienna and to participate in its frivolity, pomp, and gaiety, in a style that is reflected in the light operas of Franz Lehar. The officialdom of the government was the preserve of the oftentimes impoverished gentry.

During the later part of the nineteenth century, the feudal pattern in Hungary had begun to weaken somewhat, in part the result of the active efforts of social reformers. First came the abolition of serfdom in 1848, then the granting of civil rights to non-Magyars in the Nationality Act of 1868. These changes, which progressively weakened its privileges while mitigating the oppression of the peasant, made the ruling class uneasy. The most mobile of the non-Magyars were the Jews, whose forebears had emigrated to Hungary in the eighteenth and nineteenth centuries from regions corresponding to today's Russia, Austria, and Germany as well as other areas during periods of persecution. The careers of these families followed a predictable pattern. A Jew might at first settle in a small town, take up some kind of employment or trade, and then eventually he or his son or grandson would migrate to Budapest—if not to America—to seek his fortune. If the new arrival in Budapest did well as a businessman, his son in turn would get a good education and would be likely to turn his energies in the direction of one of the professions. This was so typical that in the period after the Nationality Act of 1868 the migration from the country to Budapest was sufficient to more than double the population of the city between 1870 and 1900. However, even by 1900, more than 95 percent of the country's population lived in rural areas or small towns; Hungary was still an agricultural land. But, notwithstanding the high rate of unemployment in the city, most of the relatively few Jews had drifted to Budapest, and they constituted half of the population of the capital. They were an upwardly mobile segment of the population.[6] Industry and commerce grew during this period—iron and metal, textile and leather works were built on the outskirts of Pest. While the typical factory worker earned bare subsistence wages for his twelve-hour day, at the same time capable men were needed to run the new commercial enterprises and to provide professional services as physicians, lawyers, and journalists. Although many failed in their efforts to move up, in such a rapidly growing economy others would succeed. Thus Budapest had, at this time, a qualified meritocracy, in which energy, intelligence, natural talent, charm, aggressiveness, wit, and cunning could lead to wealth, status, and influence. This constituted perhaps the biggest crack of all in the feudal order.[7]

The early twentieth century was a time of expansion, opportunity, vitality, optimism, and growth in Budapest. It was the period of economic takeoff in Hungary's development and was accompanied by an intellectual and cultural renaissance as well. Literature, literary criticism, theater, music, and the fine arts were vitalized through contact with the new streams of thought emerging elsewhere in Europe, and Budapest prided itself on the scholars it produced.

The upper middle class, which arose in the city of Budapest out of a highly competitive milieu, achieved a status and well-being that nevertheless remained precarious, in part because its members were by birth largely excluded from the privileged position of the nobility.[8] Two sharply contrasting political alternatives were open at the beginning of the century to members of the new professional and capitalist classes, and both were fraught with difficulty. One, chosen by some members of the Jewish minority, was to become the champions of increased respect for cultural diversity, democrats, and opponents of reactionary Hungarian chauvinism and aristocratic privilege. Those who took this view seriously were working for necessarily radical social, political, and economic reforms. However, this road, requiring some courage, was taken only by a few, some significant intellectuals among them. On the whole, the Budapest middle class took the opposite route, which entailed less personal danger and uncertainty. This route was to assimilate into the semifeudal structure and distance oneself as far as possible from the obvious victims of that structure. Some went so far as to seek ennoblement, and readily aided the leading stratum in its efforts to suppress non-Magyar cultures, reinforce the old feudal patterns, and maintain its power.

If it seems that many of the most successful and influential Jewish capitalists paid high hostage to fortune by becoming all but vassals to the aristocratic Magyar landholders, it must also be remembered that the Jews' insecurity throughout Europe was enormous, as Sartre has pointed out: "Jews are often uneasy. An Israelite is never sure of his position or of his possessions. He cannot even say that tomorrow he will still be in the country he inhabits today, for his situation, his power, and even his right to live may be placed in jeopardy from one moment to the next."[9]

The ambivalent position of Jewish capitalists in Budapest can be seen by consideration of the ennoblement of a number of them—Hungary's capitalist elite, including John von Neumann's father—during the early part of the century.[10] Members of the Jewish capitalist class had been invited by the emperor to apply for the privilege of purchasing a title. From the standpoint of the aristocracy, such an invitation compromised the very concept of nobility and conflicted with the strong anti-Semitic sentiments prevalent among them. But these considerations were far outweighed by the aristocracy's need for allies. It was beleaguered by the non-Magyar nationalities, who constituted a large majority of the population of Hungary, and it was threatened by the prospect of political organization among the impoverished lower classes. It also needed money and financial expertise so as to help bring about the modernization and economic development of Hungary while maintaining its semifeudal structure of protecting the special interests of the nobility.

Ennoblement of some of the leading Budapest bankers was a means of collaboration between the two classes, in which the bankers were coopted by the noble caste to serve its political purposes. Once a banker had become one of the nobility, once he had made that political choice, his political allegiance was alien to the interests of the urban middle class and the democratic elements in the society. He had foreclosed the option of working directly to change the social structure. One may wonder whether these Jews, no matter how closely they identified themselves with the old semifeudal tradition and the policy of Magyarization, really believed this cooperation would help to ameliorate the pervasive anti-Semitism. Historians and sociologists from all shades of the political spectrum agree in any case that this vanguard of the capitalist meritocracy, which began by breaking down the structure of feudalism, eventually became the strongest pillar holding up its remains.

On the personal level, this ambivalent progress for Jewish capitalists often involved a repudiation of their own origins and their less fortunate brothers. As one might expect, individual Jewish capitalists' success was often achieved at great psychological cost. An opportunistic outlook permitting a wholehearted pursuit of "rational" self-interest without any conflicting sense of humiliating compromise would make such a passage less stressful, but most people do not function that way. Perhaps some were able to view the acquisition of personal wealth and influence in Hungary in itself as a blow to anti-Semitism.

But why should or would a successful Jewish Budapest businessman seek the kind of pact with the government implied and symbolized by ennoblement? The capitalist, in spite of his wealth, was excluded from the social elite, and also from the political elite represented by the Imperial Court and the higher levels of the civil service. Ennoblement made him, at least nominally, an insider. Since nobility in the Austrian-Hungarian Empire carried "enormous social and political weight,"[11] a Budapest banker such as Max Neumann, John's father, might desire nobility not only for its aura but also for pragmatic purposes, to obtain political connections and influence. And in some respects the Jewish capitalists' political interests apparently coincided with those of the Hungarian government, as both the Budapest business community and the government felt threatened by the socialist movement and the workers' movement, which were growing rapidly in the years preceding World War I.

Thus the tiny Jewish middle class was instrumental in the industrialization and economic development of Hungary, and the Budapest banks especially were instrumental in financing industrial growth, obviating the need for dependence on foreign capital. Yet, although they had been the vanguard in establishing a middle class, the assimilated capitalists subsequently repudiated this class by choosing to enter the aristocracy. The Jewish middle class in Hungary, as well as in other Eastern European countries, it must be appreciated, had no mechanisms similar to what we in America consider to be normal middle-class politics with its possibility of effecting liberal reforms. Ultimately the development of a strong middle class in Hungary was aborted. Moreover, because of intermittent periods of intensely anti-Semitic government policies, and via assimilation, the Jewish community became debilitated. Although the turn of the century had been a time of economic growth for Hungary, as it moved from a feudal economy into a period of optimism and expansion and a cultural renaissance, within a decade or two the economic and cultural landscape was barren. Opportunities for young men had become highly restricted; the optimism had disappeared. A talented young man such as John von Neumann would naturally choose to seek his fortune outside the country, and many did so. As McCagg has pointed out, the ennoblement of the Budapest capitalists and its sequel, the voluntary emigration of the next generation of able, well-educated young Hungarian Jewish scientists or intellectuals, are historically two strands in the same piece of cloth.[12]

The new upper middle class of Budapest had maintained close ties to other European cities. The possibility of emigration was deeply ingrained in eastern and central European Jewish consciousness. In spite of the pressure for Magyarization, the prevailing tongue in the Hungarian upper middle class was German, also the language of Austria, and many of the Jews identified with German culture. Germany symbolized, too, an advanced, industrialized nation, in contradistinction to Slavic culture, which symbolized crude, illiterate peasants and connection with Russia. The interest in things German reflected an aspiration toward Western culture, science, industrialization, capitalism; the snobbish hostility toward Slavic culture was another form of the same sentiment. Early-twentieth-century Budapest even shared in some of the qualities of Viennese life: "[the] art of being not only rich but actually enjoying it. . . . That courteous gaiety and amused self-mockery and flickering erotic spark. . . ."[13]

Sociability, courtesy, gallantry—these as well as ambition and intensity were the normal attributes of a young man of the upper middle class growing up in Hapsburg Budapest. And of course personal charm was easily put into the service of ambition. There was great competition for the relatively small "room at the top." John von Neumann himself is sometimes credited with the joking remark that "only a man from Budapest can enter a revolving door behind you and emerge ahead of you."[14]

When it became apparent to historians of science that a disproportionately large number of leading intellectuals (especially scientists) in various parts of the world had grown up as sons of the middle class in Budapest in the early part of the present century, they sought for explanations of this so-called Hungarian phenomenon. It seemed as if Hungarian origins magically provided men with exceptional talent. John von Neumann was perhaps the most phenomenal among them, but some of his cohorts, such as Dennis Gabor, Eugene

Wigner, Leo Szilard, and Edward Teller, also constituted part of the natural-science wing of the "phenomenon." Others one might mention include Oszkar Jászi, liberal sociologist and historian, and Georg Lukács, communist philosopher. A phenomenon still more germane to the central theme of this essay, which I will examine at length in later chapters, is the avid taste for political activity that characterized some of the Hungarian scientists who had been involved with the atomic bomb in America during World War II. The roots of both phenomena, the intellectual brilliance and the political behavior, lie in the changing conditions of the Budapest in which these men grew up. A number of the gifted Hungarian-Americans have themselves speculated on the first of these phenomena in retrospective writings or comments;[15] historians, especially William McCagg, have examined it from a scholar's perspective.[16] The origins of the political behavior, however, have so far received only the most superficial attention in the literature, introspective or otherwise.

While pre-1914 Budapest, the scene of John von Neumann's childhood, was a period of increasing wealth, influence, confidence, and optimism among Jewish businessmen, the economic growth in this portion of the Hapsburg empire occurred against the backdrop of political insecurity, which generated a pressure to achieve. Always regarded and in some measure regarding themselves as aliens, the Jews were forever vulnerable to an upsurge of ultranationalism. The prosperous Jews in Budapest—businessmen, bankers, professionals—were subject to the power of the Hungarian political bureaucracy, and political powers, including the Hungarian nobility, were in turn dependent on the Hapsburg Court in Vienna. The Hapsburg throne itself was contingent on European power politics, which were about to precipitate a world war. And if the Hapsburg Empire fell, how would the assimilated middle-class Jews survive? What if the widespread discontent of the peasantry, especially the non-Hungarian majority, was to be mobilized into revolutionary action? Or what if the ever-present anti-Semitism were to burgeon into sanctioned violence against Jews? With such violent storms always on the horizon, the prosperous middle class wanted nothing to rock the boat; it wanted to make hay while the sun shone. The pressure on the sons of the Jewish upper middle class was to develop themselves and their potential talents while they could.[17] No matter what disaster overtook him, the man with extraordinary talents, competence, mother wit, and charm would always manage to survive—if not in Hungary, then somewhere else. The acquisition of these characteristics, perceived as requisite for survival in the face of deep insecurities, was supported, at least in the view of one Hungarian-American psychoanalyst, by strong, warm families with devoted mothers who eschewed all public life; in these successful families is found "the elusive but all pervasive security emanating from the mother images of that period."[18]

The desire for excellence and familial support for that desire were generally present, and the high school or gymnasium was taken seriously as a means to implement it.[19] In the classical gymnasium, academic ability was prized by students as well as faculty. The minimum curriculum of the high school was established by law.[20] The secondary schools maintained by "autonomous denominations" (Reformed Church, Lutheran, Greek Oriental, Unitarian) were the most independent from government control, though the minimum hours devoted to various subjects were still fixed. The Lutheran High School in Budapest, where John von Neumann and several other "phenomenal" Hungarians spent eight years, offered relatively personal academic training, drew some of its teachers from the "leading ranks of contemporary Magyar 'culture,' " and maintained a close liaison with the University of Budapest. It was also a fashionable school for boys, one that "inculcated an elitist outlook."[21] But many schools of this type existed in Europe, so that the "phenomenon" cannot be primarily attributed to its methods. What such a school can do is to encourage the self-assurance of a boy with exceptional natural mental ability and help him to regard himself as part of a natural-born elite.

After World War I, the dissolution of the Hapsburg empire, the dismemberment of Hungary, and the ascendancy of the counterrevolutionary, initially violently anti-Semitic Horthy regime, the optimism about economic progress had lost credibility within Hungary. A budding young scientist might keep his optimism and self-confidence intact, however, by leaving his shipwrecked land. More likely than not, a bilingual education would have already deliberately prepared him for such a step. First the tentative emigration to attend university elsewhere and then the search for a position outside Hungary. The field of opportunity was relatively large and the international community of scientists was relatively free of anti-Semitism. Furthermore, success in establishing himself in this community—and his Budapest upbringing had indeed prepared him well—could serve to reinforce a young man's optimistic confidence and self-reliance. In fact, achieving professional success early, outside his home country, after having survived various political and physical dangers as well, might turn that self-confidence into a feeling akin to magical invulnerability.

Given sufficient natural talent, the choice of a scientific career was highly desirable for a young Budapester anticipating emigration. It was a profession that could be practiced everywhere in Europe: universal, international, apolitical. It would not be expected to depend much on local political connections, painful compromises, and the uncertainties of business and finance. McCagg,[21] in his analysis of the Hungarian phenomenon, places a heavy emphasis on the observation that Hungary was unique in Europe, in that its gifted scientists emigrated not under strong political pressure but voluntarily and comfortably, in no way disrupting their careers.[22] Moreover, they emigrated without dependents, as college students, a particularly natural and propitious status in which to master the challenge of accommodating oneself, as a slightly marginal participant, to a society different from that of one's hometown or home country.

It must be recorded, however, that an entirely different option than that of leaving to seek opportunities in nonpolitical occupations was available to the bright, educated son of a Jewish businessman in Budapest. He might place himself in opposition to the semifeudal .structure of Hungarian society and work for political and cultural emancipation, accepting the personal risks involved. George Lukács and Oskar Jászi, among others, followed such a course.

Introspective comments by Hungarian emigrant-scientists not only confirm this analysis of the Hungarian phenomenon but describe motivations only they themselves could attest to. For example, Edward Teller has said that he had known as a youth "that Hungary was foundering," and *if he wanted to survive* he would "have to be better, much better than anyone else."[23] Teller appears to have experienced the necessity for superior ability as literally a matter of life and death. John von Neumann's explanation of the Hungarian phenomenon is still more interesting, especially as he was given to precision and tended to understate rather than overstate:

When asked about his own opinion on what contributed to this statistically unlikely phenomenon, he [von N.] would say that it was a coincidence of some cultural factors which he could not make precise: an external pressure on the whole society of this part of Central Europe, a subconscious *feeling of extreme insecurity in individuals*, and the necessity of *producing the unusual or facing extinction.*[24]

The external factors can be examined in retrospect through sociological and historical analysis, but it is the subjective experience that is striking. The word *extinction* suggests plainly that scientific creativity was at least in part motivated by a fear of death. Scientific innovation became for von Neumann the personal means for continuing the earlier optimism associated with his childhood in prewar Hungary. An echo of this theme is found in the writings of Dennis Gabor, who became preoccupied with the attitude, as he expressed it, of "Innovate or die!"[25] This existential theme had been given its particular form by political and socioeconomic circumstances, which were overriding facts of life in the Hungary in which John von Neumann's character was "sculpted." Not only von Neumann, but many of his "phenomenal" cohorts, even if they became involved in otherworldly occupations such as mathematics, would throughout life operate with antennae sensitive to political circumstances, especially alert to potential threats to their own well-being deriving from world politics.

I shall explore in later chapters how von Neumann responded to the life-and-death issues posed by the creation of the first atom bomb. It will be my contention that the political and technological threat to life posed by nuclear weapons revived boyhood attitudes for such phenomenal Hungarians as von Neumann, Leo Szilard, and Edward Teller. The primary response to the fear of extinction was to produce the unusual, to innovate almost obsessively. Innovation alone would stave off annihilation.[26] The optimism that marked the economic and technological progress of late Hapsburg Hungary must not be lost again. Thus Teller wrote about "How to be An Optimist in the Nuclear Age,"[27] and von Neumann also faced the threat of nuclear weapons with apparently cheerful faith in the beneficence of technological progress. The creation of nuclear weapons offered to a number of Hungarian-American scientists not only the opportunity but also (at least in the case of von Neumann) the compelling desire to serve and be part of the elite establishment, and to assume a modern American role reminiscent of the court astrologer or court engineer of a feudal military empire.

Of course it is incomplete to regard John von Neumann's origin only in terms of membership in the group of phenomenal Hungarians, to see him molded by the historical, social, and educational situation of his particular socioeconomic class without considering the unique qualities of his family, just as Norbert Wiener's unique family should not allow us to disregard the political and social setting in which he grew up. Von Neumann's maternal grandfather, the son of a poor immigrant Jew from Bohemia, had already become wealthy in the late nineteenth century, by means of a business in agricultural equipment, particularly millstones. In the 1890s the city of Pest had become the largest flour-milling center in Europe.[28] The grandfather had bought a four-story apartment house in the elegant section of town, inside the Ring on the Pest side of the Danube, at 62 Vaci-Körút.[29] The ground floor was used by Grandfather Kann for his offices, but he had given the top floor as a kind of wedding gift to his daughter Margaret when she married the hardworking and increasingly prosperous young Jewish banker Max Neumann. Similarly,

Margaret's favorite sister, when she married, came to live with her husband on the third floor. The Neumanns on the fourth or top floor and the Alczutis on the third floor were such close friends that they constituted an extended family for their children. Another indication of Grandfather Kann's financial success was the summer house he had bought in the Buda mountains, which became a vacation spot for his children and grandchildren.[30]

Already before the turn of the century, Max Neumann had been working at Jelzáloghitel Bank in Budapest. He "prospered greatly, *inter alia* because Kálmán Széll, the chief of his bank and his personal protector, became in 1899 Prime Minister of Hungary."[31] In the early part of the twentieth century, the Hungarian banking industry was instrumental in breaking the hegemony of Austrian capital as the financial basis of Hungarian industry and in finding native Hungarian capital to replace it, thereby increasing Hungary's autonomy as a nation.[32] Although the work was technical, the bankers' collective action was politically consequential. In the process Max Neumann became a rich man in his own right; he too bought a house in the country.

Max Neumann was given the "opportunity" to attain social status, his title of nobility, in 1913. Henceforth his full name and title would be Max (Miksa) Neumann de Margitta. Later his son John would use the Germanicized version of the title, "von Neumann." The ennoblement in 1913 "implied political choice. It implied both a degree of social conservatism or 'feudalization,' and a willingness to be considered Magyar as opposed to any other nationality."[33] Actually, ennoblement in the years immediately preceding World War I was a particularly opportunistic compromise for a Jew who had risen from the middle classes. While formally the title was granted by Franz Joseph, emperor of Austria-Hungary, the ennoblement of Max Neumann and a fair number of other Budapest bankers and industrialists in the years 1910–14 had been engineered by István Tisza, who had after 1910 "become the most deliberate, most brutally reactionary representative of the alliance of the ruling classes,"[34] and who used ennoblement as a tactic to build a badly needed political alliance for his regime.

Max Neumann, it is noteworthy, did not assimilate to the extent of converting to Christianity, and his closest personal friendships were with other Jewish businessmen and bankers.

A posed photograph of the Max Neumanns during the Hapsburg era shows them as a strikingly elegant couple. János (John) or Jancsi (Johnny), the eldest of their three sons, was born on December 28, 1903. The younger sons were born in 1907 and 1911. Jancsi (pronounced "Yon-shee") was also older than any of his female cousins, who lived on the third floor of the house. A Hungarian nurse became part of the Neumann household when Jancsi was born, and later there were German governesses so that Jancsi and the other children could learn German.

Margaret Neumann, Jancsi's mother, was a woman of considerable charm and native intelligence, apparently with a good grasp of human nature and human motives. Her life revolved about her husband and children, and she rarely went outside the closed family circle, which encompassed her sisters and nieces and nephews, plus the family's servants. To Jancsi she was a caring and continuously available confidante, who also provided generous emotional support for success and achievement. In spite of her exclusive commitment to the small world of the family and its well-being and success, however, she was not serene and calm but often anxious about her family. Those who came to know her during her later years, when she was living at John von Neumann's house, invariably describe her, although she was a very thin, chain-smoking, nervous woman, as kind and delightful.

The family anecdotes about Jancsi as a child, while probably not reliable evidence about the original events, do convey something about how his uniqueness was perceived by other family members. His mother recalled that once, while sitting and crocheting, she stopped and reflectively looked into space for a minute or two. Jancsi, noticing this, inquired, "What are you calculating?" Other anecdotes show him a precocious, rapid reader, fascinated especially with history, and an early prodigy at doing arithmetic in his head. They suggest not only that he identified reflection with calculation but that he would attempt to resolve emotional issues by insisting on correct form, either proper social form or statements in logical form.

Vis-à-vis his younger brothers and cousins, he was the oldest, the most aggressive one, the least sentimental, the most thoughtful. He was aloof from the games of the younger ones and the girl cousins. It was they who might be silly or sentimental. In the family group it was his brother and cousins who exhibited sensitivity for others and tender feelings; Jancsi had some contempt for all of that. His interests were reading, studying, calculating, and—especially during World War I—engaging in elaborate battles with toy soldiers, playing with brother Michael or a boy cousin. Jancsi's father took an interest in his education and made the decisions regarding it. In 1914, when he was ten, the usual age for a boy to start gymnasium, Jancsi was enrolled in the Lutheran Gymnasium for boys in Budapest.

Though Jancsi was a mathematical prodigy, and intellectually precocious in general, he never skipped a grade in school, nor did he ever become a misfit or outsider there. Quite the contrary; his intellectual ability was admired by his fellow students. Intellectual achievement was gratifying not only because of the response it elicited from teachers or parents but also because of the admiration of one's peers. He found a friend in Eugene Wigner, just one class ahead of him, and the custom of students teaching each other—which was even institutionalized in a student seminar—allowed Jancsi to make use of Wigner's knowledge.[35]

At the time that Jancsi entered the gymnasium, his parents were already aware of his exceptionally logical mind and the ease with which he mastered learning. After he had attended the gymnasium for only a few months, the mathematics teacher there, Ladislas Ratz, told Max Neumann that his ten-year-old son was an extraordinarily gifted mathematics student, a child prodigy in fact. Although Ratz was prepared to give him special treatment in his class, he felt that the boy should have individual tutoring by a professional mathematician to develop his talents fully. Max Neumann was pleased to go along with the suggestion. Ratz introduced the boy to a leading mathematician at the university, J. Kuerschak, who in turn arranged for his younger colleague, Michael Fekete, to come regularly to the von Neumann house to work with Jancsi. Thus, among the small circle of Hungarian mathematicians, the budding genius Jancsi von Neumann became well known while yet an adolescent. His father was enthusiastic about this development, managing to create for his boy a professional mathematical milieu in his own home even while he was going through the regular course of study at the Lutheran High School, which had dedicated teachers, some of whom were engaged in independent research and many of whom had a genuine devotion to the development of the talents of their students.[36]

Jancsi's father's interest in him also extended to chess games—a pastime the boy seemed to take very seriously. When he lost to his father, as was usual, he would be most unhappy. He disliked outdoor games and athletics, though he would occasionally go ice-skating in the winter. He enjoyed the science fiction of his day, and he systematically read a fifty-volume history of the world from beginning to end.

As Eugene Wigner has recalled his boyhood friend,

We often took walks and he told me about mathematics and about set theory and this and that. It was amazing. And he loved to talk about mathematics—he went on and on and I drank it in.

He was inexhaustible on such occasions in telling me about set theory, number theory, and other mathematical subjects. It was really wonderful. He never thought of going home. . . . He was phenomenal, also in his desire to talk.[37]

To Jancsi the availability of a receptive age-mate like Wigner was a stroke of good luck.

Before leaving high school, Jancsi did some original mathematical work: with Fekete he extended a theorem in analysis put forward by the Budapest mathematician Fejer.[38]

To quote Wigner again on Jancsi's demeanor in high school,

In school and among his colleagues, Jancsi was somewhat retiring. He participated in the pranks of the class, but a bit halfheartedly, just enough to avoid unpopularity. He had a few close friends and was respected by all—intellectual strength was recognized and approved of by the student body, if not always envied.[39]

On another occasion when Wigner was reminiscing about his admired boyhood pal, he somewhat ruefully admitted that Jancsi's friendships had lacked intimacy and emotional substance.[40] Yet the lifelong relationship between von Neumann and Wigner was immensely important to both of them. Wigner, the Nobel Prize–winning theoretical physicist, recalls that his own father had been right in not wanting his son to become a mathematician, although the young Eugene had remarkable mathematical talent: "At that time there were only three positions for a mathematician in Hungary. Particularly from having known Jancsi von Neumann, I realized what the difference was between a first-rate mathematician and someone [like me]."[41] Wigner's remarks again recall the strong competitive element in the Budapest professional world.

Jancsi's father, so successful in the business world, also knew the pressures and tensions, the painful compromises involved in success in the banking world. Of course, pleased at his son's talent, he did whatever he could to nurture it, through tutors and by arranging contacts with professional mathematicians. He anticipated that Jancsi would indeed become a professional mathematician[42] and that—with his talent as his amulet—he would become relatively independent of the vagaries of the business world.[43] Jancsi, it appears, at first aspired to follow in his father's footsteps and obtain success, wealth, and influence through banking or some other commercial activity. Ultimately his enthusiasm for mathematics seconded his father's urgings, and he gave up a worldly career for university life. But Jancsi was less ivory-towerish than most of his colleagues, and furthermore he retained an abiding interest in the business world and the exercise of power. In fact, even after he had become a world-renowned mathematician he continued to be "impressed by people who were successful in political and organizational activities . . . and he cultivated them."[44]

Jancsi received an additional nudge in the direction of a career in mathematics when he won the nationwide Eotvos

Prize, awarded on the basis of a high school competition requiring "excellence in mathematics and scientific reasoning."
It was a concrete confirmation to the boy that in this field none
of his peers in Hungary could surpass him.

This then had been the young John von Neumann's family
and school life. While he was in his teens, however, his education was unavoidably supplemented by first-hand acquaintance with revolution and war, occasioned by dramatic events
in Hungary itself.[45] The latent instabilities and unresolved
political, economic, and social conflicts had broken into the
open.

The first decade of Jancsi's life had been prosperous times.
There had been no revolutions and no wars; the socioeconomic status of the Neumann family had not been questioned, and the primary task of the children had been to develop their individual talents,[46] to absorb what they could of
Western culture in their classical education, and to enjoy life.
Yet behind this stable world lurked enormous conflicts between ethnic groups in Hungary (and the threatened seces-
sion of sections of the country); poverty, illiteracy, malnutri-
tion, tuberculosis, and lack of franchise for the vast majority
of the population; and international political maneuvering in
which Hungary had to depend on the uncertain support of
stronger nations, particularly Austria.

Even the ten-year-old Jancsi could not have been unaware
of or unimpressed by the great public enthusiasm, military
music, and patriotic oratory that accompanied Hungary's
entry into the First World War on the side of Germany in 1914.
The conscripted army of Hungary was made up mostly of
peasants and workers, and Max Neumann continued in his
executive banking position. Thus family life was not disrupted
by the war. By the time the war ended, when Jancsi was fifteen, Hungary had severed its connection with the Austrian
monarchy and after a bloodless revolution had become a liberal republic. It had also been on the losing side of a war and
had to suffer the consequence: loss of territory. In fact, when
the treaty was finally signed, Hungary lost two-thirds of its
territory and population to neighboring countries. Through
the peace treaty many members of the non-Magyar ethnic

"nationalities" in Hungary were freed from Magyar domination, intolerance, and abuse, but the loss of territory was a bitter pill for Hungarian patriots to swallow.

By the summer of 1919 the population of Budapest was swelled by refugees from the provinces and by returning soldiers. There was a great food shortage—all the confusion and military action had interrupted the regular supply of food from the rural areas. Not only was Budapest "turned into a gigantic debating society, wrangling all day and all night over constitutional and social problems and problems of foreign relations," but hunger and discontent among the populace led to "frequent explosions of violence and looting."[47] Nevertheless Jancsi continued to go to the Lutheran gymnasium, and Max Neumann continued to work as usual.

The liberal government had drawn support from progressive elements among the intelligentsia, the middle class, and the aristocracy; it promised to achieve a peace treaty favorable to Hungary. In the face of chaos and failure to obtain such a peace treaty, the liberal Karolyi government fell, and in March 1919 the government was taken over by a coalition of Social Democrats and Communists (Bela Kun) encouraged by the success of the Russian revolution. The escalating confusion was now centered on the attempt by the government to move rapidly away from a capitalist-feudal economy toward one in which "everyone would contribute work according to his capacity, and be provided for according to his needs." One of the first steps in this direction was the expropriation of the banks: "All the banks were speedily occupied and the higher functionaries either refused access to them altogether or, if by special favor not excluded, deprived of all authority."[48] Max Neumann was now apparently persona non grata as a banker within Hungary. Moreover, groups of toughs known as the "Lenin Boys," which included leather-jacketed sailors and escapees from prison, roamed the city; knocking on the doors of the rich, they harassed and roughed up some aristocrats and wealthy bankers. The Neumann family's physical safety was in jeopardy, and they fled the country.[49] Max Neumann was rich, and even this dislocation could be made relatively painless; yet the awareness that one is in effect exiled from one's own home can never be pleasant. The family went to Austria, spending their time either in Vienna or a resort town, Abbazia, along the Adriatic. The exiled group of bankers in Vienna began to plan the financial basis for a Hungary they expected to be recaptured by the old aristocratic establishment.[50] Indeed, the Communist government collapsed after five months when invading Romanian troops marched into Budapest. Under the Karolyi and especially under the Kun regime Jews had held many leading official government positions, a rare phenomenon in Hungarian history, and it is ironic that it was during the latter regime that Neumann and other Jewish bankers had fled.

To obtain perspective on the brief Communist reign, it is of interest to quote the personal account of another Budapest schoolboy, roughly the same age as Jancsi, whose family was neither sufficiently wealthy nor politically sufficiently conservative to feel the need to leave the country:

At school strange and exciting events were taking place. New teachers appeared who spoke to us in a new voice, and treated us as if we were adults, with an earnest, friendly seriousness. One of them was Dezsö Szabó, author of a celebrated novel about the Hungarian peasantry. He was a shy, rather tongue-tied, and absent-minded person who talked to us in a gentle voice about a subject more remote than the moon: the life of a farm hand in a village. Other new teachers were young members of the intelligentsia who had never taught in a school before. . . . On May Day, a celebration was held at our school. One of the boys from the top form, a charming and gifted youngster of seventeen who had already published several poems in a literary magazine, extolled the memory of Danton and St. Just. The speech was enthusiastically received by the boys and the new masters; the old teachers listened in acid silence. After the collapse of the Commune the young man was thrown out of school and according to rumour, killed by the White Guardists. It was my first experience of this nature; it gave the words "counter-revolutionary terror" a personal and frightening meaning.[51]

In August 1919 a counterrevolutionary regime (Horthy) took over the reins of government with support from conservative aristocrats as well as radical fascist military men and retook Budapest. Soon after, the Neumanns returned from Austria.

Jancsi took up where he had left off at school and in his studies with mathematicians at the university, and Max Neumann regained his position in the bank.

When Jancsi returned to high school, the first issue was whether or not those students (and one teacher) who had expressed sympathy toward the Communist regime should be expelled. The available evidence indicates that Jancsi was not at all attracted to idealistic leftist regimes and that his opinion was asked about how one should deal with his less reliable classmates.[52] The counterrevolutionary actions under the Horthy regime during its first years in power were vengeful and terroristic; focused on leftists and Jewish intellectuals, the vengeance was not restricted to those who had actually participated in the Kun government, and "more or less unofficial 'bands' had set out over the country meting out lynch laws . . . to many whose offence was simply that of belonging to the same race as Kun."[53] (Kun was himself a nonreligious Jew.) Approximately 5,000 men and women were killed, in effect executed, in this period of the "White Terror"; another 100,000, including many urban Jews, fled the country. The number of violent deaths produced by the White Terror was about twenty times as large as the number of violent deaths produced by the relatively moderate Red Terror of Kun's regime.[54] Oscar Jászi, a member of the Karolyi government, has described the qualitative difference between the Red and the White Terror:

The Terrorist actions of the Reds usually revealed the primitive cruelty of coarse and ignorant men; the Whites worked out a cold and refined system of vengeance and reprisal, which they applied with the cruelty of scoundrels masquerading as gentlemen. The worst atrocities . . . of the Whites were the deliberate actions of elegant officers.[55]

The Max Neumann family, having returned to Budapest during this period, remained physically and materially unscathed. As a matter of fact, men like Max Neumann were needed by the right-wing regime, and Horthy, even while establishing and implementing harshly anti-Semitic practices, at the same time established ties to Jewish bankers and industrialists, especially the ennobled elite among them, and worked with them.[56]

Just a year before Jancsi completed high school and was ready to enroll at the University of Budapest, strong discriminatory quotas were imposed on university admission by the parliament.[57] Aside from "race," the government, in a nationalistic spirit of fear and revenge against educated leftists, made the prime condition for university admission that the "student could be absolutely trusted in respect to morality and national loyalty," although "besides . . . national loyalty, the intellectual ability of applicants must be taken into consideration."[58] The fact that Jancsi was indeed admitted to the university upon completion of high school, in 1921, is supporting evidence for his "clean record" as seen by the counterrevolutionary authorities, although of course the mathematics professors were eager to have such a gifted student. But the seventeen-year-old Jancsi left the country in 1921, returning to Budapest only periodically to visit his family and take examinations at the University. The teenaged von Neumann at first attended the University of Berlin, where he listened to Albert Einstein lecture on statistical mechanics, studied with mathematician Erhard Schmidt, and had regular get-togethers with his budding fellow Hungarian scientists, Eugene Wigner, Leo Szilard, and Dennis Gabor[59]—each of whom was part of the "Hungarian phenomenon."

The ubiquitous von Neumann also was in Göttingen some of the time, especially to visit Hilbert; "the two mathematicians, more than forty years apart in age, spent long hours together in Hilbert's garden or in his study."[60] Hilbert was the champion of "axiomatics" in mathematics and in the sciences,[61] an approach that appealed to von Neumann. Even long after their period of active collaboration, von Neumann continued to develop some of the lines of investigation initiated by Hilbert. David Hilbert must be counted with the mathematicians Herman Weyl and Erhard Schmidt as having had considerable influence on von Neumann as a young mathematician. Weyl and Schmidt had both been Hilbert's students twenty years earlier.[62] After Berlin, von Neumann studied at the

Eidgenössische Technische Hochschule in Zürich in democratic Switzerland, while still enrolled at the University of
Budapest. In Zürich he had contact with Weyl and George
Pólya, another mathematician of Hungarian origin who later
emigrated to the United States.

Von Neumann took a degree in chemical engineering from
Zürich in 1925—a second string in his bow, more practical
than mathematics and useful in industry. He received his
doctorate in mathematics from the University of Budapest in
the spring of 1926. He spent the latter part of 1926 in Göttingen, and in 1927, in spite of his exceptional youth, he became a privatdozent, lecturing in mathematics, at the University of Berlin, about three hours by train from Göttingen. The
experience of early social and professional success in foreign
lands not only reinforced Jancsi's exuberance and confidence; it also helped to perpetuate his adolescent sense of
invulnerability.

Linguistically John von Neumann was amply prepared for
life in Germany, for he had often spoken German at home as a
child with his parents and with his nurse. Politically and cul-
turally, too, Germany and Hungary had some patterns in
common. Germany was a republic in the 1920s; the Kaiser
had lost the war in 1918 and the reaction to an unsuccessful
socialist revolution had strengthened the political right. Although the Weimar Republic persisted until 1933, it was continually threatened by severe economic crisis, widespread
romantic chauvinism and discontent, a nostalgia among the
upper classes for the days of the empire and special privilege, and the rising power of an angry right-wing radicalism.

In the face of these forebodings of doom, the Weimar Republic was the scene of intense cultural and intellectual activity.[63] It was the time during which the Bauhaus movement
in architecture and design thrived, as well as writers Thomas
Mann and Bertolt Brecht. In film and in art expressionism was
prominent; Berlin in particular was the center of the most
iconoclastic and nihilistic elements of the Dada movement,
and the cabarets were at the height of their popularity. Berlin
was also an active center of legitimate theater and lively

journalism. Freudian psychoanalysis had arrived on the scene. In the social sciences the Frankfurt School was important, and socialist radicals were politically active.

Professors in Germany were part of an elite; they were generally conservatives, frequently monarchists. They feared for their special status and talked of "democratization" as a threat to "higher culture."[64] Traditionally, "the status and the privileges of the universities were granted to them by the military-aristocratic ruling class."[65] Scientists and professors generally avoided identification with the middle classes, and did not take it as their mission to liberalize the society. On the contrary, it was usual to emphasize the "value neutrality" of professorial activity, while in fact the professors were eager "to maintain themselves as a highly privileged 'estate' elevated above the classes of society, not accountable to any, but protected by and spontaneously allied with a similarly privileged higher civil service."[66] Thus the universities tended to be centers of conservative opposition to the republican regime.

In spite of giving lip service to scientific internationalism, the scientists, specifically the German physics professors, tended during the Weimar period to regard German science in strongly nationalistic terms.[67] In turn the German republican government gave heavy financial support to "pure research" in physics and mathematics. The government's objectives were political and ideological: Germany sought an ascendancy in basic science as a means of international prestige, as a surrogate for the military power that according to the Versailles Treaty of 1919 Germany was prohibited from developing. The Weimar government had in addition an internal political objective in its financial largesse, namely that of persuading the academic community that democracy is not necessarily the enemy of higher culture.[68]

The major political tragedy of Germany was the erosion of the republican form of government established in 1919 until it turned into a full-blown dictatorship with the establishment of the Third Reich in 1933. The privileged class of German professors, engaged in such esoteric pursuits as "pure" mathematics and hostile to democracy, wittingly or unwittingly helped in the demise of the republic: "The universities were part of the system which made the Nazi take-over possible."[69] The complete failure of German professors and professionals (such as physicians) to provide intellectual leadership in opposition to the rise of Nazism reflects their commitment to their own status and economic interests, often expressed as apolitical professionalism. This historical situation, which has parallels in other countries and in other times, raises the whole issue of genuinely "responsible" intellectuals into prominence.

The political basis of the universities in Germany and the relation of the professors to the various strata of society had some structural similarity to the political basis of banking in Budapest and the relation of the ennobled bankers to the classes in Hungarian society. Because of these structural similarities Jancsi may have felt himself less of a stranger in Germany than he might have otherwise. They may also have led him to accept the situation as natural and normal, or at least comfortable and familiar. But von Neumann was in Germany to do mathematics, and even if as a Hungarian he could not escape some sense of his own marginality as a foreigner,[70] and in fact some threat to his academic pursuits should the radical right or the radical left win the upper hand, his enthusiastic preoccupation at that time was with the profound intellectual crises in mathematics and theoretical physics, not with politics. Recalling von Neumann at this time, a Göttingen colleague who later met him again in America wrote that "he was generally recognized as a person full of ideas, high intelligence, and with a brain that worked like greased lightning. He was friendly, not condescending, and he had a sense of humor. He liked parties, in particular if mathematical games were played."[71]

It was at Göttingen that von Neumann's and Wiener's paths first crossed. Hilbert's personality had long dominated Göttingen; he was widely regarded as the foremost mathematician of his time, and he set the tone in a number of ways. His axiomatic approach to the foundations of mathematics would be followed up by others at Göttingen, including especially von Neumann, who had become one of the circle of young mathematicians around him.[72] Hilbert had taken an active interest in atomic physics; for years he had always had a physicist for an assistant, in effect to teach him, Hilbert, the latest physics; and he was on very cordial terms with Max Born of the Göttingen physics department. Born, in turn, was the most mathematically minded of the prominent physicists of his generation, far more so than Einstein, Planck, Ehrenfest, or Bohr. Doctoral students and postdoctoral scholars, as for example the young Heisenberg and Jordan, would attend seminars in both departments, and many of Hilbert's physics assistants had received their doctorates under Born.

The exciting event at Göttingen in 1925 was Heisenberg's development of a quantum theory; although Heisenberg (age twenty-four) was a physicist, Hilbert asked him at the first opportunity to lecture about his theory at the mathematicians' seminar.[73] Von Neumann (age twenty-two) found what he had heard at Heisenberg's lecture exciting, and conveyed his enthusiasm to some of the less advanced graduate students.[74] Very soon he was working with Hilbert and Hilbert's physics assistant, Lothar Nordheim, on improving Heisenberg's mathematical formulation, with a view ultimately to putting it into axiomatic form.[75] As the three men knew, Max Born and the American Norbert Wiener—who was spending some portion of each of the years 1924, 1925, and 1926 in Göttingen—had already generalized Heisenberg's formulation in some respects. (But only von Neumann would carry the task through to its ultimate completion.) Physicists differed among themselves in the extent to which they regarded axiomatization and insistence on rigor as important. For example, Heisenberg regarded it as a misplaced emphasis,[76] a distraction from the physics itself. Born, on the other hand, valued it extremely highly. Today an "axiomatic" school of physics contends with other more "physical" schools of physics.

The other intellectual current that stimulated von Neumann's interest at that time was the foundation crisis in mathematics. The crisis arose from a branch of abstract mathematics, G. Cantor's theory of sets, in which some contradictions had unexpectedly been discovered. This led to the apparent necessity of proving mathematics generally to be free from contradiction, lest they also crop up in other branches. If such contradictions actually occurred, or even if they merely could occur, whole branches of the tree of mathematics might break off. For this reason Hilbert, and also in particular von Neumann, carried out such a program,[77] which in a sense was intended to justify mathematics. Von Neumann's works on this program were seminal contributions toward clarifying axiomatic formulations of various branches of mathematics, and in the late 1920s he was confident that all of mathematical analysis could be proved free from contradiction. A few years later, however, it was discovered by K. Gödel that the program of Hilbert and von Neumann was impossible to carry through in principle.

Von Neumann's regular position from 1927 on was in Berlin, except for a period at the University of Hamburg. Friends have recalled that at that time he enjoyed Berlin's night life and cabarets more than most of his colleagues, who saw him as a bon vivant. He is also remembered as a young man who exuded bravado and zest for life, whose involvement in mathematics was no bar to the enjoyment of worldly pleasures. When he placed the title "von" in front of his name, a title that carried prestige because it typically was used by families of the German nobility, there was some justification for it—after all, Max Neumann had obtained for himself an equivalent Hungarian title.

Not only a quick intelligence but also an urbane and cosmopolitan style, nurtured at home and in school, were aids to von Neumann's survival both inside and outside Hungary. So were political astuteness and a sensitivity to political currents. When Budapesters such as von Neumann spoke of the fear of extinction, we may ask what made the universal fear of death such an overriding motive among this group of healthy young men? Was it a displaced fear of the danger of sinking into the impoverished classes? Did it express a sense of participating in a dying culture? Or did it reflect a realistic fear of meeting violent death in that environment? I would judge that all three questions should be answered in the affirmative. The possibility of violent death was in any case not unique to Budapest. Large-scale political violence had indeed prevailed at various times and in various forms in the Hungary of Jancsi's adolescence, but it was also prevalent elsewhere in Eastern Europe. World War I had killed millions of young men throughout Europe, and soon the scale of political violence of Stalinist Russia and Nazi Germany would outdo that of the previous decades.[78]

The young Jancsi had more to protect than just life and a way of life. He needed to protect the opportunity to develop his talent in mathematics, to pursue his studies. It was not in his interest to seek social justice for the Hungarian proletariat and the peasant, or to get lost in political debate. The boy in his late teens was implicitly making a choice among alternative values. His commitment was not only to the traditional values of his family and to his own survival and well-being but also to mathematics itself. His resources for survival included his father's wealth and his own good sense, but his mathematical and logical talent did indeed become his amulet. This amulet, combined with early successes in dangerous times, could provide the emotional basis for his optimism and self-confidence in uncertain circumstances. In the belief of his family and friends in Hungary, he came closer than anyone to having the attributes that would justify such an outlook, and as a middle-aged adult, he retained an adolescent bravado that suggests his own sense of invulnerability. His notorious jovially reckless driving is a case in point. He totally wrecked numerous cars, but he was never severely hurt. "There was a special sense of invulnerability in von Neumann's make-up," an observant friend of the middle-aged von Neumann noted.[79] When I interviewed some of his family and Hungarian friends, I was surprised to hear several of them independently describe their shock at his dying of cancer (at age fifty-three) in similar terms. Each said in his own words, "Irrational as it seems, I've always regarded John as capable of surviving anything; I could not believe that *he* would die!" In any case, his creative mathematical work, as well as his innovative role in computer technology, bestow a kind of immortality on his name for as long as our civilization continues to value such things.

Comparing the origins and early years of John von Neumann and Norbert Wiener, it is evident that while Jancsi's home life may have been luxurious and pleasant, the political circumstances surrounding it were highly unpleasant. The prejudice that was experienced by a New England Jew is not comparable in harshness to the violent outbursts of anti-Semitism, pervasive exclusion and general political violence that had characterized Hungary. Wiener and von Neumann belonged to different social classes, and had different relations to the phenomenon of upward social mobility in their respective environments.

One can distinguish at least two paths of upward mobility among Jews in Western society during the periods of rapidly spreading industrialization. One path is through economic enterprise: achieving wealth and thereby also some political influence, and then using this wealth and influence to move into the social and political establishment. In this pattern the desire for and pride in upward mobility is open, unabashed, and uninhibited. This was the path of the Neumann family and was deeply ingrained in John von Neumann. For him, of course, it would naturally be his special gifts that would be his best means of reaching the top, since he lacked actual aristocratic lineage. In the class to which his family in Hungary belonged, the acquisition of sophisticated social skills was highly valued, for it too was a useful means of survival and upward mobility.

The other path involved a combination of partial acceptance of the position of outsider in relation to the economic and political elite and ambition to achieve status as an intellectual. In this pattern the emphasis is on independence, intellect, scholarly achievement, the ability to provide a critique of the status quo, and on humanitarian values rather than money, social status, and power. Such a Jew would seek recognition among scholars and intellectuals but would be uncomfortable as a member of any intellectual establishment, lest it compromise his independence. Leo Wiener's upward mobility had been by this route, and as he proudly asserted, "without any ancestry of successful pork packers or liquor dealers" or "a genealogical tree with its roots in the May-flower." This was also to be the route for Norbert Wiener. His maternal grandparents fit the other pattern, but Norbert seems not to have been attracted by it.

By 1925 both Wiener and von Neumann had become busy, productive mathematicians. Already Wiener had completed his mathematical theory of Brownian motion, a major and pregnant piece of mathematical work. His work on harmonic analysis, which he presented at Göttingen in 1925, had grown out of the earlier Brownian motion analysis. Around the same time von Neumann was introduced to the problem of for-mulating atomic physics mathematically, a problem much discussed among the physicists at Göttingen and very in-teresting to Hilbert. In 1927 von Neumann published three fundamental papers on a mathematical framework for quan-tum theory, and in the following year, just after he became an instructor at the University of Berlin, he wrote a seminal paper on an entirely different topic, the theory of games. Also, he was developing an axiomatization of set theory, which dealt with the very foundation of mathematics; this last topic at-tracted the attention of mathematicians of that time more than his work on quantum theory or game theory did.

In the following three chapters I will describe some of Wiener's and von Neumann's mathematical interests in the 1920s, to provide some notion of the kinds of problems they dealt with. To make these chapters accessible to as wide a group of readers as possible, each topic is approached in terms of the history of the underlying mathematical ideas. Since mathematics was central to the lives of these men, one could not hope to understand them without at least a good sampling of their mathematical thinking. Chapter 4 describes Wiener's theory of measure for paths and places it in the con-text of the history of ideas; the following chapters describe von Neumann's 1928 theory of games and some of his work on quantum theory in a similar way. The fact that I have chosen to discuss two topics for von Neumann reflects the diversity of his mathematical activities; Wiener's mathematical papers are more obviously connected with each other. The particular mathematical topics chosen for discussion are not arbitrary; in fact, it will be part of my purpose in later chapters to show that interest in these particular mathematical theories has some relation to Wiener's and von Neumann's respective personalities and philosophical outlooks.

Chapters 3 to 5, then, describe some substantive mathematics, the product of Wiener's and von Neumann's activities. It is natural, and not uncommon, for a creative mathematician to wish to be remembered for his impersonal mathematics, into which he has poured so much of himself, and to let his idiosyncrasies, his politics, and his personality be ignored and forgotten. Yet the process of mathematical creation itself has personal psychological and philosophical dimensions that motivate and find expression in mathematics proper. I shall focus on that topic in chapters 6 and 7. Philosophical aspects of certain applications of game theory are discussed in chapter 12.

Yet even a discussion that includes both the mathematics proper and the psychological-philosophical dimension leaves out another important determinant of mathematical activity. I have already commented on the social forces that provided prestige and established institutional settings for mathematical research in Weimar Germany and contributed to making a career in mathematics possible there. The reader is here cautioned not to forget altogether, during the discussion of creative styles and mathematics proper, that social forces lie concealed in the background even of mathematical research.

3 THE MEASURE OF AN EXTREMELY LIVELY AND WHOLLY HAPHAZARD MOTION

A mathematician, like a painter or a poet, is a maker of patterns. If his patterns are more permanent than theirs, it is because they are made with ideas. . . . The mathematician's patterns, like the painter's or the poet's, must be BEAUTIFUL. . . . Beauty is the first test: there is no permanent place in the world for ugly mathematics.
—G. H. Hardy

One of Norbert Wiener's major mathematical innovations, developed when he was still in his twenties, was the invention of what has come to be known as the "Wiener Process" or the "Wiener Measure." I am choosing this as a particularly significant example of Wiener's mathematical interests, since in it is found a remarkable confluence of ideas, ideas such as measurement, infinite set, probability, Brownian motion, and curves without tangent.

It seems that some men in each of several ancient civilizations were interested in measuring things. How far is it from city A to city B? How large is the earth? How tall is this man, and how long is his shadow in the morning, at midday, and in the evening? What is the direction in which his shadow falls? In each case the idea is to attach an appropriate number to the phenomenon measured. Usually this is accomplished by comparing one thing to another, for "big" and "small" have little meaning otherwise. The ancient Greeks in particular distinguished measuring from counting. Counting of objects always results in one of the "natural" numbers such as 1, 2, or 173, but measuring results in a "magnitude" of the thing measured. In measuring and comparing the height of persons, one might proceed by choosing a stick that is exactly equal in length to one's own foot, so that if the stick is lost another can always be fashioned. Then one marks on the stick twelve or twenty or some other number of equal divisions. The magnitude corresponding to the height of a person will usually involve a fraction of a stick length ("foot"), such as 5-7/12 or 6-1/20 feet. Thus measurement compels us to supplement natural numbers by fractions.

It is significant to observe that in making measurements such as these one is setting up a correspondence between the humanly invented numerical magnitudes and the physical heights of various persons. Also implicit is the recognition that the rule for combining partial measurements corresponds to the arithmetical rule for addition: a woman's height is measured to be 3 feet from the ground to her navel, but 2-7/12 feet from the navel to the top of her head. Her total height, our whole procedure implies, is then the sum of the two measurements. So the idea evolved that one might utilize the correspondence between the physical measurement process and the rules of arithmetic to learn all sorts of things about measurements through logical or mathematical reasoning alone. A highly original thinker in this field among the ancient Greeks was Eudoxus of Cnidos, an astronomer, mathematician, and philosopher who lived in the fourth century B.C.; subsequent to him Archimedes became the outstanding practitioner of the art. Since the kind of mathematics favored by the Greeks was geometry, the particular measurements they reasoned about were the measurements of a line (its length), the measurement of a surface (its area), and the measurement of a solid object (its volume). Here too they could test their reasoning, for they knew how to carry out physical measurements of length, area, and volume.[1]

One may wonder about the motives of the ancient Greeks for abstracting out numbers from qualitative perceptions. We have no manuscripts from Eudoxus himself, and differences among individuals may be considerable. But the fragments available from the writings of Philolaus, a significant forerunner of Eudoxus, are nevertheless suggestive:

For the nature of Number is the cause of recognition, able to give guidance and teaching to every man in what is puzzling and unknown. For none of existing things would be clear to anyone, either in themselves or in their relationship to one another, unless there existed Number and its essence. But in fact Number, fitting all things into the soul through sense-perception, makes them recognizable and comparable.[2]

Thus numbers and their relationships are means for understanding "existing things and comparing them with each other."

Some of Eudoxus's reasoning laid the foundation for aspects of modern mathematics essential to Wiener's work. Eudoxus, who had been preceded by Pythagoras, connected numbers and arithmetic with lines and curves and shapes through reasoning. He thought of all the rational numbers as arranged in ascending order on a line segment, "rational numbers" being all those that can be written as a ratio of integers, as for example one-third or four-fifths. He then devised irrefutable arguments that if each rational number is represented exactly by one point, the line would be full of gaps, and he concluded that other points on the line must correspond to "irrational" numbers, that is, numbers not expressible as a ratio of integers. The modern theory of irrational numbers was invented in the nineteenth century by Richard Dedekind, but its essential ideas and line of reasoning were those of Eudoxus's theory. Modern thinkers concede that Eudoxus's thinking about the relation of numbers to points on a line was impeccable in its logic.

In connection with the "measure" for surfaces and solid objects, Eudoxus's pioneering work has also stood the test of time. He had proved by careful logic that the measure, i.e., volume, of a pyramid is equal to one-third the volume of a prism of equal base and altitude, and similarly that the volume of a cone is one-third that of a cylinder having the same base and height. One could test experimentally that his reasoning fit physical reality. The measure of a surface is an area, and Eudoxus is also credited with inventing a systematic method, the "method of exhaustion," that allowed him to begin with a curvilinear shape and find the equivalent square, one with the same area, thereby computing the "measure" of the surface of the curvilinear figure. The method was also reinvented and greatly elaborated two thousand years later, in the seventeenth century, and in its more evolved form became known as the integral calculus.

Around 1900 a new step was taken that went definitively beyond Eudoxus. In the wake of Cantor's theory of infinite sets, mathematicians like Borel and Lebesgue began to extend the idea of obtaining a measure from volumes, lines, and surfaces to encompass all "sets of points." Suppose you have an amount of sugar that measures exactly one cup, but then you go outdoors with the cup and scatter the sugar over a ten-acre field. Still, all those dispersed grains of sugar add up to one cupful, and there ought to be a mathematical way to recognize from the numbers, distribution, and size of the grains that their proper measure is one cup. Any geometrical figure is made up of mathematical points. A point is not quite analogous to a grain of sugar, because the ideal concept known as a "point" in mathematics has no length or volume at all. True, we often represent it crudely by a dot on a piece of paper, but in principle that dot should be made to shrink until it has zero radius. Euclid had given the classical definition of a point to be "that which has no parts." Consequently even a short line must contain an infinity of points—no finite number of points with zero length each could add up to anything else but zero total length. The question answered by Borel and Lebesgue was how one might "measure" sets of scattered points, an infinity of them, and find their equivalent in terms of a length or a volume or an area. For example, imagine a line one inch long made up of points representing in sequence all the numbers from zero to one. The new mathematical tools of Cantor, Borel, and Lebesgue enabled mathematicians to assign a length (i.e., a measure), the appropriate portion of the inch, to all the rational numbers on that line, leaving the remaining portion to all the irrational numbers. It turns out that the number of irrationals constitutes a higher order of infinity than the rationals, so that the measure of all the rationals is zero, while that of all the irrationals between zero and one is one. These ideas, while building on Eudoxus's thinking, went beyond it, so that one could now assign a measure, an equivalent length or area or volume, even to dispersed sets of points.

Norbert Wiener, as a young man, mastered the Borel-Lebesgue theory of measure, a theory which at that time was still a new and exciting development in mathematics in Europe.[3] He wanted to make some new applications or devise extensions of that theory, so he adapted it to answer different kinds of questions than had so far been posed. Later on these questions will be stated carefully, but for the moment I will give only a crude illustration suggesting the types of questions Wiener asked. Imagine that a fly is seen to be on the inside of a window at a certain spot at noon, but ten minutes later that same fly is sitting on the space bar of a typewriter on a table in the room. During those ten minutes the fly was buzzing around all over the room while nobody paid attention to it. What are the odds that the fly had touched the ceiling during the intervening ten minutes? Of course to ask about "odds" is to ask a question that traditionally falls within the theory of probability. After all, we have only incomplete information about the fly's trajectory, and probability theory is designed to describe situations with incomplete information mathematically.

In the type of question he posed, Wiener was in effect proposing to express questions in the theory of probability in terms of measure theory. The connection between the measures of an area and probability was already known: for example, suppose it is raining and we are interested in what the odds are for the next raindrop to fall into a region A or into a subregion B that lies within A. Imagine a sidewalk, and suppose there are no overhanging trees or other circumstances that would favor any one spot over another. It would not do to try to estimate the number of points in each region, for that number is infinite and of the same order in regions A and B. It does make sense to measure relative probabilities by comparing the size of the areas of A and B respectively. Thus the areas serve as the suitable "measure" of the sets of points. Concerning the problem of the fly, and the odds of its having touched the ceiling, Wiener translated the question of probabilities into the question, What is a proper measure of those

myriad possible trajectories that touch the ceiling compared to all the other possible trajectories? Instead of seeking
a measure for an infinite set of points, Wiener sought a measure for an infinite set of possible trajectories. When he succeeded, he had also extended the realm of probability theory to new types of situations, and he had demonstrated the
close connection between probability theory and Lebesgue
measure.

The Lebesgue theory of measure, together with the probability concept, was the most important tool needed by Wiener
to fashion his theory of the "Wiener Measure," but other ingredients as well were necessarily contained in his theory. One
of these was the available description of the empirical phenomenon of Brownian motion, which had come to scientists'
attention much more recently than the idea of measure. It was
noticed only after the microscope had been invented, when
Leeuwenhook (1632–1723) saw the phenomenon in a drop of
water under the microscope, recorded it carefully, but misin-
terpreted its nature. If one examines a collection of small particles suspended in a liquid, such as water, one finds that
"each one, instead of sinking steadily, is quickened by an
extremely lively and wholly haphazard movement. . . . Each
particle spins hither and thither, rises, sinks, rises again without ever tending to rest."[4]

This is what has since become known as the Brownian
movement, and Leeuwenhook understandably took the spontaneous motion to be a clear indication that he was seeing
very lively little animals.

Brownian motion seems to be a very small, minor phenomenon in the physical world, one that has very little effect on
anything else. Yet, quite out of proportion to its insignificance
in the physical world, it successively caught the attention of
biologists, physicists, and finally mathematicians and became surprisingly important in the history of science. The
Scotch botanist Robert Brown (1773–1858) devoted considerable effort to the systematic study of this motion. He put
under the microscope some plant pollen, "male sexual cells"
of plants. It was Brown's first hypothesis that the spontaneous
motion, which clearly defied gravity and occurred without

external forces, was the vital motion of spermatozoa. In England the educated general public took an interest in this phenomenon. But when Brown discovered that the same type of motion occurred with powder made from other parts of plants, he formed a new hypothesis about "vital motion," a hypothesis that was widely accepted: it is the universal vital motion of the *Urmolekül,* the basic molecule of life itself, which is contained in all living matter. It was consequently as interesting a subject for biologists in Brown's day as the substance DNA was for the biologists of the 1960s. The study of "Brownian" motion, which it was at first hoped would elucidate the fertilization process in plants and sexual reproduction, was after the second hypothesis seen as offering the possibility of insight into the nature of life itself. But alas, Brown soon came to examine inanimate materials, such as the scrapings from rocks, and when he put them into a drop of liquid under the microscope, he found the perpetual motion there as well. Do even rocks contain the vital *Urmolekül?* Good empirical scientist that he was, Brown at this point refrained from hypothesis and left the nature of Brownian motion unexplained.[5] A few scientists in the late nineteenth century continued to make careful observations of this vexing little phenomenon, in which perpetual motion of a highly irregular sort occurred without any detectable forces or currents to cause it.[6]

Everyone has at one time or another seen dust particles dancing around in a beam of sunlight. The dance of the dust particles is caused by small air currents and can be stopped by preventing them. Some scientists sought to dismiss Brownian motion as a similar phenomenon, presumably owing to the currents in the water under the microscope. Careful experimentation, however, also ruled out that explanation.

Meanwhile a sharp controversy had come to dominate the practitioners and philosophers of physical science,[7] between the "atomists" and the "energeticists." Clark Maxwell and a little later Ludwig Boltzmann were the two great expositors of atomism. Maxwell especially had developed in detail earlier models of gases and their behavior in which every gas is composed of atoms or molecules in constant motion (the kinetic theory). Moreover, the speed of motion of the molecules would be manifested through measurement of the temperature of the gas, and Maxwell was able to interpret various properties of gases in terms of molecules. Boltzmann had carried these ideas much further when he derived all of thermodynamics (a theory dear to the energeticists) on the assumption of the existence of atoms in constant motion. At the same time a number of leading European scientists, Ernst Mach, Wilhelm Ostwald, and Pierre Duhem the most prominent among them, rejected the atomic hypothesis. They found such a sweeping hypothesis as that of "invisible" atomic constituents and "invisible" incessant motion in fluids as scientifically unwarranted, the product of an undisciplined imagination. The leading energeticists were philosophical positivists who sought to make only a minimum of hypotheses in describing phenomena, and they especially resisted accepting mechanical pictures of submicroscopic events. In 1898 Boltzmann seemed unhappily convinced the energeticists had won the day, but as he wrote in the introduction to part II of his Gas Theory, "It would be a great tragedy for science if the theory were temporarily thrown into oblivion because of a momentary hostile attitude toward it. . . . I am conscious of being only an individual struggling weakly against the stream of time."[8] Science historian Brush goes so far as to attribute Boltzmann's later suicide (1906) to depression engendered by his deepening pessimism concerning the future of kinetic theory.[9]

Within a few years of Boltzmann's death, however, the seemingly minor phenomenon of Brownian motion would serve to vindicate the atomic theories and convince all the leading energeticists of the legitimacy of the Maxwell-Boltzmann atomic hypothesis. Experimental physicists examining Brownian motion at that time suggested that at least qualitatively this motion was explicable as owing to the bombardment of a solid material particle in a liquid by the molecules in the liquid. Since the Brownian particles observed through a microscope are still orders of magnitude larger than the atoms or molecules themselves,[10] they would according to the kinetic theory move much more slowly than the molecules of the liquid; moreover their motion would have to be just of the zigzag type *seen* in Brownian motion, because the particles would be buffeted by their many collisions with the molecules of the liquid.

A young man named Albert Einstein knew of the great controversy between atomists and energeticists, but he did not know about Brownian motion. He independently derived thermodynamics in terms of the motion of molecules (1902–03),[11] as he recalled later:

My major aim in this was to find facts which would guarantee as much as possible the existence of atoms of definite size. In the midst of this I discovered that, according to atomistic theory, there would have to be a movement of suspended microscopic particles open to observation, without knowing that observations concerning Brownian motion were already familiar.[12]

In his 1905 paper describing his theoretical discovery Einstein wrote,

If the movement discussed here can actually be observed (together with the laws relating to it that one would expect to find) . . . then an exact determination of actual atomic dimensions is possible. On the other hand, if the prediction of this movement were to be incorrect, a weighty argument would be provided against the molecular-kinetic conception of heat.[13]

Einstein made quantitative predictions about how far a suspended microscopic particle would travel in a given time, predictions that could be tested experimentally. Before the end of the year 1905, having been in contact with experimental physicists, Einstein could report that "not only the qualitative properties of the Brownian motion, but also the order of magnitude of paths described by the particles correspond completely with the results of the theory."[14] Experimenters intensified their investigations only to further and in detail corroborate Einstein's theory. Mach, Ostwald, and other remaining holdouts against atoms thereupon capitulated. Atoms were real, and the molecular theory of heat could not be denied. Understanding of Brownian motion had come a long way from being mistaken as the activity of living organisms.

Einstein's motive was clearly to vindicate or provide a convincing test for the molecular theory of heat. His argument and his mathematical method were as straightforward and as simple as possible; mathematics was a needed tool for Einstein, and he certainly had no interest in unnecessary mathematical extravagances. Along with others who had constructed theories of Brownian motion around that time,[15] he had been able to circumvent the necessity of discussing the complicated zigzag paths of Brownian particles. They had no mathematical tools adequate for describing such trajectories, and anyway they were not necessary for making simple experimental predictions.

Once the energetics-atomism controversy was convincingly settled to everyone's satisfaction, Brownian motion was hardly of any interest to physicists. However, to mathematician Norbert Wiener, who liked to tie his mathematics to physical phenomena, it was the nature of the trajectories that seemed intriguing. An experimental physicist, Jean Perrin, had described them,[16] and Wiener would later quote him:

The trajectories are confused and complicated, changing direction so often and so rapidly that it is impossible to follow them. . . . Similarly, the apparent mean speed of a grain during a given time varies *in the wildest way* in magnitude and direction, and does not tend to a limit as the time taken for observation decreases, as may be easily shown by noting, in the camera lucida, the position occupied by a grain from minute to minute, and then every five seconds, or, better still, by photographing them every twentieth of a second, as has been done . . . it is impossible to fix a tangent, even approximately, and we are thus reminded of the continuous underived functions of the mathematicians. It would be incorrect to regard such functions as mere mathematical curiosities, since indications are found in nature of "underived" as well as "derived" processes.[17]

Wiener saw that the well-documented physical phenomenon of a Brownian particle observed first at one location and a little later at a different point is like the phenomenon of the fly moving from the window to the space bar of a typewriter. What occurs in the period between observations is not experimentally known. The phenomena could be fully described mathematically only if one could attach a measure to bunches of trajectories. Consideration of Brownian motion allowed Wiener to make concrete and specific his effort to extend the Lebesgue measure concept from sets of points to sets of trajectories. The experimental observation that the individual trajectories vary "in the wildest way" would only add piquancy to the mathematical problem.

Norbert Wiener was one of those mathematicians for whom contact with actual phenomena in physics, engineering, or biology would sometimes play a fruitful role in protecting his mathematics from becoming empty and artificial. It is in this way that the phenomenon of Brownian motion focused his mathematics. As he was to discover, he liked to work "in the middle ground where physics and mathematics meet . . . for [he later said] such work seemed to be in harmony with a basic aspect of my own personality."[18] He had begun working on the idea of extending the theory of the Lebesgue measure shortly after he had joined the mathematics department at MIT as an instructor in 1919. The connection between phenomena and mathematics in Wiener's thinking is poignantly illustrated in his musings concerning the Charles River in 1919, which have in them the makings of a minor legend but were reported by Wiener himself:

It was at MIT too that my ever-growing interest in the physical aspects of mathematics began to take definite shape. The school buildings overlook the River Charles and command an ever changing skyline of much beauty. The moods of the waters of the river were always delightful to watch. To me as a mathematician and a physicist, they had another meaning as well. How could one bring to a mathematical regularity the study of the mass or ever shifting ripples and waves, for was not the highest destiny of mathematics the discovery of order among disorder? At one time the waves ran high, flecked with patches of foam, while at another they were barely noticeable ripples. Sometimes the lengths of the waves were to be measured in inches, and again they might be many yards long. What descriptive language could I use that would portray these clearly visible facts without involving me in the inextricable complexity of a complete description of the water surface? This problem of the waves was clearly one for averaging and statistics, and in this way closely related to the

Lebesgue integral, which I was studying at the time. Thus, I came to see that the mathematical tool for which I was seeking was one suitable to the description of nature, and I grew ever more aware that it was within nature itself that I must seek the language and the problems of my mathematical investigations.[19]

Wiener did not analyze the Charles River until later, but he had hit upon the theory of Brownian motion as a suitable physical situation in connection with the extension of the Lebesgue measure he was seeking.

Aside from the "measure" concept, the idea of probability, and the available descriptions of Brownian motion, another idea entered as an important element into the invention of the Wiener process. This is the idea of a path that zig-zags so much that it is practically nothing but corners, what mathematicians refer to as a function that is everywhere "undifferentiable." The notion of such a path or function is an idealization, just as the mathematician's point, with its zero radius, smaller than any mark that can be made on a piece of paper with a pencil, is an idealization. The concepts of geometry, such as point and line, although usually visualized in terms of crude approximations like pencil marks on paper, are actually abstractions such as the point with zero radius or the line of zero width. Nevertheless, features of the real world, the subject of interest to physicists and engineers, can often be elegantly described in terms of such mathematical abstractions. Wiener would add the "nowhere differentiable" function to the list of ideal geometric concepts useful in physics and engineering.

Continuous paths were envisaged by some nineteenth-century mathematicians, and it was demonstrated explicitly that one could give recipes for constructing paths in which the number of sharp corners becomes so great that there is practically no point on the path at which it has a unique slope (mathematically a tangent)—the path is all corners and everywhere "undifferentiable."[20] We can picture a path that zigzags a great deal and has many, many corners, but a path that is all corners defies visualization as much as the point with zero radius. Nevertheless we can describe such a path mathematically, how it would be constructed in principle, and what its mathematical properties would be. It is these strange paths that would play a fundamental role in Wiener's theory. In his theory these paths describe in an idealized way the motion of Brownian particles.

From the point of view of mathematics per se, it is of interest that such paths can be constructed in principle, that there is no contradiction in the definitions and properties describing them. From the point of view of pure mathematics it does not matter in the least whether such paths resemble anything in the real world, or whether we can picture them in our mind's eye. "A mathematician . . . is a maker of patterns, patterns . . . made with ideas," Norbert Wiener's teacher G. H. Hardy had said.[21]

These nondifferentiable functions and other kinds of "freakish" functions came to interest mathematicians in the latter half of the nineteenth century.[22] Until that time mathematicians had generally relied on their geometrical intuitions to disregard the possibility of all such strange functions. The new strange functions brought about a crisis in mathematics, because it was found that the general methods of mathematical analysis existing at the time could not be applied to the newly discovered functions. This was very disturbing. In the nineteenth century mathematician Charles Hermite expressed a typical reaction in a letter to his colleague Thomas J. Stieltjes: "Je me détourne avec effroi et horreur de cette plaie lamentable des fonctions qui n'ont pas de dérivées."* Reviewing this general reaction, a later mathematician observed, "Researches dealing with non-analytic functions and with functions violating laws which one hoped were universal, were regarded almost as the propagation of anarchy and chaos where past generations had sought order and harmony."[23] The essence of the crisis consisted of the awareness that many of the tools of classical mathematical analysis could lead to erronous conclusions, although mathematicians had supposed that these methods were universally valid.[24]

---

*"I turn with fear and horror from this lamentable plague of functions that have no derivatives."

Henri Lebesgue recalls his first awareness of a paradox generated by the classical methods, in effect a proof that 2=1 (see figure 1):

Formerly, when I was a schoolboy, the teachers and pupils had been satisfied by passage to the limit. However, it ceased to satisfy me when some of my schoolmates showed me, along about my fifteenth year (1890), that one side of a triangle is equal to the sum of the other two, and that $\pi=2$. Suppose that ABC is an equilateral triangle and that D, E, and F are the midpoints of BA, BC and CA. The length of the broken line BDEFC is AB + AC. If we repeat this procedure with the triangles DBE and FED, we get a broken line of the same length made up of eight segments, etc. Now these broken lines have BC as their limit, and hence the limit of their lengths, that is, their common length, AB + AC, is equal to BC. The reasoning with regard to π is analogous. . . .

This result has been extremely instructive to me. . . . The preceding example shows that passing to the limit for lengths, areas, or volumes requires justification, and . . . it is enough to arouse all one's suspicions.[25]

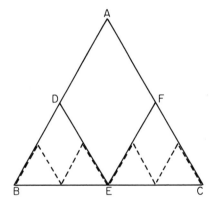

Figure 1

The crisis in mathematical analysis led to a new insistence on greater clarity and rigor in mathematics and a radically new formulation of the foundations of the subject. The two intimately related major innovations that emerged to help clarify analysis and provide the tools to end this particular crisis were the previously mentioned general theory of measure of sets and the Lebesgue theory of integration (1907), which was based on the theory of measure of sets (see note 3).

A mathematician is a maker of patterns, patterns made with ideas. We have noted the background of ideas themselves: the idea of assigning a number (i.e., a measure) to things as it evolved from Eudoxus to Lebesgue; the idea of incomplete information about an event translated into probabilities and numerical odds; the knowledge of the phenomenon of Brownian motion from the work of Robert Brown to that of Albert Einstein; and the nineteenth-century invention of the concept of a trajectory that nowhere has a unique tangent. How now did Wiener proceed to devise a significant theory from this group of ideas?

He began with the notion that it would be mathematically interesting to develop a probability measure for sets of trajectories, an extension of probability theory to such problems as that of the fly on the window. Moreover, it was clear to Wiener that some variant of Lebesgue measure would provide the appropriate mathematical tool. Several other mathematicians had already been exploring such variants.[26] A prototype of the kind of problem Wiener considered is that of the drunkard's walk: a drunk man is at first leaning against a lamp post; he then takes a step in some direction—it may be a short step or a long step; then he either stands still maintaining his balance or takes another step in some direction; and so on. The path he takes will in general be a complicated path with many changes in direction. If we made a graph of his position along the east-west direction against the time elapsed since he started at the lamp post, we would expect a very irregular zigzagging graph describing the man's trajectory. Assuming the man has no a priori preference for any particular direction or particular step size and may move fast or slowly according to his whim, is there some way to assign a probability measure to any particular set of trajectories? If so, how can we then use the probabilities to compute averages, to answer a typical question such as, How far away from the lamp post is the man likely to get in one minute? Here a probability measure will have to be assigned not only to sets of points or to areas but to sets of alternative paths.

If one made a graph of the man's *various possible* distances from the post in the north-south direction at various times, one would generate many such zigzag lines describing trajectories. (See figure 2 for a sketch of a few sample trajectories.) Each zigzag line is a geometrical picture of what in mathematical jargon is called a function—it relates a point on the horizontal axis to a point on the vertical axis, in this case given by the height of the zigzag line at that point. There exist quantities that are different for different trajectories, such as the length of the trajectory, the number of times it crosses the horizontal axis, or how far north it goes in thirty seconds. Since these quantities depend on the whole zigzag line or function one chooses, they are functions of trajectories or "functions of functions," not merely functions of a point. The mathematical term for a function of a function is a functional. Wiener proposed to formulate the idea of the "average of a functional" and set up a mathematically correct method of computing this average.

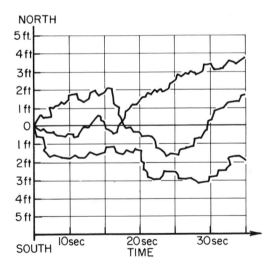

Figure 2

In the older probability theory one had rules for computing averages of functions, but there were no ways to compute the average of a functional. To take an average of a functional meant that in some way one would have to average over all possible trajectories; this required that a probability be associated in some way with sets of trajectories. Anyway, how was one to give a precise meaning to the notion of "all possible trajectories" that start from zero? This was the problem Wiener set himself, to formulate a method of calculating the average of a functional. In his first article on the subject in 1920 Wiener generalized the concept of measure sufficiently so that he could define an average over possible paths.[27] Although he was apparently successful in achieving this generalization in a formal way, it was not clear to what extent his generalization was really meaningful. It seemed to Wiener himself at the time that the work was "artificial and unsatisfying."[28] The trouble was that he had no criteria for assigning probabilities to particular sets of paths, other than that the assignments must not lead to mathematical contradiction; it was to avoid contradiction that in his formulation all trajectories that doubled back on themselves were given negligibly small probability (see figure 3, which shows two examples from all possible paths).

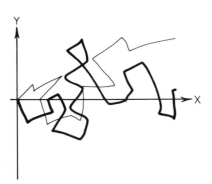

Figure 3

How was one to find a natural and satisfying formulation? For Wiener this came about through his appreciation of the possibility that mathematics could describe natural phenomena. He might have chosen the "trajectories" in the time series generated by the varying price of a stock in the stock market, or he might have selected the drunkard's walk or the fly on the window as the instance to make his mathematics more concrete. But he wisely chose to deal with the phenomenon of Brownian motion and the rather convenient description of it that Einstein had provided. Through this choice Wiener established a link between Lebesgue measure and the whole field of statistical physics.[29]

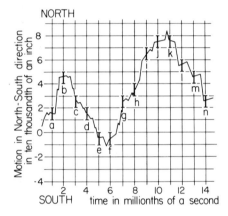

Figure 4

To enumerate the possible trajectories, Wiener in effect considered a graph of the position of the particle in the north-south direction versus the time, as it might be drawn on a piece of graph paper (fig. 4). The particular trajectory drawn on the figure goes through gate a at time one (millionth of a second), through gate b at time two, through gate c at time three, and so on. We can lump together as one group of trajectories all those one might draw that go through all the same gates. This provides a systematic way of dividing the possible trajectories into groups. If the subdivisions in the graph are made finer and finer in both the vertical and horizontal directions, the enumeration of the specific gates through which a trajectory passes would seem to specify a trajectory practically completely. The probability or measure Wiener assigned to a particle having a trajectory going through a particular set of gates was chosen by him to be just the probability that would result if one used the same physical picture of Brownian motion that Einstein had used.[30] This is a good beginning, but the burden of proof was on Wiener to show that this definition of probability constituted a self-consistent definition, that it satisfied the rules of probability (i.e., the same rules for combining probabilities that apply to throwing dice and to the raindrops on the sidewalk considered earlier). The number of possible trajectories is beyond counting; in this respect it is similar to the number of points on a line (or the number of points in the area in the raindrop problem). Wiener showed that it is possible to label each of the trajectories that contribute to the process by a different number, and that the total number of these trajectories is equal to the number of points on a finite line; he then used the length of the equivalent finite line as a "measure." By basing his measure on the physical model of Einstein, Wiener implicitly prevented many conceivable kinds of paths from contributing to the total. This was crucial for, as was proved later, if these paths are not eliminated, it is no longer possible to define a measure for the paths without self-contradiction.

What kinds of trajectories do contribute to the Wiener (Brownian motion) process? Which ones are not eliminated? This may come as a bit of a surprise: Wiener proved that the whole contribution (except for a set of trajectories of measure zero) comes from trajectories that are continuous but *nondifferentiable*. In other words, the trajectories are those freaky mathematical curves that are nothing but corners. Moreover, with all the zigs and zags the trajectories become infinitely long in a finite time.[31]

The nondifferentiable trajectories are an idealization of whatever the actual trajectories are; they are sufficiently close to the truth to yield results in agreement with experimental observation of Brownian particles. To understand physical nature, idealized models are often more helpful than more complicated exact descriptions, although the latter are also interesting. For example, if he wishes to compute the orbit of the earth around the sun, a physicist is likely to use the idealized model of the sun and the earth in which each is concentrated at one point instead of having a finite size. For other calculations he will regard the earth as a perfect homogeneous sphere, again an idealization that for many purposes leads to correct results. If one were sufficiently far away from the earth and the sun, these bodies would in fact appear indistinguishable from a point or a homogeneous sphere. On the other hand, if one is interested in the shape and variation of density of the earth itself, a much more elaborate description will be required. The nondifferentiable trajectories of the Wiener process leads to correct results for all the averages describing Brownian motion. These trajectories are the "point" or the "perfect sphere" of the Brownian motion. The time scale for observing Brownian particles makes their motion indistinguishable from the nondifferentiable functions. It is possible to refine the Einstein-Wiener physical picture, so that it becomes correct also for much briefer intervals between observations than are physically possible. If one does so,[32] one finds that the trajectories that contribute are differentiable but nevertheless highly irregular and mathematically strange curves.[33]

Wiener's work (1920–23) did not teach the physicists much about Brownian motion that they didn't know already.[34] It did, however, open up a whole new field of investigation in mathematics, which ultimately would find applications in many branches of physics, engineering, and biology.[35] It initiated the mathematical theory of stochastic processes and solved its central problem, and it developed the fundamental new technique of functional integration. It also set one important precedent in that it was one of the first instances in which a problem in the molecular theory of heat was put in terms of the measure theory of Lebesgue[36] (a precedent to be followed up in the 1930s by the development of ergodic theory by Birkhoff, von Neumann, and others) and set another precedent in formulating a problem in probability theory in these terms (a precedent to be followed up by the formalization of probability theory in terms of measure theory in 1933 by Kolmogoroff).

The fate of Wiener's work was to be rederived by other mathematicians and also by himself, by different and sometimes shorter and more elegant methods; to be rewritten in very different notation; to be generalized and to be imbedded in a formal scheme of postulates. The probability for groups of Brownian trajectories came to be known as the Wiener measure, and the expressions for averages over Brownian trajectories became known as Wiener integrals. It came to be realized that the Wiener process applied to many problems other than Brownian motion. Recently it has become a favorite tool of quantum field theorists. The averages for many different kinds of quantities, i.e., different Wiener integrals, were later calculated in detail by other mathematicians. One of the mathematicians who extended Wiener's work on Brownian motion pays tribute to that element in Wiener's work that most impressed him:

In retrospect one can have nothing but admiration for the vision which Wiener had shown when, almost half a century ago, he had chosen Brownian motion as a subject to study from the point of view of integration. To have foreseen at that time that an impressive edifice could be erected in such an esoteric corner of mathematics was a feat of intuition not easily equalled now or ever. . . . It gave us not only a new way of looking at problems but actually a new way of *thinking* about them.[37]

But then, perhaps, it was more than mathematical intuition that moved him to choose the problem.

We can mechanize insofar as we can make a formal rule.
—M. Taube

The games people play in any particular civilization have
something to do with the realities of life in that civilization. In
some instances one finds particular elements of the realities
of life picked out, mimicked, translated, and reduced into the
form of a game. Thus the ancient Chinese game of Go contains features of a model of war. The much later game of
chess,[1] which is believed to have originated in India where its
pieces had names such as "elephant," "horse," "chariot," and
"foot soldier," was apparently also a game of military strategy.
The festive Olympic Games of ancient Greece, described by
Herodotus, constituted a veritable microcosm manifesting
not only the high achievements of that civilization but also
the competition between nations—some at war with each
other—in spite of the vision of a united Greece. They also
displayed the Greeks' love of beautiful men.

Games can be grist for the mathematician's mill. Whereas
games already mimic some elements of social reality, the
mathematician will carry abstraction from reality one step
further when he reduces elements of a game to precise mathematical formulation. From the point of view of mathematics,
the connection to reality will have served its purpose if it has
stimulated the creation of interesting new mathematics. For
the practical man it of course has other purposes.

In the instance of the French mathematician Blaise Pascal
(1623–1662), a pioneer in the mathematical study of games of

chance, the stimulus from practical gambling came through a friend. As the philosopher Leibniz described it,

Chevalier de Méré . . .—a man of penetrating mind who was both a gambler and philosopher—gave the mathematicians a timely opening by putting some questions about betting in order to find out how much a stake in a game would be worth, if the game were interrupted at a given stage in the proceedings. He got his friend Pascal to look into these things. . . . Other learned men took up the subject. Some axioms became fixed.[2]

So it was that the game of dice became an intellectual preoccupation in the middle of the seventeenth century among men of the caliber of Pascal, Fermat, Leibniz, and Huygens, but in the history of the mathematical theory of probability it is Pascal who is particularly significant.

Pascal is also remembered for his oft-quoted remarkable saying that "the heart has its reasons whereof reason knoweth not,"[3] which asserts that passions and feelings are not irrational, but possess a rationality of their own.

He is a forerunner of von Neumann in that he was, at the age of 19, all but the first modern to invent and patent a calculating machine to carry out arithmetical operations;[4] again it was Leibniz who improved upon and extended the operations of the Pascal computer.

The style of reasoning characteristic of the modern game theory associated with von Neumann is somewhat different from that of merely calculating odds in gambling. As Ian Hacking points out, Pascal had also anticipated the twentieth-century game-theoretic style of reasoning, but in connection with a philosophical argument rather than the gaming table. In fact, "Pascal's wager is the name given to some game-theoretic considerations that concern belief in God."[5] Pascal considered the decision problem, given that we have no way of knowing whether God exists, is it more advantageous to act on the premise that God exists, or on the opposite premise that God does not exist? He suggested a method of reasoning to arrive at a wise decision in the face of an agnostic position concerning God's existence and thereby invented a pattern for structuring decision problems of all kinds in which one has

incomplete information. In his analysis Pascal makes various
assumptions: in particular he assumes that if one acts on the premise that God exists, one will eventually come to believe in God; and that the nature of God, if he exists, is that described by the Catholic Church. He considered the advantages of a libertine's life as opposed to a pious life, the libertine's life being perhaps preferable if there is no God. This had to be balanced against the infinite benefit of the salvation that would result from a pious life if God exists. On balance, Pascal, the apologist for the Church, concluded by his game-theoretical reasoning that in the face of human uncertainty as to God's existence the pious life is the more advantageous. Of course one can seriously question Pascal's whole approach, viewing it as "a little indecent and puerile," as Voltaire did: "The idea of a game, and of loss and gain, does not befit the gravity of the subject."[6]

One crucial aspect of modern game theory missing from Pascal's wager is the presence of a competitor, an intelligent opponent. Modern game theory is remarkable in that in its description of a game, for example the game of checkers, all possible moves of the opponent are taken into consideration in making one's own move. Competitive games, whether athletic games or board games, mimic the competitive elements of social reality. In the late eighteenth and the nineteenth century, the notion of competition was central to the ideas that dominated the intellectual formulations of the nature of life and social reality,[7] and unmitigated seeking of one's genuine self-interest and advantage received intellectual sanction. When the seminal political economist Adam Smith identified economic gain, or more generally self-interest, as the universal motive in civilized society and put forth his theory of the "invisible hand" according to which the free competition among individuals each pursuing his own gain leads to the maximum benefit to society, he had provided the intellectual foundation for a capitalist economic system.[8] Moreover, free rivalry among competitors had been extolled. Charles Darwin, who was well acquainted with Adam Smith's ideas, pointed to the universality of competition for food and territory among organisms, a competition most severe between individuals of the same species. While the struggle for existence of which he spoke may involve complex interrelations, Darwin gave examples of the kind of prototype on which that metaphor is based—"Two canine animals, in a time of dearth, may be truly said to struggle with each other which shall get food and live."[9] Here is a case of "pure competition" within the context of a struggle for survival. Although Marxian political economy would provide an alternative to Adam Smith's view of economic competition, the basic outlooks of Darwin on the struggle for existence and of Adam Smith on economic competition have continued to be prominent well into the twentieth century. In the military field the choice of tactics and strategy in the face of an enemy or rival nation had in the nineteenth century been given sharp although not mathematical formulation by the Prussian military officer, Karl von Clausewitz,[10] a formulation that dominated military thinking until World War II. Thus the universality of competition was widely accepted as a fundamental element of social reality.

We have spoken of Pascal's wager as a distant ancestor of von Neumann's game theory and incidentally noted Pascal's invention of a computer. In the nineteenth century we find a more immediate forerunner of von Neumann in connection with both computers and game theory in the person of Charles Babbage, an Englishman. Babbage is primarily remembered today for his design of two computing machines far ahead of their time. The one he actually built was designed as an aid for calculating tables, but the one he envisaged, planned out in detail, and tried very hard and unsuccessfully to finance, was a mechanical counterpart of the kind of general-purpose electronic computer that von Neumann had helped to design a century later. It is noteworthy that the kinship between Babbage's and von Neumann's ideas extend to making models of games, but there is a minor difference: Babbage thought in terms of a mechanical device that would automatically make good moves in a game, while von Neumann thought in terms of an abstract mathematical formalism that would inform either the human player or a machine of the best moves. In his own words, Babbage was interested in "the contrivance of a machine that should be able to play a game of purely intellectual skill successfully; such as tit-tat-to, drafts, chess, etc."[11] It would test some of the principles he had laid down for his general-purpose computer and might provide a convincing demonstration for the purpose of fund-raising.

He first examined the possibility of constructing such a game-playing machine in principle and came to the conclusion that "every game of skill is susceptible of being played by an automaton. Further consideration showed that if *any position* of the men upon the board were assumed (whether that position were possible or impossible), then if the automaton could make the first move rightly, he (*sic*) must be able to win the game, always supposing that, under the given position of the men, that conclusion were possible."[12] This is an early assertion of the kinds of general results that von Neumann would later formulate and prove with mathematical precision. Note that Babbage was concerned with strictly determined competitive games, in contrast to mere games of chance. Of course in mechanizing game playing the focus is on winning, not on the pleasure of playing the game. For the reader who plays games for fun rather than to win, there is something ludicrous about inventing a robot to play in your place. This very absurdity highlights the observation that in devising mathematical (or mechanical) formulations only certain elements of reality are caught—the objective of winning, but not the pleasure of playing. As consolation, we note that the robots may be fun, or have fun, too.

The first step in a proper mathematical theory of games is to provide a suitable description of games in mathematical language. Such a description must contain all the necessary information concerning any game but should contain no irrelevant information. Irrelevant information would impede insight into the mathematical problem. But one can devise such a suitable description only after one is very clear about the mathematical problems that one wishes to pose in connection with games. This was first done by the French mathematician Émile Borel in 1921: For particular classes of games, does a winning (or definitely nonlosing) strategy exist? If it exists, how do I discover the strategy? With these questions clearly in mind he knew what information is required about each game, and he could and did describe games in mathematical language.[13] But was it mathematically interesting? For Borel it was only a mathematician's hunch that a profound mathematical theory would emerge from tackling these questions. He did not know. However, already in his first article on games in 1921 Borel suggested that game theory has certain analogies with problems of military strategy and economics, although admittedly these are far more complex than parlor games. But he warned that the application of game theory to "the art of war" would lead to the recommendation of unsatisfactory strategies that ought to be modified to take into account "the knowledge of the psychology of the adversary." Incidentally, Borel had been a legislator (member of the Chamber of Deputies) in France and in 1925 became minister of the navy.

Von Neumann, to his chagrin, sometimes had to rely on others for the first intuition and suggestion of a problem, although he would outshine others in its systematic formulation and solution. He later claimed that he had quite independently of Borel arrived at a similar stage in his thinking about parlor games, and he well may have, though he published nothing on the subject until 1928.[14] The great contribution of his 1928 paper, "Zur Theorie der Gesellschaftspiele" ("Theory of Parlor Games") is that he provided answers to the questions Borel had posed, answers that went beyond the questions.[15] Moreover he obtained these answers by proving a difficult new theorem, the minimax theorem, which turned out to be so profound that it opened up new areas and manifested new connections within mathematics.

In the remainder of this chapter I shall describe something of the Borel–von Neumann game theory, beginning with the language through which an actual game is represented formally in game theory, after which I shall proceed to a verbal statement and explanation of the content of von Neumann's minimax theorem. Finally I will discuss von Neumann's advice on how to win at stud poker. Von Neumann's proof of the minimax theorem is highly technical and will not be discussed here.[16]

Which features in describing a game are relevant for the
purpose of finding a good strategy of play? Which features should the mathematical description pick out and which should it ignore? Consider a card game: The cards are shuffled; the deck is cut; the cards are dealt face down, each player receiving so many cards; then each player picks up his cards and looks at his hand, he looks at the other players' faces, and so on. This is how a chronologist would describe the course of events in a card game, but for the mathematician such a physical, chronological description obscures the essentials. What the mathematician finds essential is that each player can decide on any of a large variety of policies in playing his cards. Which policy he chooses and which policies each of the other players chooses will determine how much he is likely to win or lose, on the average.[17] Therefore the theory must list all the strategic options open to each of the players, and what we need to know is what the net gain or loss is on the average for each player when a particular policy is followed by each. That is all we need to know about the game; the rest is frills, dross that would only get in the way.

Suppose we wish to describe a game for two players, and the net result after each round of the game is that the loser pays some money or gives points to the winner. This is the situation especially considered by Borel and by von Neumann. Von Neumann labeled it the "zero-sum two-person" type of game. The phrase *zero-sum* means that the sum of the assets of all the players does not change, because one person's loss is the other's gain. It is a paradigm for a strictly competitive situation, which is in fact a frequently encountered situation in games. Darwin, in the remark earlier quoted, describes a zero-sum two-dog "game"; "Matching Pennies" is an example of a zero-sum two-person game. One can also count such games as ticktacktoe, checkers, chess, and nine men's morris as zero-sum two-person games by stipulating that the winner receives a point for each round won and the loser loses a point (or gets a negative point) for each round lost. For a zero-sum two-person game *all* the needed information can be conveniently arranged in a rectangle, sometimes a square, and the first step in analyzing any zero-sum two-person game is to arrange all the information in such a rectangle, which is called the "normal form" of the game.

I will illustrate the normal form of a game by a simple example. Suppose we have two players, Abigail and Boniface. We suppose that within the rules of the game each party can follow only three possible strategies, or game plans. We systematically designate the three options open to Abigail as A_1, A_2, and A_3. Similarly, the three options available to Boniface will be labeled B_1, B_2, and B_3. We then work out, from the rules of the game, the net gain or payoff to Abigail for each possible case, and insert the value of the payoff into the chart.

	B_1	B_2	B_3
A_1	−8	0	−4
A_2	4	10	−2
A_3	0	−5	−3

Thus from the chart we see that if Abigail adopts strategy A_1 and Boniface has decided on strategy B_1, Abigail is going to lose $8.00, or eight points. Since it is a zero-sum game, this implies that Boniface wins $8.00 or eight points. If Abigail chooses A_1 but Boniface opts for B_2, then both come out even, and so on. Except for mostly requiring a larger number of rows and columns, other zero-sum two-person games are represented in an entirely analogous way.

Having reduced the game to its normal form, we identify ourselves with one of the players and ponder the merits of alternative strategies. The normal form permits us to see the possible consequences of any particular choice at a glance. Note that in the example if Boniface follows strategy B_1 then the best choice for Abigail is A_2, because she gains four points, while otherwise she only comes out even or loses eight points. If Boniface chooses policy B_2 or B_3, inspection of the chart shows that for Abigail the strategy A_2 is again the most advantageous. Consequently the advice to Abigail is clear: Follow strategy A_2—it is in every case the best choice. The situation for Boniface is not quite so obvious, because his choice should depend on what policy Abigail follows, but he has no way of knowing which policy that is going to be. Suppose he follows strategy B_1; he might be lucky and win eight points or be unlucky and lose four—it is a gamble. Strategy B_2 is definitely a worse choice, because he risks losing ten points and can win at most five. Finally, strategy B_3 is low-risk: he wins something no matter what Abigail does but only modest amounts. While he doesn't know what Abigail will do, he reasons, "Since Abigail is a smart lady with a healthy self-interest, chances are she realizes A_2 is her best strategy, and I'd better be prepared for it." Thus Boniface confidently and cautiously opts for strategy B_3. With this reasoning and with Abigail following her best choice, the result is that Abigail loses two points to Boniface.

Game theory is more interesting if it comes out with definite answers. In the example just given, Boniface arrived at a choice of policy by assuming his opponent had considered all her options and selected the one that would most benefit her, and in fact we assumed that Abigail did just that. This led to a sound choice for both parties and a definite result for the payoff. To make two-person, zero-sum game theory yield definite solutions, von Neumann also posited that both players act on the premise that the opponent has examined the payoffs in each case and acts in his or her self-interest. It led to a definite result in our example. Will it do so in every case, or are there exceptions? This is a mathematical question of considerable generality.

Before presenting von Neumann's answer to this general question, it is first necessary to state more precisely what is meant by the concept of a strategy. What is meant is a complete plan of action for all contingencies that may arise. For example, in a game of ticktacktoe the first player might describe one strategy as: "I begin by putting my X in the upper right-hand corner; then if the opponent on his first move puts his O in the central square I put my next X in the middle square in the top row; if the opponent on his first move puts his O in the opposite corner from my X, I put my second X into the other corner in the top row; if the opponent puts his first O anywhere else, I put my second X into the central square; on my third move, if the opponent has put his first O into the central square and his second O into the lower left-hand corner, I will . . . ," and so on. In game theory it is assumed that both players have examined all the strategies before the game begins and have also decided which strategy to follow. Consequently the playing itself is completely mechanical and predetermined. All contingencies have been anticipated.

In the general zero-sum two-person game that interested von Neumann, we can imagine the normal form written down, with each of the strategies for one player (player A) corresponding to a particular row and each strategy of the opponent (player B) corresponding to a column. Each row has a minimum value, which measures the worst that can happen to A if he chooses that row. In my little example, the first row has the minimum value −8, the second row the minimum value −2, and the third row is −3. If Abigail chooses the second row, as she does in the example, the worst that can happen to her is the least bad outcome. If she chooses either of the other rows the worst would be worse. In mathematical language one would say that the second row has the "largest minimum" of the three rows. For Boniface the choice of column three corresponds to the column in which the worst that could happen to him is least bad. This expresses itself numerically in the fact that it is the column that has the "smallest maximum." The net payoff in our solution, −2, is simultaneously that smallest maximum of any column and the largest minimum of any row. They coincide, and their point of coincidence marks the solution of the game. In the language of game theory, −2 is the "value of the game."

Von Neumann proved in 1928 that for a large class but not all zero-sum two-person games, the largest minimum of any row coincides with the smallest maximum of any column, and that then the result of two cautious, self-interested, calculating, rational opponents engaging in a game will result in a *unique "value of the game" which is in fact both the smallest maximum of any column and the largest minimum of any row.* This is an informal statement of von Neumann's first minimax theorem. It also implies one or more best sound strategies for each party to follow. The class of games to which von Neumann's first theorem applies is all those games in which both players have at each moment full information about the position reached in the game so far. Ticktacktoe is such a game. In a game of chess or checkers, both players have complete information unless some portion of the playing board is hidden from the view of one of the players. For such complicated games as chess, it would be an enormous task even for a computer to enumerate all comprehensive strategies and write down the game in normal form, to look for the coincidence of smallest maximum and largest minimum and thus pick out the sound strategies. If this task were ever to be carried out for the game of chess, it would make chess tournaments a thing of the past. Nevertheless the theorem applies even to chess.

Von Neumann's second theorem is a generalization of the first. It deals with all the zero-sum two-person games not covered by the first theorem, namely the games with incomplete information. His second theorem shows that for these games a unique value of the game exists *on the average*, and that a sound policy can be recommended to each player who plays many rounds of the game, but no best strategy is available for playing only one round.

A simple example of a game without perfect information is "paper, scissors, stone." Here neither player knows what the other is going to do. Both players simultaneously show either a fist (stone), a flat hand (paper), or two fingers (scissors). If both parties show the same symbol, it is a draw; otherwise scissors beats paper, paper beats stone, and stone beats scissors. To make it numerical, let us say each time the loser gives the winner one point. The normal form for a single round of the game looks like this:

	B (scissors)	B (paper)	B (stone)
A (scissors)	0	1	−1
A (paper)	−1	0	1
A (stone)	1	−1	0

Analyzing this normal form, it is seen that the largest minimum of any row is −1, while the smallest maximum of any column is 1. So in this case it would appear that the two are not equal.

This is typical of what may happen in games without complete information. In this example it seems obvious from common sense that if one plays only one round of this game there is no reason to prefer any of the three alternatives to any other. But suppose A and B play many, many rounds. If A is cool, conservative, and eager to win, he'll want to play in a way that shows no pattern that B could easily recognize: what policy should he follow? Von Neumann's approach to this type of game was, following Borel, to use a "mixed strategy" or "random strategy," sometimes stone, sometimes paper, sometimes scissors. To determine which to choose in a particular round, von Neumann recommends using some random device. In view of the symmetry of the game, it is clear that if B adopts a sound strategy, A cannot on the average expect to win. At best he can expect to come out even in the long run, and the sound strategy for assuring that is to mix up the three choices in equal proportion. Of course one might do better if one is sensitive to the psychology of one's opponent, and if the latter is not using the sound strategy, but such psychological considerations are excluded from game theory.

Von Neumann's general minimax theorem, applicable to every zero-sum two-person game, whether one has perfect information or not, states that the largest minimum of the average payoff to A coincides with the smallest maximum of the average payoff from B, and this value is the unique average value of the game. The theorem differs from the first theorem by the necessarily repeated insertion of the word *average*, which reflects the importance of the element of mixing up various strategies randomly in the case of incomplete information.

There is another difference; since in games of perfect information, the normal form indicates that if one player deviates from the sound strategy but the other sticks to it, the one who deviates will definitely lose. Not so in games of incomplete information. For example, for stone, paper, scissors, if A deviates from the sound strategy (e.g., always plays "stone"), while B sticks to his sound strategy, they will still come out even on the average, so that A is not at all punished for deviating. However, B might possibly take advantage of A's

policy by himself changing from the sound policy to one
adapted to A's pattern (e.g., always play "paper"). At that moment, of course, B will have given up his conservative style of play, and A might do well to take advantage of it. This would take them outside the realm of game theory.

Von Neumann's general minimax theorem is noteworthy for its great generality and simplicity. His proof was a tour de force. The minimax theorem is a statement belonging to algebra, dealing with the existence of a solution for a certain set of equations and inequalities. However, von Neumann did not find the means to prove the theorem altogether within that branch of mathematics. Instead he made use of topology, a branch of mathematics dealing with the mapping of one shape or set of points onto another. Von Neumann's proof is complicated; many years later other mathematicians as well as von Neumann himself found different and simpler proofs, as is usual for important theorems.[18] An interesting feature of the von Neumann proof is that it showed a close connection between the algebraic minimax theorem and certain theories in topology (namely, the theory of fixed points). In fact the minimax theorem was later used by von Neumann to extend the important Brouwer fixed-point theorem in topology.[19]

While the original stimulus for even considering the minimax theorem had come from thinking about zero-sum two-person games, the resulting theorem is of general mathematical interest quite aside from its interpretation as a means of finding sound strategies for games. Many another mathematician might have left matters there, with the proof that a unique value of the game and sound strategies always exist, but not von Neumann. He set out actually to mathematically discover a "sound strategy" for a simplified version of stud poker—a far more interesting game than any we have described so far.[20] The first step for such an analysis is the invention of a game that is simpler to analyze than stud poker but nevertheless retains its main elements, especially the possibility of bluffing. Von Neumann invented the following variation on stud poker for two players: The same number of cards are dealt to each player face down; after each player looks at his own cards he decides on his bid, *without* knowing what the other person is going to bid; to keep things simple, only two alternative bids are allowed—high (H) or low (L)—and each player must choose one or the other; if both players have independently bid H or L, the outcome of the game is that the player with the better hand gains the amount H or L respectively from the other player; if one player has bid H and the other L, then the second player gets a chance either to "pass"—in which case the first player collects the amount L—or to raise his bid to H and "see"—in which case the two players compare hands, and the one with the better hand collects the amount H. Different suits are given different rankings, so that one hand is always better than the other.

The next step in the analysis is to convert this verbal description into mathematical language. First of all, to compare hands, it is convenient to assign to each possible hand a number, the better the hand the bigger the number. In five-card stud, with no joker, there are 52!/47!5! or approximately two-and-a-half million (exactly 2,598,960) different possible hands. From the point of view of game theory, the first step in the game, dealing the cards, is equivalent to each player's picking an integer between 0 and 2,598,961 out of a hat. Each number corresponds to exactly one possible hand. When it comes to comparing cards (or numbers), the player with the higher number will win.

What are the alternatives available to player A, once he has looked at his cards? There are three choices: (1) he can bid high; (2) he can bid low, and if B bids high, pass; (3) he can bid low, and if B bids high, he sees. Of course B has entirely similar choices. A complete strategy for A would specify not only which of the three choices he will make for every hand he might get but—since in a game without perfect information one expects to use a mixed strategy—also how often in the long run he will choose each of the three choices for a given hand.

Having enumerated the possible deals and the possible choices, one must finally tabulate possible payoffs or outcomes. This last step is illustrated by the small chart below, which gives the outcomes for the case that A's hand is better (has a higher number) than B's; in a similar way one can construct a chart for the case that A and B have hands of equal value or the case that B's is better than A's. (The reader can do this, if he likes.)

	B(hi)	B(lo-see)	B(lo-pass)
A(hi)	H	H	L
A(lo-see)	H	L	L
A(lo-pass)	−L	L	L

Chart of alternative choices for A and B resulting net payoff from B to A if A's cards are better than B's. (The choices are: bid high; bid low and then "see" opponent; bid low and then pass.)

If we had spelled out in dollars the two permissible bids, H and L, the chart would have contained the dollar amounts rather than the letters H and L. As in other problems in algebra, the letters are preferred, because the chart and all results will apply to the game no matter what amount the players decide on for the allowed high and low bids. (The reader who prefers numbers may everywhere replace L by $1.00 and H by $2.00 to make things concrete.)

From the minimax theorem it is known a sound strategy (i.e., minimax strategy) of the mixed kind must exist for both players. It is a complicated problem of algebra and combinatorials to discover a sound strategy even for this simplified stud poker, but von Neumann successfully worked it out.[21] One feature of the invented game that helped simplify the analysis is that it is exactly symmetrical between the two players. The usual game of stud poker is not quite symmetrical because one player bids before the other, so that the latter knows the other's bid when he bids himself. Von Neumann's algebraic manipulations are not suitable for presentation here, but we can state his conclusion qualitatively: the sound strategy for either player consists of *always* bidding high when he has a strong hand; of *never* bidding low and then seeing; of *sometimes* bidding low and then passing when he has a weak hand; and *other times* bidding high when he has a weak hand. The fact that it is a part of the sound strategy sometimes to bid high on a weak hand is a way of saying that it is part of the sound strategy sometimes to bluff, as all poker players know. Von Neumann notes two reasons for bluffing in this simplified stud poker. One is to keep the opponent guessing about your hand—this was the element so important in the stone, paper, scissors game; the other is to be protected against the opponent's bluffing.

Von Neumann's conclusions are quantitative, and so far I have presented only the qualitative result. Of course to bid high or low always means to bid the allowed values H or L respectively, and the words never and always are similarly precise. The words that need to be made quantitative are strong hand, weak hand, sometimes, and other times. How strong? How weak? How often? Those who are interested may consult my summary of von Neumann's quantitative results.[22]

If player A follows the minimax policy in this variant of poker, he can at least expect to break even, on the average. However, if player B fails consistently to bid high on his strong hands, on the average A will come out ahead. If A follows the sound strategy strictly but B deviates from it somewhat by changing how often he bluffs, it makes little difference. The reason is similar to what happens in stone, paper, scissors. Player A can take proper advantage of player B's failure to follow the sound strategy in respect to frequency of bluffing only by changing his own policy. The minimax strategy, this reminds us, is a conservative one, suited best to playing against a knowledgable and conservative opponent. It is not a strategy for exploiting the psychology of the opponent or his lack of knowledge. To get a feeling for the sound strategy, I urge the reader to spend an evening testing von Neumann's theory for the simplified version of stud poker, strictly to follow the sound strategy, while his opponent plays whatever way he pleases. The expectation on the basis of the von Neumann theory is that on the average (over several hours' play), the reader will win or at least break even. If the reader, at the end of the evening, then explains the sound strategy to his opponent, they might spend another evening with both playing the sound strategy, in the expectation that both will just about break even. Finally, if they can spare a third evening, they can test a fine point of theory, by having one of them follow the sound strategy exactly while the other follows the sound strategy except that he may vary the frequency of bluffing to suit his own taste. The expectation, according to von Neumann's theory, is that they will again break even on the third evening.

Aside from the general minimax theorem, and the explicit analysis of simplified variants of poker, von Neumann during the years 1926–28 was already considering the zero-sum three-person game. The essential new feature arising with three persons, and which is absent in the zero-sum two-person game, is that coalitions may form. If the three players are A, B, and C, then A and B for example may join forces and in effect play a two-person zero-sum game where C is on one side and the team (A + B) is on the other; generally both A and B benefit by their collusion (or can arrange to share the benefit), but since it is a zero-sum game it must be to C's detriment. Von Neumann focused on the question of which of the various possible coalitions among three persons are likely to form and the benefit a player can achieve by forming a coalition. All this is contained in the same 1928 article in which the minimax theorem is proved.

What, then, became of game theory after the 1928 paper? During the 1930s, although some new proofs and applications of the minimax theorem were given, it was not an active field of research. In the 1940s Abraham Wald initiated a new and fruitful approach to statistics in which he focused on the making of decisions. His theory, known as statistical decision theory, relies heavily on the minimax theorem because he envisaged decision making as in effect playing a kind of two-person game against nature. Thus the minimax theorem, developed first in connection with game theory, was instrumental in the development of a different field of mathematics than game theory itself. Since the 1940s game theory has been an active field of research in mathematics. Other extensions of the theory—namely its application to economics and military decisions—will be discussed later when the philosophy underlying the application of game theory to the social sciences is critically examined. Who could have guessed in 1928 that in the 1950s military and political strategists would take game theory as seriously in advising rulers as the advisers to the kings of an earlier era took astrology?

5 AXIOMS AND ATOMS

Some of the best inspirations of modern mathematics clearly originated in the natural sciences . . . the multiple phenomenon which is mathematics. . . . I think it is correct to say that his (the mathematician's) criteria of selection, and also those of success, are mainly aesthetical.
—J. von Neumann, 1947

In giving even a very brief sketch of the development of some of the ideas prominent in the mathematical physics of the twentieth century, it is necessary to go back to the ancient Greek thinkers whose mathematics and natural philosophy set the mold for modern developments. Whether we prefer to regard the discovered "truths" of modern mathematics and physics as among the highest achievements of human civilization, or whether our perspective is that of an anthropologist who finds them to be interesting myths that European culture lives by, or whether we accept both of these perspectives or neither, their roots unmistakably lead to ancient Greece.

Democritus (ca. 460–360 B.C.) put forth a simple unifying concept encompassing the diversity and complexity of the world we know: "By convention color, by convention sweet, by convention bitter; in reality nothing but atoms and the void." According to Democritus, the reality of the world is that of a great variety of atoms with differing sizes, shapes, locations, and motions, together with the empty space within which atoms move about. All the physical events and changes in the world are nothing more or less than rearrangements and interactions among the atoms of which matter is constituted. Although Democritus was also a mathematician, he and his followers made no effort to describe atomic reality by means of mathematics and perhaps never imagined that possibility. But he was able to interpret a host of familiar phenomena in terms of atoms and their motions. By relegating sensations to the merely conventional and illusory and abstracting out the concept of atoms, he had anticipated the route toward a unified mathematical description of the physical world that after various vicissitudes came into its own with the kinetic theory of the nineteenth century and the quantum theory of the 1920s, both subsumed under modern quantum statistical mechanics. Atomism represented one serious school of thought in Greek philosophy. Aristotle, who did not accept atomism, nevertheless gave an exposition of Democritus's views and put forward his own reasoned criticism of them.[1]

A view radically different from Democritus's implicit philosophical materialism is found in Euclid's *Elements of Geometry*. Although the original inspiration must have derived from observation, the *Elements* deal with mathematical truths, truths pertaining to the relation among ideas rather than with the material world. Euclid's geometry (ca. 300 B.C.) begins with definitions and proceeds with postulates and axioms on the basis of which numerous theorems can be deduced by logical steps. It is the classic instance of the axiomatic or postulational method in mathematics. It is Euclid who defined a point as "that which has no parts" and a line as "a length without breadth." The postulates, axioms, and theorems deal with relationships among elements such as points, lines, and similar formally defined constructions of the human mind. The theorems are derived by rigorously logical reasoning from the givens and consequently are "necessary truths." Within the framework of the Euclidean system they are essentially tautological.[2]

Facts taken from observation of physical events are truths of an entirely different kind from mathematical tautologies. Nevertheless, it has been the aspiration of modern mathematical physicists to devise formal axiomatic systems in the tradition initiated by Euclid which can be made to describe the empirical world with its flux of events. There is tension if not incompatibility between the requirements of axiomatic form and faithful description of physical phenomena. After all, there is no reason why it should be possible to fit empirical theories based on generalizations from observed events into any rigid scheme of definitions and postulates in the spirit of Euclidean geometry with its invented definitions. But one can try to find a suitable set of definitions and self-consistent postulates. In any intellectual history of Western civilization the interplay between axiomatic mathematics and empirical physics deserves a prominent place. I say "Western" because the Oriental civilizations, especially those of the Far East, did not seem to share that particular preoccupation in spite of their considerable mathematical and scientific activity.[3]

Around 1926 von Neumann became acquainted with empirical theories that adequately described the host of observed phenomena of atomic physics, yet he himself was imbued with the spirit of axiomatics. Hilbert challenged him to reconcile the two. Clearly atomic theory had to have come a long way from Democritus and axiomatic mathematics would have to deal with far more complex structures than Euclidean geometry before one could conceive of incorporating modern atomic physics within any formal mathematical structure by means of definitions and axioms that would resemble Euclidean geometry in their logical clarity, self-consistency, and completeness. Von Neumann's famous work on the mathematical foundations of quantum mechanics was just such an effort at reconciliation.

Of course, a number of major innovations in physics and mathematics intervened in the 2,200 years that passed between Euclid and von Neumann, but even disregarding innovations, the later history of Euclid's *Elements* is remarkable.

The *Elements* form, next to the Bible, probably the most reproduced and studied book in the history of the Western World. More than a thousand editions have appeared since the invention of printing, and before that time manuscript copies dominated much of the teaching of geometry. Most of our school geometry is taken, often literally, from eight or nine of the thirteen books.[4]

The axiomatic method of Euclid's geometric demonstrations has represented the model for deductive reasoning. Its popularity may have reflected the fascination with a method for finding indubitable and eternal truths. Yet a critical reader might well have noticed that the definitions are ambiguous, and if one should want all terms defined one would seek definitions of the very words used in Euclid's definitions, which would require definition in turn, ad infinitum. Moreover, Euclid's postulates themselves were unproved propositions. Nonetheless the pattern of reasoning exemplified by the deduction of the theorems of geometry in the *Elements* was emulated by the scientists who in the scientific Renaissance in Europe devised deductive systems, most especially and importantly by Isaac Newton himself.

The appearance of Isaac Newton's *Principia* (1687) is perhaps the moment in the history of Western thought, clearly anticipated by Galileo and others, when physics and mathematics become irreversibly, indissolubly joined. In the book Newton asserts as postulates or propositions general mathematical equations for the motion of objects, be they terrestrial or celestial, namely his "laws of motion." As the deductions from the equations were repeatedly verified by experience in the most diverse circumstances, the single most fundamental and truly remarkable tenet of modern physics was established: the motion of objects follows a strict mathematical pattern and can be predicted by calculation. The dynamics of the universe, from pebbles to planets, is appropriately described by the language of mathematics. Moreover, Newton gave the theory an axiomatic formulation and proved theorems by geometric demonstrations, entirely in accord with the methodological model provided by Euclid's *Elements*.[5]

As Newton appreciated, insofar as he was describing the physical world he could not achieve the certainty of the purely tautological statements of Euclidean geometry:

The proofs in natural philosophy cannot be so absolutely conclusive, as in the mathematics. For the subjects of that science are purely the ideas of our own minds . . . ; as the mind can have a full and adequate knowledge of its own ideas, the reasoning in geometry can be rendered perfect. But in natural knowledge the subject of our contemplation is without us, and not so completely to be known.[6]

100
John von
Neumann
and
Norbert
Wiener
Newton had introduced postulates about the nature of space, time, and "absolute" motion and some general propositions based on observations. It had required a process of abstraction and induction, so there was no absolute certainty here. Nevertheless, once all the postulates and propositions were granted, Newton's deductions were fully as rigorous as Euclid's geometric deductions.

Newton was an heir not only to Euclid's method but also to Democritus's atomism. He acquiesced in the conception that matter is made of hard and impenetrable atoms, although his ideas about the structure of matter were not crucial for either his gravitational theory or his laws of motion. He had made a specific atomic model to explain the pressure of a gas, and in his optics corpuscles played an explicit role. In Newton's day there was very little empirical basis for assigning many specific properties to postulated atoms. More than a century later experimental data of chemists on the proportions in which different substances combine chemically, to which Dalton, Gay-Lussac, Prout, and Avogadro called attention early in the nineteenth century, gave rise to research programs that would make the atomic hypothesis more specific and complex.[7]

Newton's mechanics forms a complete and deductive theory that contains all the needed physical principles and statements to permit the mathematical analysis of any specific concrete situation arising in mechanics. Yet its formulation and its mode of presentation in the *Principia* are cumbersome, especially as Newton relies on complicated geometric proofs, where the calculus, his own invention, would have provided much simpler derivations. The book does not provide a systematic and conveniently automatic approach to most applications of mechanics. In the development of a science a merely deductive theory is not yet the final statement, but "the deductive is followed by the formal development."[8] After Newton other mathematicians found a variety of different mathematical forms for his principles. One formulation seemed theologically more appealing,[9] another exhibited a deep internal symmetry and elegance and permitted a formal unification of mechanics and optics,[10] while Lagrange's

Mécanique analytique, published a hundred years after Newton's *Principia,* provided the exemplary formal treatment. Not only is this formulation particularly clear in its logical structure, but it brings more powerful mathematical tools to bear than Newton did, avoiding entirely his tedious geometric demonstrations. Lagrange had himself contributed to inventing the mathematical tools he used, particularly the calculus of variations.

One historian of mechanics has described the function of Lagrange's formal development in the following terms:

Lagrange strives to think through all needed considerations once and for all and to represent as much as possible in terms of a formula. Every case which occurs can be treated according to a very simple, symmetric and surveyable scheme. What thinking remains to be done, can then be carried out through purely mechanical mental processes.[11]

The formal development of a major physical theory is no mean task. One strives at the same time for an impeccable axiomatic or logical structure, comprehensiveness, and choice of a type of mathematics that provides a "simple, symmetric and surveyable scheme" and reduces the solution of specific problems to following a prescription. It is a task distinct from originating the physical theory in the first place, as Newton did. Von Neumann's place in the history of modern quantum mechanics (i.e., atomic physics) corresponds to Lagrange's position in relation to Newtonian mechanics, in that both men formalized an original physical mechanics.

By 1925 or 1926, when von Neumann appeared on the scene at Göttingen, a scientific revolution had already occurred in connection with atomic theory. Not only had Maxwell's and Boltzmann's atomic theories of heat been vindicated, but an increasing number of experiments had given detailed information about the internal structure and dynamics of individual atoms. This detailed information had generated a crisis in physics because the data were incompatible with the Newtonian laws of motion, which by the time the twentieth century arrived had become part of the dogma of

102

John von
Neumann
and
Norbert
Wiener

science. Of course if Newton's principles were violated, then the Lagrangian formulation of Newton's principles and every other equivalent formulation would be equally violated.

One experimental observation apparently contradicting Newtonian physics had to do with the color of the light that atoms absorb and emit. Each particular species of atom emits or absorbs very specific wavelengths, i.e., very specific colors of light and no others. The wavelengths of visible light that hydrogen atoms absorb, for example, are precisely 656.3 nm (nanometers) or 486.1 nm or 434.0 nm or 410.2 nm, but not the wavelengths in between.[12] If Newton's physics and the nineteenth-century theory of light (Maxwell) were correct, then it was incomprehensible why the hydrogen atom absorbs and emits some colors but not others. It was known from scattering experiments (Rutherford) and other evidence that atoms consisted of a heavy nucleus with a positive electric charge and lightweight electrons traveling around the nucleus. According to the principles of physics universally accepted at the end of the nineteenth century, the electron would have to lose energy by radiating *all* the colors of the rainbow and itself eventually fall into the nucleus. Experiments showed clearly that what actually happened was very different.

During the first quarter of the twentieth century it became increasingly obvious to some physicists that no way of reconciling the experiments on atoms and radiation with classical physics was possible. New principles of mechanics had to be found for the microworld of atoms, molecules, and their constituents. A "revolution" in physics was stirring. In 1925–26 two independently formulated new sets of principles of micromechanics appeared in print, two sets of principles that at first appeared to be quite different from each other. It was in this respect unlike the historical circumstance of Newton's mechanics. The analogous situation would have been if two Newtons had arisen simultaneously, each setting forth laws of motion apparently very different from those proposed by the other Newton.[13]

Throughout the period of searching for new principles in the atomic case the suggestion had been made by various physicists including Einstein and de Broglie that submicroscopic particles, specifically electrons, had waves associated with them, and experimental evidence appeared to confirm the suggestion. Schrödinger, then in Zürich, hit upon the following idea: Suppose the electron in the hydrogen atom is analogous to a string—tied at both ends—in a musical instrument. Such a string emits a very definite tone together with its overtones, but not the wavelengths in between. With this idea in mind, Schrödinger proceeded to generalize one of Hamilton's formulations of Newton's principles and set up a wave equation for the electron. Lo and behold, when he solved the equation for the case of the hydrogen atom, it gave exactly the right wavelengths for the emitted colors of light for hydrogen! This was a happy and intuitively plausible solution to the prototype scientific puzzle of the microworld. Schrödinger's theory hinted at a kind of symmetry between the nature of material particles and radiation.

Meanwhile a more algebra-minded group of physicists in Göttingen had approached the problem in a different way. Werner Heisenberg, Max Born, and Pascual Jordan reasoned that the basic variables in the Hamiltonian formulation of Newton's mechanics were the position of a particle and its momentum; simultaneous measurement of these two quantities are just the information needed to predict the motion of a particle in Newtonian physics. But neither the position nor the momentum of an electron inside an atom had been observed directly by experiments. What one observes are the colors and intensities of the light the atoms emit, they asserted. So Heisenberg and his colleagues decided that one could treat the "unobservables," position and momentum, quite differently from Newtonian physics, but that the significant quantities were the observables—the colors and intensities of the light. Perhaps the rules of algebra for positions and momenta are changed, they thought. If they are not observed, they may not be definite numbers, but some other kind of mathematical object. It should be noted that mathematicians play certain kinds of games, one of which consists in envisaging mathematical objects (other than ordinary numbers) and then inventing rules of algebra for these objects. Such rules could be anything as long as all the rules and definitions are consis-

104

*John von
Neumann
and
Norbert
Wiener*

tent, i.e., they don't lead to contradictions. So Heisenberg and his coworkers introduced a new rule for the multiplication of the position (one component) with the momentum (same component) of an electron:

$$(\text{position}) \times (\text{momentum}) - (\text{momentum}) \times (\text{position})$$
$$= \sqrt{-1} \times 1.05 \times 10^{-27} \text{ erg sec.}$$

The magnitude of the number on the right side is known as Planck's constant ($\hbar$). If position and momentum had been ordinary numbers, the right-hand side would have been zero; they found mathematical objects (called matrices) already known to mathematicians which could represent position and momenta and would satisfy the at first strange and unfamiliar rule of multiplication. The Göttingen physicists were amazed at what they had wrought. However, the mathematicians reassured their physicist colleagues that matrices were perfectly acceptable and interesting mathematical objects. One Göttingen graduate student in mathematics, Maria Goeppert-Mayer, found the fact that such interesting mathematics would play a role in physics "wonderful" and soon decided to become a physicist instead of a mathematician.[14] With this multiplication rule they were able to obtain agreement with experiments in atomic physics. Objects for which the order of multiplication matters are in mathematical parlance said to be "noncommuting." Numbers always commute, but matrices are generally noncommuting. The two operations "putting on your socks" and "putting on your shoes" do not commute; the order in which you do them matters. Matrices might represent these two operations.

The Heisenberg algebraic approach to atomic physics and the Schrödinger wave approach are very different.[15] The two men did not like each other's methods. Heisenberg referred to Schrödinger's physical picture of atoms as "disgusting," and Schrödinger said he was "frightened away, if not repelled" by Heisenberg's algebraic approach.[16] But very quickly it became apparent that the two approaches did *not* constitute two different principles but were merely two different forms of the same principle. One could be derived as a mathematical consequence of the other.

So by the spring of 1926 two forms of the same quantum mechanics were available, namely wave mechanics and matrix mechanics. For practical purposes of calculating the properties of atoms, the Schrödinger wave equation formulation and the Heisenberg matrix mechanics formulation were adequate. For some calculations the Schrödinger formulation was the more convenient, while for others the Heisenberg formulation was. One chose whichever was more convenient, knowing that mathematically the two can be proved always to agree. It is as if one had two different languages to describe each experiment, and it was generally possible to translate one language into the other.

But von Neumann, among others, found this unsatisfactory. He asked himself the fine question, *What features do these two formulations have in common? Can we abstract out the essence of quantum theory by identifying the characteristics common to both formulations and formalize this abstract essence?* Then the two known formulations will appear only as two special representations of the abstract formulation, which would then be more fundamental than either of the two known formulations and would bring unity to the theory. Moreover, with possible future extensions in mind, there is ample reason to examine the mathematical structure of the theory in depth. Being a mathematician and not a physicist, von Neumann would also insist on a higher standard of mathematical proof than the physicists.[17]

Such was the state of atomic physics in 1926. What developments had meanwhile occurred in the field of axiomatics? By the beginning of the nineteenth century Euclid's geometry had come to be viewed as the one and only true geometry, whose postulates are "self-evident" truths, but mathematicians found that it was possible to alter one of Euclid's postulates without causing internal contradictions, thereby producing "non-Euclidean" geometries with theorems different from Euclid's. Empirical science would have to decide which if any of these fit the physical world. Late in the nineteenth century several mathematicians, and in particular Hilbert, reexamined the logical structure of definitions, postulates, and axioms of the sacrosanct Euclidean *Elements,* and Hilbert's own resulting formulation of the logical foundations of geometry initiated the modern postulational method in mathematics.[18] It differed from Euclid's ancient method, since Hilbert left the primitive concepts such as point and line completely undefined. The relationships between the primitive concepts emerge from the axioms, but no other definition is required. Hilbert required of the axioms he devised for geometry that they be strictly consistent with each other, independent of each other, and that they form a complete set of axioms for geometry. Moreover, these logical relationships among axioms required proof.[19]

Hilbert had in his studies of the foundation of geometry inaugurated a method of critically examining the nature of mathematics itself, a field of inquiry now known as metamathematics. In his formalistic approach any meaning or interpretation of the symbols used is irrelevant. Mathematics is seen as a purely formal, mechanical manipulation of meaningless symbols according to specified rules. Hilbert had hoped that not only geometry but all of mathematics would be described in these terms. He had said of the axiomatic method that with it one thinks consciously, rather than believing mathematical formulae naively and dogmatically, that it gives one the advantage of faith in the formulae but without naiveté.[20] But Hilbert went further: he believed in the desirability of using the axiomatic method not only in mathematics but also in physics. In the year 1900, when an international centennial celebration and meeting of mathematicians was held in Paris, he gave a historically influential lecture in which he enumerated what he believed to be the outstanding problems in mathematics for the new century. The sixth task he listed was to treat the axioms of portions of the science of physics analogously to his own axiomatic treatment of geometry. He himself subsequently worked on the mathematical statement and axiomatic foundations of various areas of physics (kinetic theory of gases, theory of radiation, general relativity theory).[21] He repeatedly expressed his optimistic faith in the method,[22] and when the new quantum theory arrived (1925–26), he examined the mathematical structure of this new development in physics. This was the period when John von Neumann came to Göttingen and worked with Hilbert and his assistant Nordheim on the formal structure of quantum theory; eventually he inherited the problem from Hilbert, as well as some of his faith. (I observe parenthetically, for whatever conclusions one cares to draw from it, that Hilbert's optimism concerning the power of the axiomatic method was unjustified, just as Boltzmann's pessimism concerning the acceptance of the kinetic theory had been unjustified.)

As it turned out, the mathematical idea that von Neumann found best suited to a formalization of the empirical theories of quantum mechanics is also one that had been particularly developed by David Hilbert. This mathematical concept has to do with an interplay between algebra and geometry, so that its genealogy can be traced back to the analytical geometry devised by Fermat and Descartes in the seventeenth century. The concept to which I am referring, however, may seem surprising to the nonmathematical reader. It is a kind of extension of the mathematical concept of space from three to an infinity of dimensions, an extension that von Neumann in honor of his older colleague called "Hilbert space" or "abstract Hilbert space."

In order to understand this idea one should recall the Cartesian connection between algebra and geometry: it is possible to make a graph of an equation. Suppose one wishes to solve the two simultaneous algebraic questions, $x^2 + y^2 = 1$ and $x - 3y = 0$. The reader can easily find by elementary algebra that one solution is $x = 3/\sqrt{10}$, $y = 1/\sqrt{10}$ and the other is $x = -3/\sqrt{10}$, $y = -1/\sqrt{10}$. One can interpret the algebra problem geometrically if one makes a graph of the two equations, marking off values of x along the horizontal axis and y along the vertical axis (fig. 5). The first equation graphs into a circle with its center at $x = 0$, $y = 0$ (the "origin") and a radius of unity. The second equation becomes a straight line that goes through the origin. The two solutions of the simultaneous algebraic equations correspond to the two points of intersection of the line and the circle. Both solutions are a distance of unity from the origin and a distance of two units from each other. Thus we can visualize and discuss an algebra problem in two variables in terms of geometric concepts like point

108

John von
Neumann
and
Norbert
Wiener

of intersection, distance, radius, and line. We can use the Pythagorean theorem for the hypotenuse of a triangle to express the distance of any point from the origin, or what would be the same, the length of the line from the origin to that point. Suppose the point is a horizontal distance A and a vertical distance B from the origin; then its distance from the origin is $\sqrt{A^2 + B^2}$.

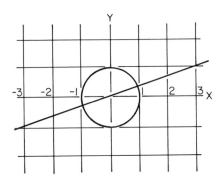

Figure 5

But suppose one has an algebra problem involving three variables. It still helps to visualize it if one views it in terms of lines and shapes in three-dimensional space. If a point is located at distance A in the east-west direction and a distance B from the origin in the north-south direction and a distance C in the vertical direction, then the distance of the point from the origin is with the help of the Pythagorean theorem found to be $\sqrt{A^2 + B^2 + C^2}$. A sphere of unit radius with its center at the origin is by definition the locus of all points with distance unity from the origin. Thus the algebraic equation $x^2 + y^2 + z^2 = 1$ can be thought of as such a sphere.[23] In fact, the problems of algebra involving three variables or unknowns can be depicted nicely in three-dimensional space. However, in principle an algebra problem might present many more than three variables. Suppose one has to deal with the five variables v, w, x, y, and z, and one is given the equation $v^2 + w^2 + x^2 + y^2 + z^2 = 1$. Noting the similarity to the circle in two dimensions or the sphere in three dimensions, one is tempted to use geometric language. It describes a five-dimensional sphere (hypersphere) with unit radius centered about the origin. If one found a solution to a set of equations with five variables to be $v = A, w = B, x = C, y = D, z = E$, one would be tempted to interpret it by analogy as a point in a five-dimensional space, the point being a distance $\sqrt{A^2 + B^2 + C^2 + D^2 + E^2}$ from the origin. In this geometric language the five-dimensional sphere would indeed correspond to the locus of all points a unit distance from the origin.

Such analogies between the geometry of two or three dimensions and higher dimensions turns out to be a valuable heuristic device for the mathematician and to provide a convenient language. However, even when the language is that of geometry, all proofs and derivations are algebraic. Analogies may suggest ideas and relationships in mathematics, but validity can be established only by rigorous proof. It is possible to extend the theory of algebraic equations to the case of an infinite number of variables and even in that case to interpret it geometrically. Especially in this case one would be inclined to be cautious about using analogies, preferring to begin one's reasoning with mutually consistent axioms and then accepting geometric language only where each step has been proved and every analogy justified by rigorous deduction from the axioms. David Hilbert had pioneered the study of equations with infinitely many variables, proving the validity of the extension of various results from the study of equations with a finite number of variables. In the process he introduced geometric notions describing a "space" with infinitely many dimensions. His articles on the subject published in 1906 stimulated further studies of spaces with infinitely many dimensions, especially by E. Schmidt, E. Fischer, and F. Riesz.[24]

It was von Neumann's deep insight in 1926 that if he was to be the Lagrange of quantum mechanics, he might "abstract out the essence of quantum theory" by use of the language and algebra of an infinitely dimensional space but make the abstract formulation itself independent of the choice of coordinate system, just as the relation between figures in Euclidean geometry is independent of any coordinate system. Such an abstract formulation would have to contain the Heisenberg and Schrödinger descriptions of quantum mechanics as particular representations. Moreover, it would be desirable to found the mathematical theory on strict axioms in the spirit of Hilbert's philosophy. These are the three streams that von Neumann joined in his *Mathematical Foundations of Quantum Mechanics*: physical quantum theory itself, axiomatics, and a powerful mathematical tool, the theory of infinite-dimensional spaces. The last was to reconcile the first two.[25]

Already in 1925, even before Schrödinger's paper had appeared, Norbert Wiener had taken the first steps toward a general formulation of quantum mechanics.[26] Also independently of von Neumann, P. A. M. Dirac worked out a complete and elegant abstract formalism for quantum mechanics, but certain steps in his deductions he could not or did not prove rigorously at that time.[27] Only von Neumann evolved a complete formulation that was fully rigorous. It ultimately proved itself to be a superb formalism for quantum mechanics and could in fact encompass later extensions of quantum theory. Von Neumann's original work appeared in a series of three articles in 1927, in which his theory was formulated,[28] although some crucial mathematical points remained unproved until von Neumann's later articles in 1929.[29] Finally in 1932 von Neumann prepared an exposition of the whole theory in the form of a book, an elaboration on the earlier papers. The book has become a classic in mathematical physics,[30] and I will attempt to give the reader some small, nontechnical indication of part of the content of this elaborate and admittedly highly technical and abstruse work. In doing so I will have to shuffle back and forth between the physical theories, the mathematical formalism, and the axiomatics.

Fundamental in quantum theory is the mathematical description of the "state" an atom is in.[31] Now the state of an atom is one of those undefined primitive terms in von Neumann's formulation,[32] analogous to point or line in Hilbert's axiomatic version of Euclidean geometry. The theory gives a description or characterization of the state but contains no definitions of the concept "state" itself. The state of an atom is, according to von Neumann, characterized by "a vector of unit length in the Hilbert space for the atom." This is

a mouthful and requires explaining: In figure 5, which shows
a circle with unit radius centered on the origin, any point on the circle could be characterized by two numbers, the values of x and y respectively at that point. If one drew a directed line segment from the origin to that point, it would also be characterized by these two numbers. Such a directed line segment is a vector of unit length in two-dimensional Euclidean space. In three-dimensional Euclidean space three numbers are needed to describe a point on the sphere with unit radius or its associated vector of unit length. In the Hilbert space it requires infinitely many numbers to characterize a specific vector of unit length, which in turn corresponds to one specific state of the atom.[33] Indeed, in the Heisenberg form of quantum theory the state of an atom is to be characterized by a denumerable infinity of numbers, and moreover the representation of each state as a vector of unit length in a Hilbert space is perfectly compatible with the Heisenberg theory. So this is how von Neumann chose to conceptualize the description of the state of an atom.

But von Neumann purported to give a formal characterization of the state of an atom that would unify the Heisenberg and the Schrödinger forms of the theory. How could he do that? He argued that even though the Schrödinger theory characterized the state of an atom by a (wave) function of a continuous variable, which would seem to correspond to a vector of unit length in a Hilbert space with a *non*denumerable number of dimensions, it can be proved that a perfect one-to-one correspondence (more precisely, an isomorphism) can be set up between the states of an atom in the Heisenberg and Schrödinger theories.[34] Furthermore, the formal rules of algebra, for example the rules for addition and multiplication, are the same for the Schrödinger wave functions as they are for the Heisenberg vectors. To encompass both theories von Neumann invented the "abstract" Hilbert space. He *defined* the abstract Hilbert space by certain rules he laid down as axioms, rules about the way in which the "elements" (vectors, wave functions) of the space combine, rules that apply equally to the Heisenberg vectors and the Schrödinger func-

112

John von
Neumann
and
Norbert
Wiener

tions. Von Neumann wrote down five axioms in all to specify everything that needed to be specified about abstract Hilbert space. The first two axioms deal with the rules for addition and multiplication. The third merely states that the number of dimensions of the space is infinite. The fourth and fifth axioms are rather technical and will not be described here.[35] After von Neumann had successfully proved that both the Schrödinger states and the Heisenberg states of atoms indeed satisfy all the axioms of von Neumann's abstract Hilbert space, he had achieved the formal unification—at least insofar as a description of the state of an atom is concerned—of the two theories. Moreover, he could prove many theorems about the vectors in abstract Hilbert space from the axioms and could interpret some of them in terms of the geometry of abstract Hilbert space.

However, quantum mechanics, whether in the Heisenberg form or the Schrödinger form, involves a mathematical entity quite different from a state vector or state function. This other mathematical entity, known as an "operator," acts on one state to change it to another. We can picture for example in three dimensions that an operator rotates a unit-length vector about the origin so that it will point in a different direction than it did before the operator acted on it. It is an interesting and peculiar fact about rotations that if two rotations are carried out in succession it often matters which of the two is carried out first, i.e., the net result depends on the order in which things are done. To convince yourself that the sequence of operations can matter, do the following experiment: Place a book flat on the table in front of you with only a sheet of paper between the book and the table, and the back edge of the book toward you. Draw a line on the paper along the front edge of the book. Let the first operation, symbolized by operator Q, be to "rotate book 90° clockwise about this line in a direction away from you." Now imagine a vertical axis and carry out the second operation, symbolized by operator P, namely to "rotate book 90° clockwise about the vertical axis." Take careful note of the final position of the book (or any "vector" pasted on it) after the two operations. Now begin all

over again but reverse the order, i.e., first rotate the book
clockwise 90° about a vertical axis and then 90° about the axis you had originally drawn on the piece of paper. You will see that the final position of the book depends on the order in which you carried out the two operations. Consequently, the rules for the algebra of the "operators" P and Q, if they are to correspond to reality, must be such that PQ does not equal QP. This is typical of physical operations—we mentioned earlier how the net result of the two operations "putting on one's socks" and "putting on one's shoes" depends on the sequence in which they are performed. An operator in Hilbert space generally rotates vectors in that infinitely dimensional space, and in many instances the order of operations does matter. In this way the Hilbert space formalism for quantum mechanics will reproduce Heisenberg's result mentioned earlier concerning the noncommuting character of a position operator and its corresponding momentum operator.

The suitable formal language of quantum mechanics, von Neumann proved, consists of vectors in abstract Hilbert space and certain classes of operators, with rules of algebra of their own, acting on the vectors. The von Neumann formulation in its abstract form is independent of any particular choice of representation or choice of coordinate system. In this respect it resembles the geometry of Euclid more than the analytic geometry of Descartes. Since the abstract equations are "invariant" to changes in representation, they contain only the essentials, rather than incidental features that depend on special representations. By contrast, the Schrödinger and Heisenberg schemes reflect special representations.

The formal edifice of vectors and operators in abstract Hilbert space, held together by axioms and strict theorems, was hailed in 1927, as it fully contained the facts and theories of atomic physics. In the view of Göttingen mathematician Richard Courant, "It was the great merit of von Neumann, guided by [Erhardt] Schmidt, to see that what matters really is the structural situation in abstract space."[36] Since that time many experimental discoveries and theoretical developments have taken place in physics. Experimental and theoretical nuclear physics were developed; new particles were discovered by physicists working with cosmic rays, natural radioactivity, cyclotrons, and linear accelerators; relativistic quantum theory developed; and quantum field theories were constructed to describe all the properties, including their creation and annihilation, of the ever-increasing variety of known particles. Throughout this period of growth in physics, the edifice of abstract Hilbert space has proved viable. Today von Neumann's work is still seen as laying bare the essential mathematical structure of quantum theory, and—with some reservations—also that of the major extension of quantum theory, namely quantum field theory.

In 1952 it was found that von Neumann's rules for addition of vectors ignored some exceptions arising in quantum field theory from the fact that the total electric charge (i.e., the sum of positive and negative charges) can never change in an experiment.[37] A second type of limitation of abstract Hilbert space has also been found in the domain of quantum field theory. In fact, some of the most useful formulations of quantum field theories violate one or another of the five axioms of abstract Hilbert space.[38] This state of affairs has led to formulations of abstract algebras in which the formalism for elements of abstract Hilbert space appears as only one special case. Such abstract algebras are not in disagreement with abstract Hilbert space but are more comprehensive. Von Neumann himself in 1936 devised in collaboration with G. Birkhoff a formal "quantum logic" more general than the abstract Hilbert space formalism.[39] However, it remains a question for the future to decide the extent of the comprehensiveness of abstract Hilbert space in the description of the diverse kinds of particles and their interactions that are the subject matter of quantum field theories.[40]

After devising the formalism of abstract Hilbert space for atomic physics in 1927, von Neumann continued to work on other aspects of quantum theory. In particular he constructed within the framework of his new formalism a theory of measurement that involved the interaction of the observer and the instruments or objects he is observing. Although primarily interested in the formalism, von Neumann found himself dealing with problems in epistemology. He found that his formalism strangely predicted that consciousness, awareness, can affect electrons in the laboratory! I will return to this philosophical aspect of his theory in a later chapter.

It is von Neumann who, following Hilbert, had viewed the philosophy of quantum mechanics as a problem in finding a suitable logical and axiomatic structure for atomic physics. To this day a significant school of mathematicians and physicists agree with that point of view.

The pursuit of mathematics is a divine madness of the human spirit, a refuge from the goading urgency of contingent happenings.
—A. N. Whitehead

The book [the poetry of Vergil] shows me clearly what I fled from when I sold myself body and soul to Science—the flight from the I and WE to the IT.
—A. Einstein

The previous three chapters have described a sample of Wiener's and von Neumann's mathematical work during the period between the two world wars. The theorems proved are impersonal; their validity is independent of their authors, independent of their personalities, their motivations, or their views of the world. The stark impersonality and universal validity of mathematical statements is the very reason for cultural interest in them. Nevertheless, the motives leading to the creation of mathematics, and even the style in which it is done, are individual and personal. Links between highly private emotions or personal psychology and universal mathematical results can be found in the philosophies of individual mathematicians, in their attitudes, in their explicit or implicit premises, in their styles of work, and even in their relations with the larger mathematical community. The philosophical level of discourse and the description of style mediate between personal psychology and impersonal mathematics. Wiener was an articulate philosopher—his degree from Harvard had been in philosophy, not mathematics—and his own writings made explicit connections between his personal viewpoint and his mathematical work. Von Neumann was not so interested in or articulate about his philosophical views, so that discernible connections are only partly explicit in his writings, and remain partly implicit.

I shall link mathematics to personal elements through a description of bits and pieces of premises, philosophies, and work habits, which after all, do form some pattern. In placing their mathematical work in this context, I should like first to mention several possible ways of regarding their creative mathematics. The classical approach to mathematical creation is the introspective one, pioneered by H. Poincaré and J. Hadamard.[1] While this approach leads to some interesting insights into styles of mathematical creation, it is imbued with a romantic individualism that blinds it to the impact that a man's colleagues and everyday life might have on his work. On the other hand, a psychoanalytic biography might tend to err in the opposite direction. A third way is to examine the mathematical style of each mathematician's published works, just as a music critic might compare the styles of different composers' works. This approach is also revealing, but the more subtle and technical aspects of such an analysis of style are of interest primarily to other mathematicians. Another approach, the tracing of influences received from teachers or books or colleagues, would again be primarily of interest to other mathematicians. Alternatively, one might emphasize the Zeitgeist in attempting to characterize the styles of mathematicians;[2] finally, one could stress the social forces acting on the individual. This array of possible approaches to the topic at hand is intended to remind the reader of the incompleteness and modest aims of the present discussion.

As the first point of comparison between Wiener and von Neumann, consider their respective choices of topics of research in relation to the importance the leaders in the mathematical community attached to the problems. As the mathematician S. Ulam has described it:

Mathematicians, at the outset of their creative work, are often confronted by two conflicting motivations: the first is to contribute to the edifice of existing work—it is there that one can be sure of gaining recognition quickly by solving outstanding

118
John von
Neumann
and
Norbert
Wiener

problems—the second is the desire to blaze new trails and to create new syntheses. This latter course is a more risky undertaking, the final judgement of value or success appearing only in the future.[3]

As Ulam points out, von Neumann "in his early work chose the first of these alternatives; it was only toward the end of his life that he felt sure enough of himself to engage freely and yet painstakingly in the creation of a possible new mathematical discipline," namely automata theory. By contrast Wiener, in particular in his work on Lebesgue measure and Brownian motion, took the second route early in his life. Recognition for his achievement was consequently delayed some years, a fact that Wiener, eager to be recognized, resented.

Von Neumann had unhesitatingly jumped into highly competitive areas of research in the mainstream, mostly within the framework of problems posed by the great David Hilbert or his students. His unfailing craftsmanship in their solution assured him of immediate acclaim, and from the first he moved rapidly to the top of the profession. Wiener was never so rational about his own career and moved up more slowly. He had worked on topics such as quantum theory and especially the formulation of axioms for vector spaces that were regarded as important,[4] but when the atmosphere became highly competitive, he moved to other topics. He primarily chose to concentrate on a problem in "an esoteric corner of mathematics"[5] well outside the mainstream.[6] The problem, a profound one, required high originality; the choice of topic required independence, especially from the judgment of the American mathematical establishment.[7]

Wiener had been an outsider, and an exceptional outsider, before. He knew that his work, the ideas and the craftsmanship, was good. If the mathematical establishment was slow to recognize it, making him unable to get an appointment at any of the leading universities, however frustrating and irksome that may be, it was evidence of the limitations of that establishment. It did not change the direction of his work, even if it magnified his self-doubts. In all his early upbringing he had become predisposed to a mixture of contempt for and helplessness toward the political and economic establishment. He would henceforth always be skeptical of the opinions of the mathematical establishment, whose blessings he nonetheless needed and wanted.[8] One of Wiener's heroes would always be Oliver Heaviside, the British mathematician who had developed the highly original operational calculus late in the nineteenth century but had been treated with contempt by the mathematical establishment of his time. The most outstanding native American theoretical physicist, Willard Gibbs, who was little understood in his own country at first, was another of Wiener's scientific heroes. In the view of one mathematician, "Norbert Wiener was a true eccentric and von Neumann was, if anything, the opposite—a really solid person."[9]

Much of mathematics deals with imagined truths created by definitions, far removed from any empirical science, but both von Neumann and Wiener belonged to that minority among mathematicians who take a serious interest in empirics. In the history of science from ancient times to the present, the periods when the impulse to bring the most advanced mathematics to bear on the material world predominated, an impulse linking the esoteric to the senses, and even to the practical, have usually also been periods of technological advance and social change.[10] I believe that generally a mathematician's predilection for mathematics related to empirics and applications expresses an impulse to have an impact on a wider world than that of the mathematical cognoscenti. Von Neumann's and Wiener's later careers gave direct expression to such an impulse.

However, von Neumann's early work on the mathematical foundation of quantum theory and Wiener's on the mathematics of Brownian motion already belong to mathematical physics and are directly related to empirics. On one level, one might see here the influence of their respective teachers. In particular, it had been Wiener's mentor, Bertrand Russell, who had called his attention to mathematical physics. Russell had urged Wiener, when he was studying with him, to read the papers of Einstein and take an interest in the new physics, including the work on Brownian motion. Russell would have liked to have been a physicist himself, but having gone off in a

120
John von
Neumann
and
Norbert
Wiener

different direction, the best he could do was to urge Wiener to take an interest in it.[11]

Von Neumann had become involved in theoretical physics in his student days, when he visited Berlin, listening to Einstein lecture on statistical mechanics and engaging in lengthy discussions with the budding physicists Wigner, Gabor, and Szilard. He encountered mathematical physics especially at Göttingen. There von Neumann learned from Hilbert, who, though a mathematician, was very eager to keep up with and contribute to the axiomatic formulation of the new physics, something he himself could do only to a limited extent,[12] but which von Neumann in fact did. So it seems that when working in mathematical physics, both Wiener and von Neumann were not only pursuing their own interests but also inheriting the unfulfilled desires of an earlier generation.

Yet why should the two younger men respond to just that particular facet of their mentors' interests? Wiener's and von Neumann's persistent interest in empirics became so much part of their mathematical style that the motives for it must have come from within. This does not necessarily mean that the motivations were the same, although both men were conscious of a predilection for relating esoteric mathematics to the material world. For example, speaking of his work describing Brownian motion, Wiener said that it "stood approximately in the middle ground where physics and mathematics meet . . . such work seemed to be in harmony with a basic aspect of my own personality."[13] How is one to understand this?

A tension prevails between the ideas of pure mathematics, which are freely invented mental constructs, and the facts of the material world. As we shall see later, Wiener experienced and identified his private world with such tensions; some of his mathematical work was energized by efforts to create harmony, in his personal world as in the world of mathematics, in the face of apparently incompatible elements. Moreover, at the MIT of the 1920s, an institution oriented to practical engineers, Wiener's ability to relate sophisticated mathematics to physics and engineering made him a more highly valued and appreciated member of that community. It helped him to belong at MIT; it resolved a discord.

The tension between pure mathematics and the practical world is an old one. Alchemists in medieval Europe and in ancient China saw in their craft a process symbolizing both the cosmos and themselves.[14] Although it is rare among moderns, Wiener appears to have been attracted to problems that could and did serve a similar kind of function for him. In his autobiography Wiener has described instances in which an unsolved mathematical problem came to be a metaphor for a painful personal conflict.[15] He sometimes experienced his mathematical research as a means for attaining harmony, for resolving inner discords that may have had their origin in the manner of his upbringing.

Von Neumann described his interest in empirical topics in very different terms:

At a great distance from its empirical source, or after much "abstract" inbreeding, a mathematical subject is in danger of degeneration. At the inception the style is usually classical; when it shows signs of becoming baroque, then the danger signal is up. . . . In any event, whenever this stage is reached, the only remedy seems to me to be the rejuvenating return to the source: the reinjection of more or less directly empirical ideas. I am convinced that this was a necessary condition to conserve the freshness and vitality of the subject and that this will remain equally true in the future.[16]

This brief quotation evokes two persistent Neumannian strains. The first is the commitment to "progress" in mathematics. Such "progress" is the very raison d'être of the mathematical community, so that the justification for his interest in empirical ideas is an appeal to a value central to the mathematical community. But for von Neumann it appears to be intimately linked to the second theme, the theme of rejuvenation, as a counterpart or palliative to the "fear of extinction," a governing concern in von Neumann's life, especially his later years. On a practical level, von Neumann's work in quantum theory gained him the appreciation of physicists and not only the mathematicians in the 1920s. It made him part of the broader intellectual stream of that time. Contacts with the empirical help to prevent barrenness in mathematics, and consequently von Neumann valued these contacts. But empirical

122
John von
Neumann
and
Norbert
Wiener

theories, and especially engineering devices, held an attraction of their own for him, so that later we find him becoming so enthralled by developments in technology that he could barely find the time to work out his highly innovative mathematical ideas.

While the language used is not so unusual for mathematicians, the above quotations from Wiener and von Neumann seem to hint at personal concerns which may have provided a stimulus to each man's work. These same personal concerns did indeed reappear persistently in various contexts in their respective activities. These emotional undercurrents were transformed through their gifts for original creation and free play, and lifted out of the range of compulsion or anxiety to the realm of freedom in mathematical innovation. This psychological process eludes clear analysis, and besides, however important, it is only one strand in their makeup, inextricably interwoven with other objectives: success and achievement in their mathematical endeavors, aesthetic motives,[17] craftsmanship, and simple pleasure in the doing.

Further definition of the two men's differences and similarities can be obtained from a description of their working habits. For example, both Wiener and von Neumann trusted their "subconscious"; both were in the habit, at least during some periods of their lives, of sleeping with paper and pencil near their beds; and both explicitly reported making use in their mathematical work of dreams, imagery, and initially inexplicable impulses—impulses they followed.[18]

Mathematical creation, like every other human activity, has its share of pain and suffering as well as pleasure. This is reported by mathematicians or manifested in their working habits. Deep differences existed between von Neumann and Wiener, I think, in the characteristic ways in which pain and pleasure entered into their work—differences that reflect their differing personalities.

Wiener has been described by his close friend Dirk Struik as usually "entweder Himmelhoch jauchzend oder zu Tode betruebt,"[19] and similarly by another friend, Norman Levinson: "His mood could shift quickly from a state of euphoria to the depth of dark despair."[20] At times these moods visibly had something to do with his own self-doubts or with the recognition accorded his work. At other times they had to do with the mathematics itself. Wiener has recalled an instance of his work as a young man (1915):

The idea occurred to me to generalize the notions of transitivity and permutability . . . to systems of larger number of dimensions. I lived with this idea for a week. . . . I soon became aware that I had something good; but *the unresolved ideas were a positive torture to me until I had finally written them down and got them out of my system.*[21]

His relief at unburdening himself of the painful "unresolved ideas" is in part a reaffirmation of his ability to do so. I surmise that he experienced moments of pain, helplessness, frustration, and self-doubt until he had succeeded in working out such ideas, or at least saw clearly how he might do so. Later, in 1929, speaking of mathematical research, he points to "the thrill, when the several threads of thought in a mathematical theory are seen to gather themselves together into one perfect fabric. *It is an inner unity and purpose* masked by a superficial diversity."[22] The pleasure appears in the harmony and unity created—"the elation as the work approaches its final form." Armand Siegel, who became Wiener's close younger collaborator around 1950, described to me informally Wiener's characteristic pattern of working, as Siegel had encountered it, which was

to go into a depression, and to goad himself by this depression, to goad himself emotionally into a higher level or deeper level of probing and examining and a kind of desperation. . . . He would show [his depression], for he had a very expressive face and expressive voice, he would be mournful and lugubrious. He would say things like "It isn't going at all," "I can't find it." Sometimes he said, "We are wasting our time," but only if he said it cheerfully, did it mean he was giving it up. When he said it with a strong affect of misery and hopelessness, he was in the midst of it, and suddenly would become enthusiastic—"Aha, I have it"—and write it on the board—he couldn't stand to be depressed too long. Then he might discover that his idea didn't work after all, so back into the de-

124
John von
Neumann
and
Norbert
Wiener

pression. And then one of the times his "I have it, I've got it!" would really work out. And that would be it.[23]

Wiener's working pattern probably changed from decade to decade, and perhaps from one type of problem to another. But it appears that from his early to his later work, moments of painful depression served to trigger creative innovations for him, and the mathematical innovations, if successful, reaffirmed him and led him out of depression toward a happy mood.

Von Neumann's demeanor did not show the mood swings that characterized Wiener's; although occasionally angry, he had a sunny disposition, nearly always cheerful and positive. Whereas Wiener was impulsive, planning and forethought characterized all von Neumann's responses. His closest associates agree that von Neumann profoundly enjoyed mathematical thinking, in fact all kinds of thinking.[24] One of them ventured that of all activities thinking was what von Neumann enjoyed most in life.[25] Another notes that while groans might be heard coming from the rooms of some mathematicians when they are in the throes of working something out, von Neumann was never like that. He worked

with ease and pleasure. . . . If he would press on a problem, and he couldn't push it through, it would be put aside. It might be two years later when suddenly you'd get a phone call—it might be from God knows where at two in the morning, and there was von Neumann on the wire: "I now understand how to do such and such."[26]

Von Neumann himself noted that if a mathematician gets bogged down seriously, however challenging or "important" the research, he is much freer than an engineer or a physicist to turn to another topic.[27] Von Neumann's talents were such that he was always aware of many topics on which he could work productively, and he did not easily get bogged down. Thus it seems that von Neumann had adopted a conscious strategy that required only patience, trust in "time," to obviate frustration and feelings of helplessness and discouragement. It is of interest that in his later extensions of game theory it was von Neumann who first formalized the mundane problem of finding strategies to preclude undesired events in everyday life and increase the likelihood of desired ones.

Von Neumann's enjoyment of mathematics is noteworthy in itself. He described the mathematician's motivation as one of achieving beauty, a particular kind of mathematical beauty. In this relationship to the beautiful are surely some of the roots of enjoyment, although for von Neumann his own sense of mastery in solving mathematical problems also appears to have been a primary source of pleasure. When von Neumann identified the criteria of "beauty" in any mathematical theorem, he particularly mentioned

"Elegance" in its "architectural," structural make-up: Ease in stating the problem, great difficulty in getting hold of it and in all attempts at approaching it, then again some very surprising twist by which the approach, or some part of the approach becomes easy, etc. Also, if the deductions are lengthy or complicated, there should be some simple general principle involved, which "explains" the complications and detours, reduces the apparent arbitrariness to a few simple guiding motivations, etc.[28]

This criterion and the emphasis on "difficulty" suggests some commonality between mathematical work and a virtuoso performance, such as by a violinist or figure skater. As these similes suggest, the extensive employment and display of technical virtuosity apparently gave von Neumann much pleasure,[29] over and above the creation of elegance in the structural makeup of a proof. Ulam confirms this impression:

Von Neumann was the master of, but also a little bit the slave to, his own technique. When he saw that something could be done, he let himself be carried away on tangents. My own feeling is that some of his mathematical work on classes of operators or on quasi-periodic functions, for example, is very interesting technically, but to my taste not terribly important; he could not resist doing it because of his facility.[30]

Akin to mastery is the sense of freedom found in mental play with concepts and symbols. Operating simultaneously on many levels, from the empirical and the simple numerical to high abstraction, or moving back and forth easily, involves a sense of play and freedom that can be highly pleasurable and

126
John von
Neumann
and
Norbert
Wiener

creative, and von Neumann seemed to have been continually engaged in play of this sort. It could be either highly sophisticated or very mundane; a striking example is the range of his work with computers, in which he moved continuously back and forth between the highly formal automata theory, neurophysiological data, and the engineering characteristics of available components, as well as the politics and financing needed to get computer projects started. His formal lectures on mathematics were frequently interlaced with hidden jokes designed for a small group of associates. (Incidentally, his lectures were usually given without any notes.) Even his normal conversations often involved play with double meanings and shifts of viewpoint. While most of his mental preoccupations suggest strongly what Einstein described as "the flight from the *I* and *we* to the *it*," von Neumann's humor, including the dirty limericks that were his specialty, was also a tool of sociability.

Von Neumann's mastery often had a particular quality. Whatever the problem, he began by holding and organizing the multitude of needed elements in his mind, in this way gaining insight into their structural interrelation, before he wrote or even spoke about it. This period of concentration and preparation lasted a few minutes if the problem presented to him was relatively easy; it could extend over a long period of time if it was very difficult. The subsequent explicit working out on paper or blackboard often had the quality of spontaneity available to a mind absorbed in and fully attuned to a problem. This state of mind, in which absorption in the problem is followed by its spontaneous working out, has the makings of a relatively passive form of enjoyment over and above that of mastery.[31]

A still different kind of gratification or motivation is indicated by von Neumann's approach to the philosophical issues underlying mathematics, the sciences, life, and mind, which he dealt with by using the tools of formal logic and abstract mathematics. To approach this realm of arbitrariness and uncertainty by means of the abstractions of logic reveals an amazing confidence or optimism that existential anxiety can be assuaged or avoided through formal logical structure.

This confidence in the power of logic, which after all is mechanical, to resolve the problems of human life is merely another form of the optimism and faith in technology that von Neumann inherited from his early years during Hungary's economic takeoff. He was beholden and captive to this heritage, which he never felt obliged to discard.

The engineer Julian Bigelow had worked closely with Wiener during World War II and after the war worked closely with von Neumann.[32] No two working situations are entirely comparable, but Bigelow's descriptions have special merit, coming from the same participant-observer.

Wiener thought as a physicist really. He thought in terms of process. He thought in terms of some kinds of physical models or some kinds of intuitive models. Von Neumann was a fantastic craftsman of theory. He was also immensely imaginative. . . . The craftsmanship was so strong that you weren't even aware of it, but he also had the kind of imagination that could see through the problem. . . . He could write a problem out the first time he heard it, with a very good notation to express the problem. . . . He was very careful that what he said and what he proved and what he wrote down was really exactly what he meant . . . Wiener didn't give a damn. For Wiener, mathematical notation and language was an encumbrance. He could get an insight and he would be wanting to say it and he would fumble with the notation . . . because he was not thinking about the notation he was working in, but he was thinking about the problem behind it. . . . Von Neumann would get a new problem and analyze it all the way through to the end, and end up with a computation. He would write down the numerical computation and the order of magnitude of how big it would be, you see, and if it then didn't come out correctly, he would go back and check his steps and see where he made a mistake. Now Wiener never did that. Wiener would erase the blackboard and start and do it again, a different way, in a new notation, which didn't fit either, but he was trying to say something, which after he got all through was correct.[33]

In agreement with Bigelow's observation, Wiener commented on his own tendency to think in terms of physical models: "In many problems which we [Paley and Wiener, early 1930s] undertook, I saw, as was my habit, a physical and even an en-

128
John von
Neumann
and
Norbert
Wiener

gineering application, and my sense of this often determined the image I formed and the tools by which I sought to solve my problems."[34]

Many of Wiener's mathematical articles, such as his classic ones on differential space and on generalized harmonic analysis, make frequent reference to physical processes.[35] His use of apparently visual (and possibly tactile) images derived from the apparatus or phenomena of physics offers another insight into his preference for mathematics related to empirics, although later I shall show that at times Wiener derived visual images from entirely different sources. Wiener, whose eyesight was deficient, saw vivid images at certain moments during the creative process of mathematics. By contrast, visual or geometric visualization seemed not to play an important role in von Neumann's thinking. This has been especially noted by Ulam:

Von Neumann gave the impression of operating sequentially by purely formal deductions. . . . If one has to divide mathematicians, as Poincaré proposed, into two types—those with visual and those with auditory intuition—Johnny perhaps belonged to the latter. In him, the "auditory sense," however, probably was very abstract. It involved, rather, a complementarity between the formal appearance of a collection of symbols and the game played with them on the one hand, and an interpretation of their meanings on the other. The foregoing distinction is somewhat like that between a mental picture of the physical chess board and a mental picture of a sequence of moves on it, written down in algebraic notation.[36]

What, precisely, were the mental characteristics that led his colleagues to regard von Neumann as a phenomenon? Bigelow's observation that von Neumann "would write a problem out the first time he heard it, with a very good notation to express the problem" is telling. It indicates more than only that von Neumann, regardless of the content of a problem, was immediately concerned with form. For to choose a good notation the first time implies that mentally von Neumann had anticipated correctly all the considerations that would enter into the solution of the problem. Colleagues were always impressed with the fantastic speed with which he could give the appropriate mathematical formulation as well as the solution of a problem. A colleague at Göttingen, where at the time Europe's most brilliant mathematicians were either teaching or visiting, described von Neumann as equipped with "the fastest mind I ever met."[37] Decades later, a colleague at the Princeton Institute of Advanced Study, then one of the world's centers of mathematics, noted that when presented with a mathematical difficulty von Neumann "beyond anyone else, could *almost instantly* understand what was involved and show how to prove the theorem in question or to replace it by what was the true theorem."[38] Innumerable recollections and anecdotes testify to the great speed of his mental processes; a typical comment is, "I discussed the problem with von Neumann, and in five minutes he advanced the subject as much as I would have in perhaps five months."[39] Another of his mental powers was an effectively photographic memory. His coworker Goldstine wrote, "As far as I could tell, von Neumann was able on once reading a book or article to quote it back verbatim; moreover he could do it years later without hesitation. . . . I tested his ability."[40]

Von Neumann's enormous working speed and his remarkable power of memory enabled him to defy, as it were, the ordinary limitations imposed by time, since for him these limitations were less severe than for normal persons. His memory seemed unfazed by the transitoriness of events, and his ability to carry out logical reasoning with unequaled rapidity and accuracy seemed to exceed the reach of ordinary human faculties. In this way von Neumann's talents are likely to have conditioned his personality and his view of himself in relation to the world in the sense that he had some power over the exigencies of time.

This logical mastery also may have affected von Neumann's views and premises concerning the world. It became his mathematical and scientific style to push the use of formal logic and mathematics to the very limit, even into domains others felt to be beyond their reach. He seemed to regard the empirical world, probably even life and mind, as comprehensible in terms of abstract formal structure. This premise, insofar as it is shared by others, gives some of his work its very great importance for the select who can understand it. His

130
John von
Neumann
and
Norbert
Wiener

work on the foundations of mathematics was conducted on the highest level of abstraction, and he tended to embed empirical theories (game theory, quantum theory, organization of computers) into formal, axiomatic structures. He seems to fall under that tradition in Western thought in which it is believed that only rigorous logics will ever succeed in containing the timeless, universal truths that govern everything.

Von Neumann is doubly interesting because he also functioned effectively on a practical, political plane, where rather humble levels of abstraction are called for, where finite time durations and practical facts need to be taken into account, and where even the most lucid reasoning falls far short of rigorous logic. Thus when von Neumann was a member of the US government's Atomic Energy Commission, its chairman, Admiral Lewis Strauss, observed, "If he [von Neumann] analyzed a problem, it was not necessary to discuss it any further. It was clear what had to be done."[41] Lucid, logical thinking—his boyhood amulet—joined with a respect for concrete and detailed facts had become his habitual response to complex situations of all kinds, his means for survival. Other people have other kinds of habitual characteristic responses to complex situations, as for example a sensitivity to moods and feelings; for von Neumann lucid, rapid reasoning based on a thorough familiarity with facts was by far his strongest suit for dealing with anything. With the technical, scientific problems like weather prediction, computer design, or explosions, von Neumann was very good at making quick, approximate numerical engineering estimates when they were called for. Von Neumann could and did function on several levels of abstraction simultaneously.

An extreme view in one area of life generates in the human personality some compensatory elements. Thus we have von Neumann's capacity for abstraction on the one hand and his practicality on the other. Similarly, his commitment to timeless truths is counterpointed by his avid interest in history: "His knowledge of ancient history was unbelievably detailed. He remembered, for instance, all the anecdotal material in Gibbon's *Decline and Fall* and liked to engage after dinner in historical discussions."[42]

The search for logical rigor and the wish to defy the exigencies of time are closely related. Like Plato's world of essences, a perfectly rigorous system of logic is independent of any special place or time. The very language used by von Neumann in formulating quantum theory, namely his description of the process as identifying the "abstract essence" of empirical theories, has the flavor of philosophical idealism, of the Platonic Idea. In fact, when von Neumann actually did give an interpretation of his quantum-mechanical formalism, it was close to an idealistic one. By contrast, Wiener's interpretation was explicitly materialistic. The interpretation of quantum theory is one useful topic providing a concrete means of comparing von Neumann's and Wiener's orientations; the difference between their attitudes is more interesting and complex than the opposition between materialism and idealism. It has to do with both time and formal logical structure, as well as with the relation of the human observer to the world.

To appreciate properly von Neumann's and Wiener's individual outlooks on quantum theory, it is necessary to see them in relation to that of the leading interpreter of quantum theory, Niels Bohr, whose institute in Copenhagen was in the 1920s and 1930s the center for the most exhaustive and searching discussion of the problem of interpretation. In prequantum physics, there was universal agreement among scientists about how to understand the relation of an observer to the object or external events he observed or measured. A measuring instrument might perturb the object in question, but one could always correct for any error introduced thereby; and a properly designed measuring instrument gave an unambiguous reading, quite independent of the person or the mind of the observing scientist. There was no fundamental problem of measurement. But in the quantum theory, which dealt especially with small objects such as atoms, a problem arose: When a measuring instrument is coupled to a small object (to be referred to as an "atom"), an interaction occurs that according to quantum theory is to a certain extent indeterminate; this creates an intrinsic uncertainty in the state of the atom, and no way exists to remove the uncertainty. Consequently any prediction about the future of the atom can only be a statistical one. The intrinsic uncertainty in measurements of atoms is, according to quantum theory, of a special kind; in particular, it frustrates efforts at exact prediction of an atom's subsequent motion. To predict its future from the present, one should have simultaneously to measure its exact location and its rate of motion (momentum). But according to quantum theory it is just this that is impossible: the more accurately one determines the atom's location, the more uncertain its momentum becomes, and vice versa.

Bohr termed position and momentum a pair of complementary variables. There are also other pairs of complementary variables in quantum theory, as for example the energy of an atom and the precise time at which it has that energy, which cannot be determined simultaneously without intrinsic uncertainty. Electrons in some experiments show wavelike properties and in others particlelike properties. The more wavelike the electron, the less particlelike it appears. Light also has at some times wave properties and at other times particle properties. Bohr termed the wave properties and the particle properties "complementary aspects." The philosophically minded Bohr conjectured that nature is characterized throughout by a general "principle of complementarity" which defines the limits of knowledge; he expected that in biology and psychology one would also find complementary facets of phenomena, facets whose observation is in some sense mutually exclusive.

Bohr's complementarity principle was one pillar of his interpretation of quantum theory. Another was that the reading of the measuring instrument must be unambiguous, independent of the mental characteristics of the observing scientist; it must be an "objective" fact. In other words, the relation of the measuring instrument to the human observer is the same as in classical physics; the pointers or counters are not of atomic dimensions but sufficiently large that they can be described by classical physics. Without this requirement the measurements on atoms would not be objective public facts at all, but only private subjective impressions, different for different observers. A third pillar in Bohr's interpretation is a clarification of the complementarity concept in physics. To measure the location of an atom, one brings it into some kind of contact with a position-measuring device, and the human observer only inspects the pointer reading of the "position-measuring device in contact with the atom." The human observer does not observe the isolated atom but only the combination of "atom plus measuring device." Every measurement device creates a special environment for the atom; a momentum-measuring device creates a different environment from a position-measuring device. The essential limitation of the human observer is that he can have no objective observation of the isolated atom but only the atom in the special environment of one measuring device or another.[43]

Von Neumann's approach was more formal and less directly philosophical. Thus the general philosophical concept of "complementarity" did not interest him; it was merely a matter of formulating appropriate, self-consistent rules for the algebra of operators describing location or momentum.[44] The algebra interested von Neumann, but not the metaphysics. The formalism von Neumann had devised for quantum theory incorporated these rules of algebra. Von Neumann's mathematical formulation of quantum theory in terms of Hilbert space, described in chapter 4, differed from other formulations in that it was fully axiomatic and was logically and mathematically more rigorous. For purposes of interpretation it only needed to be supplemented by Bohr's general ideas concerning the coupling of "atom" to measuring instrument and the relation of measuring device to human observer. Instead of doing so, von Neumann characteristically extended the mathematical formalism to incorporate the process of measurement as well as the human observer. The formalism, which had to be free of logical contradiction, was for von Neumann the determinant of philosophical meaning, the relation of observer to measuring instrument, and so on. In the derivation of his mathematical measurement theory von Neumann found he had to make the awkward assumption that the measurement process is instantaneous, instead of requiring a finite time. He makes the assumption, acknowledging that it cannot be correct in principle, but argues that it

134

John von
Neumann
and
Norbert
Wiener

should not make much difference in the end result. This is the only blemish on an otherwise logically satisfactory formalism.

Accepting von Neumann's formalism, however, entails a philosophical view different from Bohr's. Bohr had insisted that the element of the measuring instrument observed must be an ordinary "classical" quantity (such as a pointer reading), so that the observation is entirely objective as in classical physics, and different observers will consequently agree on the result. Von Neumann, however, makes no such stipulation, and in his interpretation an observer *cannot* make assertions about any "objective" facts. In von Neumann's words, one "only makes statements of this type: an observer has made a certain (subjective) observation; and never any like this: a physical quantity has a certain value."[45] All observations are "subjective"! The point of view is clearly positivistic, but rather than simply and hastily labeling it also solipsistic, it is worthwhile describing it and its implications more specifically.

Von Neumann mentally divides the world into "three parts: I, II, III. Let I be the system actually observed, II the measuring instrument, and III the actual observer."[46] He argues that it is arbitrary just where one draws the line between II and III, the measuring instrument and the observer. The "observer" includes a person's "abstract ego," or consciousness. It may also be taken to include his eyes, his nerve paths, and his brain, and some laboratory equipment as well, but it need not include anything other than his consciousness. Although von Neumann described systems I and II in terms of quantum theory, the "observer" remains outside of any mathematical description. It is a crucial and novel part of von Neumann's measurement theory that *the "observer's" merely taking cognizance of (I) and (II) changes the state of the physical system!* In physicists' jargon this change in the vector in Hilbert space describing (I) and (II) is referred to as the "reduction of a superposition." In the von Neumann theory "this takes place whenever the result of an observation enters the consciousness of the observer—or, to be even more painfully precise, my own consciousness, since I am the only observer, all other people being only subjects of my observations."[47]

The notion that a person's mere consciousness of a physical phenomenon changes that phenomenon in the way required for "reduction of a superposition" is a novelty in physics; that the mind has such powers is an implicit assumption in the von Neumann model.[48] If one observer's "awareness" influences the physical event being observed, the question arises whether a second observer who becomes aware of the same physical event will agree in his observations with the first. To obtain "intersubjective agreement" von Neumann offers only the possibility that the second observer can regard the first observer as an object, a measuring instrument, and by proper use of this human measuring instrument will arrive at results for systems (I) and (II) that agree with those of the first observer. However, he has no mechanism for obtaining agreement between two observers, each of whom has an independent "consciousness." In the words of one philosopher,

Unless "pre-established harmony" is accepted, the only *Weltbild* in which quantum theory is rigorously maintained appears to be one in which there is a single ultimate subject. This is a different conclusion from the common skeptical position regarding the existence of other minds, and in one respect it is stronger. It does not say that I am unsure other people have the same kind of subjective characteristics as myself, but rather that if they have minds then certainly their minds lack some of my powers, particularly the ability to reduce a superposition.[49]

In von Neumann's bold formal theory the clean separation of subject (observer's consciousness) and object has been sacrificed; a type of philosophical idealism has replaced the usual materialistic assumptions of science. Mind only we know, not matter. The philosophical framework of the theory is incompletely spelled out. To von Neumann formalism was supreme; the elegance and completeness of the formalism devised and its positivistic formulation are decisive for the theory, let the metaphysical, epistemological, and ontological chips fall where they may.[50] While attractive to many mathematicians, this approach is unsatisfactory to philosophers of science:

Modern science has developed mathematical structures which exceed anything that has existed so far in coherence and generality. But in order to achieve this miracle all the existing troubles had to be pushed into the *relation* between theory and fact, and . . . . Von Neumann's work in quantum mechanics is an especially instructive example of this procedure . . . the theory becomes a veritable monster of rigor and precision while its relation to experience is more obscure than ever.[51]

However problematic the mathematical supremacy may be for philosophy, von Neumann took a similar stance on topics other than quantum theory.[52] Von Neumann's work is an attempt to confront and master through an axiomatic and mathematical formalism certain paradoxes and mysteries. If his formulation appears to have novel epistemological and ontological implications, it also has the characteristic that deep mysteries of knower and known, of subject and object, of mind and body, appear to be securely contained in a mechanical formalism that is logically satisfactory except for the one awkward assumption of instantaneousness in the measurement process. The guiding characteristic that informs so much of von Neumann's work is the effort to devise as far as possible—and even further—a formal or mathematical structure within which to contain the mysteries and complexities of life. It is a naive and optimistic faith in mathematical machinery. He is pushing out the bounds of the subject matter amenable to logic. His compression of the finite duration of the physical measurement process to a mere infinitesimal instant is perhaps a quite natural assumption to his mind, and the philosophical implication of a single ultimate subject, whether it was congenial to him or not, did not deter him. This does not belittle von Neumann's accomplishment, but only reveals his style. After all, this was the first serious, viable mathematical model for the measurement process, and it demonstrated that such an entirely self-consistent formal model could be constructed. From von Neumann's viewpoint, if there were some serious difficulties of interpretation, it did not mean he would abandon the idea of a mathematical (i.e., mechanical) model for incorporating the interaction of the observer and the object via the measurement instrument. In-

stead, he would look for a new set of self-consistent formal
axioms as the basis for a different mathematical structure, one that might remove the difficulties of interpretation. To mathematize, to formalize, he must.

When von Neumann was older, several decades after his quantum theory of measurement, he developed the theory of automata, of which more will be said later. But here a comment is in order about its possible relation to epistemology. The idea that the process of perceiving and knowing can be described in terms of "incorporal (logical) automata," somehow associated with corporal entities, goes back to Leibniz.[53] Von Neumann's automata theory contains rudimentary logical descriptions of at least analogues of the processes of perceiving and knowing. The "ultimate mathematical vision" is suggested by the availability of a logical description of quantum mechanics and the idea of an "automata theory" describing the organization and dynamics of cognitive processes. It then requires only the discovery of the proper description of the links (which need not be causal, nor have a one-to-one correspondence) connecting the quantum logics and the logics of the perceiving automata to encompass the object known and the knower in a coherent logical description. (The concepts of information transfer and coding, later so much emphasized by Wiener, suggest possible types of links.) Leibniz, in the seventeenth century, explicitly championed this kind of "ultimate mathematical vision," and John von Neumann in the twentieth century implicitly came to share that vision. Leibniz is the philosopher who also aspired to the invention of a logical calculus that could make even philosophical thinking mechanical. "If controversies were to arise," he said, "there would be no more need of disputation between two philosophers than between two accountants. For it would suffice to take their pencils in their hands, to sit down to their slates, and to say to each other (with a friend as witness, if they liked): Let us calculate."[54] Von Neumann also surely had great sympathy for this rationalistic aspiration.[55]

Norbert Wiener's approach to the interpretation of the mathematics of quantum theory was different and in certain respects diametrically opposed to von Neumann's.[56] Like Bohr,

138
John von
Neumann
and
Norbert
Wiener

Wiener viewed the complementarity principle as a valid philosophical idea, concrete and precise in the context of quantum physics but with much wider applicability as a general principle. In Wiener's hands the principle was made to describe the limits of logical reasoning and the centrality of finite duration, not only in atomic physics but in all of science. Consider for example the Wienerian statement concerning the complementarity of the wave and particle aspects of electrons and of light (1934): "The present state of physics represents a Hegelian synthesis between Newton's thesis of particle and Huyghens' antithesis of wave." In this way of putting it, wave and particle are not only seen as physical or logical contraries but are put in a historical context, in which one concept (partial truth) predominates during one period of history to be followed by ascendance of the contrary concept; ultimately the two are reconciled as part of the process of the evolution of ideas. It was Wiener's view not merely that the "truths" of empirical science are true relative only to a particular historical era, but that the same can be said of the truths of mathematics and logic! He buttresses this viewpoint by examining the history of logic and mathematics and pointing to the many changes in the concepts of mathematical correctness, of logical rigor, and of the foundations of mathematics that have taken place. This is one kind of temporal bound, which in Wiener's mind precludes absolutely "timeless" mathematics.

Besides reinterpreting complementarity in a historical way, he also found an analogy to complementary variables in the characteristic process of scientific generalization from concrete experiments. He regarded experimental concreteness and generality as complementary attributes, so that any improvement in one is at the expense of our ability to apply the other in the same situation: "Rationalism and empiricism are alike accounts of asymptotic modes of thought toward which we tend in one mood or another, rather than adequate epistemologies of our normal experience."[57] Wiener concludes from the conflict between concreteness and generality that "any really useful logic must concern itself with Ideas with a fringe of vagueness and a Truth that is a matter of degree." In turn, he noted that every observation, particularly introspec-

tions, take some finite time, thereby introducing uncertainty.
Science, which to Wiener in general meant "explanation of process" and incidentally was itself a process, required a suitable compromise between rationalism and empiricism. Wiener's prescription for at least one kind of useful physical-mathematical theory is to incorporate the imprecision of measurements in the mathematical description, for otherwise the theory would be inappropriately definite. Yet such a mathematical description incorporating imprecision could be completely rigorous. This line of reasoning led Wiener to become an advocate of a statistical approach to the "explanation of process," and specifically the general statistical approach introduced into physics by Willard Gibbs, the late nineteenth-century American physicist. Wiener's theory of Brownian motion was indeed that kind of theory, and he thought that quantum theory could and should be understood in similar terms. Here, though, Bohr disagreed with Wiener,[58] although Einstein for his own reasons fully agreed.[59] In Wiener's way of regarding it, quantum theory was more nearly of a piece with other physical theories, those describing processes involving incomplete information.[60] In agreement with Bohr, and in disagreement with von Neumann, the relation between the human observer and the measuring device is for Wiener the usual classical one, so that subject and object are cleanly separated.[61] He found no need to include the consciousness of the observer, as von Neumann had. J. B. S. Haldane, a biologist friend of Wiener's, had published some detailed speculations concerning the application of quantum theory to organisms, putting quantum theory in the framework of dialectical materialism.[62] Without endorsing Haldane's detailed conjectures, Wiener explicitly noted that his conclusions did not violate his own general ideas concerning the proper interpretation of quantum theory.

I have called attention to the difference between Wiener's and von Neumann's attitudes toward the limitations of logic and finite time durations in the special context of the interpretation of quantum theory. To seek to deal with subjects of great generality and high levels of abstraction, Wiener had recourse to a philosophical mode of thought, using the tools of language; on this level he specified limits on the comprehensiveness of useful, precise logic, as well as temporal boundaries. For Wiener mathematical theories were special instances in which general philosophical ideas are made concrete. By contrast, von Neumann, skeptical of limitations imposed by a priori philosophical considerations, sought to deal with the most general and abstract in terms of formal logic. These are differences in scientific style that probably reflect differences in emotional makeup. Wiener's approach leads to a unified view of the world, including its human inhabitants, where everything is connected to everything else, but in which the most general principles have an element of vagueness. Von Neumann's attitude leads to a duality, the timeless truths contained in otherworldly, rigorously axiomatic, abstract logic versus the mundane empirical descriptions of the world of time and phenomena. However, every phenomenon partakes of both, because abstract logics describe the formal structures that contain and constrain all material and human events, all historical and dynamic descriptions.[63] To which of the two sides von Neumann assigned priority is not clear, but it is unlikely that any man is consistent in that respect over a lifetime. Christian theology contains a similar duality, but one in which descriptions in time and history are supplemented by descriptions in terms of timeless myths, not formal logic.[64] To put it succinctly, von Neumann's formal logics were his myths.

The significance of von Neumann's approach, in respect to extending mathematical and logical description, does not lie in its being idiosyncratic. On the contrary—it epitomizes the point of view characterizing a major portion of the modern scientific community, and specifically a major portion of the quantum theorists in the period between the two world wars.[65] Von Neumann differed only in that he was thoroughgoing, exceptionally effective, and uncompromising about an approach generally accepted as desirable by the community of hard scientists, and in that he made it seem an utterly natural, effortless expression of his personality. The perfection of his logical mind, that "well-oiled machine," had something almost nonhuman about it.

The attitudes adopted by von Neumann and Wiener toward the efforts to prove that mathematics itself is complete in its axioms, definitions, and rules and free of contradiction give an additional indication of their respective styles. Wiener encountered the issue as a young man when he worked with Bertrand Russell, but his reflections on the general topic of the completeness of knowledge date back to his childhood. The first essay Wiener mentions in his autobiography was written at age ten, while he was a student at Ayer High School. It was a philosophical paper entitled "The Theory of Igno rance" in which he gave a "philosophical demonstration of the incompleteness of all knowledge." In his autobiography Wiener writes, "My father liked it, and as a reward for this paper he took a long trolley ride with me to spend a few days at Greenacre, Maine, near Portsmouth, New Hampshire, among the mists of the Piscataqua River."[1]

The theme that human knowledge and also our capacity to control events must be incomplete was a palliative against any illusion of omniscience which somebody so extraordinarily knowledgeable and curious about nature as Norbert Wiener might otherwise be subject to. The awareness of limits to our knowledge and understanding was a persistent and basic theme with Wiener, who throughout his life was at the same time enthusiastically expanding his knowledge. He makes the remarkable comment, "I cannot find in my own intellectual history any brusque change between the striving of childhood after childish knowledge and the power and striving of my grown life after the new and unknown."[2]

142
*John von
Neumann
and
Norbert
Wiener*
Wiener worked with Bertrand Russell when he was nineteen or twenty. Russell had already formulated a formal logical system that was supposed to include mathematics and logical reasoning. Wiener, again, was a skeptic. In 1915 he had already come to the view that it is "highly probable that we can get no certainty that is absolute in the propositions of logic and mathematics, at any rate in those that derive their validity from the postulates of logic."[3] Looking back on that period, he later wrote, "I already then felt that an attempt to state all the assumptions of a logical system, including the assumptions by which these could be put together to produce new conclusions, was bound to be incomplete. . . . My heresies of that time have been confirmed by the later work of Gödel."[4] Gödel's famous proof, given in 1931, demonstrated conclusively that all complex mathematical and logical systems must be in a certain sense incomplete.[5] The proof came as an unexpected shock to most mathematicians of the time. "as an astounding and melancholy revelation."[6] To Wiener the Gödel proof apparently came as a comforting reassurance that things were as he had felt them to be. It confirmed his intuition of the incompleteness of all tight logical systems; it helped him to reconcile the nineteenth-century romanticism so dear to him with his work in mathematics. To find comprehensiveness, however open-ended, Wiener had recourse to discursive philosophical reflections, in which mathematics and logic played the role of suggesting ideas or of making vague general ideas precise within a restricted context. But a closed logic itself was to Wiener's mind necessarily insufficient; although he had worked for several years in the framework of the Russell-Whitehead *Principia*, he did not become an advocate of Russell's analytical philosophy, which was based on formal symbolic logic.

Von Neumann's relation to these events in the history of logic and mathematics was very different. Notwithstanding his appreciation of facts and empirical research, von Neumann devoted himself to extending logicalization or mathematization as far as possible. A premise underlying this commitment is that logical systems have universality and consequently comprehensiveness; formal logical structure in some way captures the abstract essence of things. This view gave von Neumann an affinity with the seventeenth-century philosophers, especially Leibniz. He was not particularly interested in identifying a priori limitations of logic, but if a limitation appeared, it was taken as an obstacle to be surmounted. The problem was then how it might be surmounted in the course of the endless progress of formalization.

Von Neumann became involved with formal structure in the 1920s at Göttingen. Serious doubts had been raised by other mathematicians concerning proofs of the internal consistency of mathematics. In the face of these threats to mathematics, von Neumann joined Hilbert in the program he initiated to prove all mathematics to be free from contradiction. The approach of Hilbert and von Neumann was a purely formal one. Von Neumann succeeded in setting up a proper axiomatic formulation of the fundamental branch of mathematics, set theory, going as far as he could in proving it to be free from contradiction. At that time von Neumann optimistically believed that one could prove more, that in fact it was possible to prove all of mathematical analysis to be consistent. Other colleagues too thought that the foundations problem was now all but settled. "Then came a catastrophe: assuming that consistency is established, K. Gödel showed how to construct arithmetical propositions which are evidently true and yet not deducible with the formalism."[7] Von Neumann has retrospectively described Gödel's work:

Gödel was the first man to demonstrate that certain mathematical theorems can neither be proved nor disproved with the accepted, rigorous methods of mathematics. In other words, he demonstrated the existence of *undecidable* mathematical propositions. He proved furthermore that a very important specific proposition belonged to this class of undecidable problems: The question, as to whether mathematics is free of inner contradictions. The result is remarkable in its quasi-paradoxical "self-denial": It will never be possible to acquire *with mathematical means* the certainty that mathematics does not contain contradictions. It must be emphasized that the important point is, that this is not a

144
John von
Neumann
and
Norbert
Wiener

philosophical principle or a plausible intellectual attitude, but the result of a rigorous mathematical proof of an extremely sophisticated kind.[8]

When von Neumann first read Gödel's proof, which appeared in print in 1931, he, unlike others, quickly understood and accepted its conclusion and decided that his own and others' efforts at proving all of set theory (and with it classical mathematics) to be free of contradiction were hopeless.[9] In the course on mathematical logic he was teaching at the time at the University of Berlin, he changed from discussing proofs establishing the foundations of mathematics to discussing Gödel's theorem. He had boundless respect for Gödel's genius and later spoke of him as the "greatest logician since Aristotle."[10] In sum, von Neumann unhesitatingly accepted a serious limitation on "rigor" or "completeness" in mathematics, but only because the limitation itself was imposed by superior rigorous logic.[11]

It is clear that von Neumann and Wiener had encountered the foundation crisis of mathematics in different ways and responded to it differently. Wiener, highly intuitive, with a mind rich in every kind of imagery, was careful to avoid the disturbing presumption of omniscience. In his method he often appeared unsystematic, hit or miss, yet he generally had a sure sense of where he was headed. When he sought intellectual harmony, generality, and a high level of abstraction, he was inclined to move beyond the restricted language of mathematics to that of literary English and philosophical discourse. His rigorous mathematical thinking was usually contained within a framework of general concepts, as special instances valid in limited domains. Thus the highest level of abstraction was for Wiener neither logic nor mathematics but discursive philosophy. He accepted the imprecision as in the nature of things, the price exacted by generalization. His philosophic attitude militated against the narrow specialization required of twentieth-century scientist-mathematicians, and in this respect he was an anomaly in the professional community.

Von Neumann, ultralogical, fast thinking and fast talking, was a mathematical craftsman, a true virtuoso. He avoided philosophizing but loved formal mathematical and logical structure which transcends time and place. For von Neumann the highest levels of abstraction, such as the bases for logics and mathematics, would have to be found again through rigorous formal logics. His commitment to the tools of reason and his prowess in their use, together with his apparent disinterest in philosophical issues beyond the assertion of the primacy of formal structure and faith in scientific progress, helped make him a paragon among early-twentieth-century scientist-mathematicians.

A relatively technical distinction between Wiener's and von Neumann's mathematical styles is contained in the observation that Wiener had a strong liking for the "constructional method" in mathematics while von Neumann seemed to be particularly fond of the "postulational" approach.[12] I shall not explore this particular difference in taste any further here.

For Wiener the content of at least some of his mathematics and science, his articulated philosophical views, and his private experience all appear to be intimately connected. It was no accident that some of the physical phenomena he described mathematically could be taken as metaphors for his psychological concerns. In creating mathematics, Wiener regarded himself[13] as the mythological sculptor Pygmalion creating his own image, just as he had regarded his own father, Leo Wiener, as a Pygmalion in relation to him.

To appreciate Wiener's attitude more fully, consider the wider context of his work on the application of Lebesgue integration to Brownian motion, described in chapter 3. He had chosen the topic in 1919, a topic that in retrospect is seen to have been a very fruitful one. One of his colleagues attributes Wiener's choice to "a lucky accident,"[14] while another sees it as a great "feat of intuition."[15] There are other topics he might have chosen in connection with Lebesgue integration. I am inclined to the view that over and above accident or intuition, the fact that the subject matter was particularly congenial, "in harmony with his personality," also contributed to his choice. This problem again contains the constellation of incomplete information and partial ignorance; in this respect the theme of the ten-year-old's essay reappears. The task that Wiener

146
*John von
Neumann
and
Norbert
Wiener*

undertook was to give rigorous mathematical form to the description of a sequence of events, even though the events are derived from an underlying chaos ("random process"). In a loose conceptual way the mathematical problem paralleled what he regarded his own task to be as a human being. In the face of our partial ignorance of the nature of the forces from the external world that impinge on us and the involuntary ("unconscious") impulses that arise within us, the task is to identify patterns, to create order, to prevent inundation. In the Brownian motion problem he had found in mathematical form a modus operandi for situations in which information is incomplete. He was delighted with this. There are quite a number of topics he might have chosen to study in connection with the theory of Lebesgue integration, but he chose one that had to do with probabilities and uncertainties, what now we call stochastic processes, where random events form the basis for orderly and statistically predictable events, where one can at least predict average motions and compute the expected fluctuations about the average.

Application of the most rigorous mathematics to situations of chaos or little information leads to interesting results. Although in his early papers Wiener did not mention his predecessor, the theoretical physicist Willard Gibbs, he later wrote that it was when he was working on the Brownian motion problem that he first became aware of Gibbs's work. Twenty years earlier Gibbs had founded the subject of statistical mechanics in its general form. Gibbs had derived a statistical algorithm for describing the properties and behavior of physical substances that did not require any detailed knowledge about the state of the substance. To put it crudely, he had shown by the use of statistical averaging how orderly macroscopic physical events result from random patterns of molecular motion.[16]

As I studied Wiener's own writings and interviewed his friends, I came to feel more and more that problems involving an underlying random process and incomplete information not only held mathematical interest but also had personal meaning for Wiener. The mathematical problem characteristically involved finding predictability through chaos or signal

through noise. Could it be that when working on such problems he was simultaneously working out, in symbolic form,
the parallel problem of his own modus operandi in the face
of unknown external forces and unconscious forces within
himself, whose existence he never forgot? To put it differently, is it possible that the symbolic imagery that he used
in the mathematical innovation in part derived from or coincided with the imagery of his personal life?

Recall the picture of the young Norbert Wiener: his strong
sense of vulnerability to other people, his difficulty in master-
ing personal relations, his volatility and impulsiveness—
which he sometimes viewed as the bane of his existence and
at other times as the source of his creative power—and his
profoundly sensed need to impose discipline, that is, order,
on himself when he could no longer rely on his parents to do
that for him. Perhaps his myopia and physical clumsiness ac-
centuated the sense of randomness, of being buffeted, and
the lack of precise control. Mastery of everyday affairs required conscious effort; it had been intellectual interests that
had been stressed in his early years. Moreover he was
strong-willed, and in spite of the unpredictability of events
he pursued his purposes, even if to some extent they were his
father's purposes become his own.

To illustrate the quality of Wiener's style of dealing with
spontaneous impulses and ideas, consider his style of mathematical creation. Wiener has reported that his creative mathematical thinking coalesced at times with his own emotional,
psychological, unconscious processes. Here is how he describes one such occasion:

I was down with a first-rate case of bronchopneumonia. All
through the pneumonia, my delirium assumed the form of a
peculiar mixture of depression and worry concerning my row
with the Harvard mathematicians and of an anxiety about the
logical status of my mathematical work. It was impossible for
me to distinguish among my pain and difficulty in breathing,
the flapping of the window curtain, and certain as yet unresolved parts of the potential problem on which I was working. I
cannot say merely that the pain revealed itself as a mathematical tension, or that the mathematical tension symbolized
itself as a pain: for the two were united too closely to make

such a separation significant. However, when I reflected on this matter later, I became aware of the possibility that almost any experience may act as a temporary symbol for a mathematical situation which has not yet been organized and cleared up. I also came to see more definitely than I had before that one of the chief motives driving me to mathematics was the discomfort or even the pain of an unresolved mathematical discord. I became more and more conscious of the need to reduce such a discord to semipermanent and recognizable terms before I could release it and pass on to something else.[17]

This self-description refers to the same period in his life when Wiener developed the theory for Brownian motion, although the specific reference is to his work on a different problem, in the branch of mathematics known as "potential theory."

At that time he was still living at home and was dependent on his mother in many ways, most especially to help him with social contacts. At MIT he was taken to be a budding genius, an image that generated in colleagues and students a generally good-natured indulgence of his social immaturity and idiosyncrasy. If his dependence on his parents was frustrating to Norbert, he was learning that a different kind of dependence was essential for his most creative mathematical work: dependence on the subconscious processes out of which his intuitions arose. For an adult and even for a creative mathematician he was in close touch with this wellspring of insight and nonsense. Often when he encountered a colleague or a student, or even a grocery clerk, he would impulsively talk with an intensity that suggests that he was in the process of releasing an inner pressure of ideas and imagery, a pressure to sort them out, express them verbally, expose them, give them rational form. A number of people, having listened to Wiener on occasion when he was enveloped in his own ideas, blissfully unaware of the interest or concern of his listeners, were independently reminded of a four-year-old child with a new toy.

The imagery and ideas that came to him constituted the passive element in his mathematical work and his talk with colleagues or other listeners another seemingly necessary element; the third crucial element was of course the lone task of working through a problem or proof in every detail and testing his hunches mathematically. Nor was Wiener remiss in this respect. He had the mathematical tools, and the habit of working hard and diligently had been inculcated in him by his father early in his life. The mathematical work he carried through in the early 1920s required much careful, detailed analysis, much trial and error, and doubtless a number of false starts. The task that Wiener had to master was first to fashion clear concepts out of the chaotic and uncontrolled elements "given to him" in blessed abundance and finally to create rigorous mathematical statements. The erratic quality and the crucial role of the hit-or-miss aspect that characterized Wiener working has been described by some of his coworkers, together with the tasks of making hunches concrete, and testing conclusions mathematically. He himself commented on it (1936):

When I want an auxiliary function to do a definite job, I try one after another, finding the first too big here, the second too small there, until by grace of luck and a familiarity with the habits of the species, I come on an exact fit. Nine-tenths of the possibilities are eliminated on the basis of a general feeling for the situation before it comes to a matter of any real deductive logic whatever. . . . The final deductive finish is important for a pretty job, but may be left in the hands of any competent apprentice.[18]

Since Wiener was familiar with the introspective descriptions by other mathematicians of the process of mathematical invention, as collected and analyzed by Hadamard,[19] his description of his own mental process, as he gave it in 1952, is another contribution to the "varieties and similarities among mathematicians in respect to the psychological concomitants of mathematical inventions."

With me, the particular assets that I have found useful are a memory of rather wide scope and great permanence and a free-flowing, kaleidoscope-like train of imagination which more or less by itself gives me a consecutive view of the possibilities of a fairly complicated intellectual situation. The great strain on the memory in mathematical work is for me not so much the retention of a vast mass of facts in the literature

150
*John von
Neumann
and
Norbert
Wiener*

as of the simultaneous aspects of the particular problem on which I have been working and of the conversion of my fleeting impressions into something permanent enough to have a place in memory. For I have found that if I have been able to cram all my past ideas of what the problem really involves into a single comprehensive impression, the problem is more than half solved. What remains to be done is very often the casting aside of those aspects of the group of ideas that are not germane to the solution of the problem. This rejection of the irrelevant and purification of the relevant I can do best at moments in which I have a minimum of outside impressions. Very often these moments seem to arise on waking; but probably this really means that sometime during the night I have undergone the process of deconfusion which is necessary to establish my ideas. I am quite certain that at least a part of this process can take place during what one would ordinarily describe as sleep, and in the form of a dream. It is probably more usual for it to take place in the so-called hypnoidal state in which one is awaiting sleep, and it is closely associated with those hypnagogic images which have some of the sensory solidity of hallucinations but which, unlike hallucinations, may be manipulated more or less at the will of the subject. The main ideas are not yet sufficiently differentiated to make recourse to symbolism easy and natural, they furnish a sort of improvised symbolism which may carry one through the stages until an ordinary symbolism becomes possible and appropriate.[20]

The "hypnagogic images which have some of the sensory solidity of hallucinations" emerge of their own accord. To manipulate them implies a degree of control over the powerful forces that produce the images; to find a mathematical notation that expresses and communicates their content, to draw clear mathematical sense from them, requires conscious mathematical work. The "free-flowing, kaleidoscope-like train of imagination" is a kind of qualitative experiential random sequence in time, for which the quantitative random sequences of the Brownian motion theory could be a metaphor. For Wiener science is "description of process in time," but for him scientific-mathematical activity is in itself a process in time;[21] thus research activity and the phenomena described by mathematical-empirical theories share a common ground.

I have so far drawn some analogies between the random process of Brownian motion as it appeared in Wiener's mathematical research and his style of carrying out mathematical research. Wiener was very partial to formulating scientific problems in terms of a time-varying statistical process of which the Brownian motion is a special prototype, and at times he came close to disregarding any other kinds of mathematical descriptions. His reasoning was based on the propositions that "science is an explanation of process in time" and that it is useful to give explicit recognition to ignorance (e.g., imprecision in observation) in mathematical descriptions. These two propositions practically suffice to make a time-dependent statistical description of events seem the most natural mathematical form. In this spirit he had interpreted quantum theory as similar to the Brownian process.[22] And in the same spirit he suggested that insofar as time-varying statistical theories are inadequate to describe the course of events in the subject of economics,[23] one should be skeptical of any and all sophisticated, rigorous, mathematical economics. Meanwhile John von Neumann was busy devising a sophisticated mathematical model (nowhere involving probabilities) in theoretical economics, one that described a uniformly expanding economy.[24]

Wiener's tendency was to see the theory of time-series (Brownian motion, harmonic analysis, stochastic processes), which he had largely created in the first place, as more all-embracing and fundamental than others saw it. His own mathematical theories, in which he perhaps incorporated his own image, may have come to influence his philosophy. Vanity, infatuation with his beautiful creation, may have given him special glasses through which to see the world and to reinforce one element in his earlier perceptions. Here the interplay between scientific style, personality, philosophy, and substantive content does not have the character of cause and effect but is more complex. What is notable and unmistakable is that all these elements seem to mirror each other. Even though Wiener's experience and the content of his mathematical work shared common ground, even if one might be seen as a metaphor for the other, even if in the process of mathematical creation personal concerns symbolized mathematical problems in their initial stages, one should like to have asked Wiener directly whether there was a close connection between the two in his mind. It is too late for that, but he wrote and talked to his colleagues about these parallels in the 1940s and 1950s.

In describing scientific styles and viewpoints, it is an oversimplification to assume they are invariant from youth through age. For von Neumann and Wiener, World War II was a watershed in the sense that it brought about a change in their accustomed activities; it changed the political setting of science in the United States. Yet von Neumann's and Wiener's attitudes toward formal logic, probabilities, time and process, limits and errors all appear to evolve gradually if at all—in any case without any abrupt changes. The most prominent change in these attitudes is that they appear to become more clearly defined, more obvious. They are expressed in new contexts after World War II.

Take for example the subject of weather prediction, a practical problem that in the 1940s had come to interest both Wiener and von Neumann. Von Neumann's approach was basically to solve with a high-speed computer the dynamical equations describing the motion of the air. In this practical problem we see von Neumann dealing not with formal perfection but with useful approximations and numerical techniques, using mathematics as a practical tool. Where others had doubts about the computer's capacity for sufficient speed and accuracy, von Neumann had a justified optimism in such a program. Of course he knew that the incompleteness of the records for today's weather will introduce some error in predicting tomorrow's weather, and considerable uncertainty in predicting that of the day after tomorrow. Nonetheless, von Neumann's attitude was to push the pure-dynamics approach to the limit by carrying the computation through in the best possible way.[25] At first the computation involved what seemed a race against time: The computer had to make a twenty-four-hour prediction in less than

twenty-four hours, or it would only afford hindsight. This race
was won by von Neumann's weather prediction group. Out of the work of this group grew a method of weather prediction that was fairly reliable for the period of a day or two. The decaying accuracy of the prediction, the inevitable decay with time, was fought or ignored insofar as possible. Wiener was highly critical of von Neumann's approach. Wiener's method was to take into account from the beginning the fact that one has only very incomplete information on the state of the air from all the available weather reports.[26] He believed it was crucial to formulate all problems of prediction in terms of statistics, that is, in terms of the statistical properties of past weather, and not in terms of deterministic dynamics. "It is philosophically right," he insisted.[27] "Time" and "error" play a different role in von Neumann's approach to weather prediction than they do in Wiener's. Today long-range weather prediction depends on a combination of Wiener's and von Neumann's approaches.

A second illustration of persistent differences is offered by the problem of automating the playing of a competitive game. Of course von Neumann's fully axiomatic game theory provides a most elegant approach, but the game player in that theory is omniscient from the start, anticipating every contingency, and learns nothing while playing. Wiener, although an admirer of the mathematical theory of games, thought of his automated player as relatively ignorant at the beginning, but gaining sophistication by trial and error, by learning from "experience" with other competitors.[28] The differences between von Neumann's and Wiener's automated game players is reminiscent of Bigelow's description, cited earlier, of the differences between the two men's working habits. Time does not enter into von Neumann's game theory but is a measure of the "learning" of Wiener's automated player. When later both men became interested in parallels between modern high technology and the human nervous system, each did so in his characteristic style. Von Neumann devised automata theory, a fully axiomatic formal logical theory. Wiener put the general problem in a broad philosophic context, which he defined as

154
John von
Neumann
and
Norbert
Wiener

"cybernetics," and presented strict mathematical descriptions only for particular topics that had empirical content. The ideas of chaos, the Gibbsian viewpoint, and time series were an integral part of Wiener's cybernetics. Whereas Wiener viewed random process, chaos, as fundamental, von Neumann saw mechanisms and logics underlying all scientific phenomena. Von Neumann's feeling about random process was articulated in the 1950s in a letter to physicist George Gamow, who had developed a theory for the way proteins are built up in nature, a theory resting on random process. Von Neumann wrote, "I shudder at the thought that highly efficient purposive organizational elements, like the protein, should originate in a random process."[29]

In later chapters I will deal with Wiener and von Neumann in the 1940s and after, but here some of Wiener's own verbal or written comments, dating from the 1940s and 1950s, relating his mathematical-scientific interest in the statistical approach to phenomena (the "Gibbsian" approach) to his personal philosophy and personality will be noted.

One does not want to read too much philosophy into a mathematical problem, because certainly a man can do a mathematical problem that has little or no philosophical significance to him. But Wiener was strongly and passionately involved with the substantive, physical, and conceptual content of some of the theorems he proved, and it was a characteristic peculiar to his open, discursive style to connect, in conversation, fact and image, object and metaphor in a literary rather than logical way, thus exasperating some of his scientific colleagues and at the same time apparently revealing what in his own mind was connected to what. A close friend of Wiener, a physicist colleague, was receptive to my conjecture about him: "Wiener, when he talked about 'chaos,' which he loved to do, was very close to talking about the psychic and instinctual chaos inside himself, which both scared and intrigued him all his life."[30]

In his purely mathematical papers Wiener frequently was concerned with chaos, but there the concept was very strictly and mathematically defined. Wiener himself writes about "chance":

[Through Gibbs's work] chance has been admitted, not merely as a mathematical tool for physics, but as part of its warp and weft. This recognition of an element of incomplete determinism, almost irrationality in the world, is in a certain way parallel to Freud's admission of a deep irrational component in human conduct and thought. In the present world of political as well as intellectual confusion, there is a natural tendency to class Gibbs, Freud, and the proponents of the modern theory of probability together as representatives of a single tendency; yet I do not wish to press this point. The gap between the Gibbs-Lebesgue way of thinking and Freud's intuitive but somewhat discursive method is too large. *Yet in their recognition of a fundamental element of chance in the texture of the universe itself, these men are close to one another* and close to the tradition of St. Augustine.[31]

In Gibbs's work the "element of chance" means a practical limit to the capacity of the experimenter to control or observe the behavior of individual molecules; in Freud's work it implies a limit to a person's ability to control, through his will or reason, the "deep irrational component" in his own thought and conduct. But even if the intellectual connection between Freud's and Gibbs's theories is somewhat farfetched, the fact remains that Wiener in conversation spoke about chance, irrational impulses, stochastic processes, as if to him these phenomena were close to each other, and it is only natural to suppose that in his psyche they were.[32]

After his early reaction against the logical atomism of Bertrand Russell, according to which the world has the structure of mathematical logics,[33] and a similar opposition to the logical positivism popular in the 1930s, Wiener came to define his own outlook as closer to existentialism, embodying a general attitude toward life very different from logical atomism. Wiener refers to Kierkegaard, and I am reminded of Camus's famous essay on the myth of Sisyphus, but Wiener's language differs from theirs in that his idiom is that of the second law of thermodynamics and the statistical theories of Gibbs:

We are swimming upstream against a great torrent of disorganization, which tends to reduce everything to the heat death of equilibrium and sameness described in the second law of thermodynamics. What Maxwell, Boltzmann, and Gibbs

156

John von
Neumann
and
Norbert
Wiener

meant by this heat death in physics has a counterpart in the ethics of Kierkegaard, who pointed out that we live in a chaotic moral universe. In this, our main obligation is to establish arbitrary enclaves of order and system. These enclaves will not remain there indefinitely by any momentum of their own after we have once established them. Like the Red Queen we cannot stay where we are without running as fast as we can.

We are not fighting for a definitive victory in the indefinite future. It is the greatest possible victory to be, to continue to be, and to have been. No defeat can deprive us of the success of having existed for some moment of time in a universe that seems indifferent to us.

This is no defeatism, it is rather a sense of tragedy in a world in which necessity is represented by an inevitable disappearance of differentiation. The declaration of our own nature and the attempt to build up an enclave of organization in the face of nature's overwhelming tendency to disorder is an insolence against the gods and the iron necessity that they impose. Here lies tragedy, but here lies glory too.[34]

Here the "chaotic" is cast in the role of enemy; in other passages Wiener identifies chaos with the devil ("incompleteness" in St. Augustine's theology).[35]

I see the confluence of what is psychologically relevant for Wiener with his articulated philosophy and his choice of mathematical problems as indicative of a coherence and unity in his world. In this sense the topic of Brownian motion theory belonged to him, or he belonged to it.

Wiener was unusual for his time because of his integration of the content of his mathematical work with his philosophy and his personal psychology. Those undifferentiated kinds of symbols, such as Wiener had described in connection with the problem in potential theory, and which apparently facilitated his thinking, were surely one means of integration. For an early or mid-twentieth-century scientist, his outlook was peculiarly holistic: everything is connected to everything else. The combination of the holistic viewpoint and intellectual comprehensiveness with the highly technical specialized ability that enabled him to make major contributions to mathematics and engineering was and is a rarity among members of his profession. It seems as if he had not totally rejected some of the values of a different (pre-Cartesian) era, in which

interconnectedness was part of the usual outlook, and symbolic parallels between science (alchemy) and personal experience (spiritual transformation) were considered respectable. He not only connected the two by means of philosophical ideas, metaphors, and literary allusions but could deal
fairly adequately with a wide range of topics by these means.
At the same time Wiener had productively applied his mind to
the most advanced and highly specialized technical topics.

His style disturbed colleagues. His use of precise scientific
concepts as loose metaphors and his holism seemed unsci-
entific, and consequently not to be taken seriously. The ethos
of the scientific community, built on the requirements of scientific progress,[36] had elevated the value of specialized
technical proficiency while belittling the values of the integrity
of personal psychological processes and a holistic conception of life. The scientific ethos seemed to require that the two
last-mentioned values be sacrificed, suppressed, or com-
partmentalized by scientists for the sake of their craft. As the
sociologist Max Weber observed in 1922, "Science has entered a phase of specialization previously unknown. . . . A
really definitive and good accomplishment is today always a
specialized accomplishment. And whoever lacks the capacity to put on blinders, so to speak, . . . may as well stay away
from science."[37] As specialization increasingly became the
earmark of our technological civilization, so has the disciplined wearing of blinders.

But human proclivities suppressed and sacrificed at one
time have a way of reasserting themselves later.[38] Wiener's
style poses the issue of conflicting simultaneous demands:
the professional necessity of specialized proficiency, the integrity of personal psychological processes, the desire for a
holistic understanding. Wiener's style, or perhaps his lucky
circumstances, his strength of character, or his genius, enabled him to reconcile the three and make his professional
work a meaningful response not only to his private condition
but to the conditions of modern society as he perceived them.
His means for this achievement included his omnivorous
interests as well as his selective and integrative philosophical
reflections.

158
John von
Neumann
and
Norbert
Wiener

Perhaps we will always have among scientifically oriented people some who tend toward a Neumannian orientation favoring the dualism of formal logics and empirics and others who tend toward the conceptual synthesis of a Wienerian outlook. Von Neumann represents a peak of achievement for the attitudes that have evolved through professionalization since the seventeenth-century Newtonian-Leibnizian revolution. Ordinary scientific thought, in agreement with the Neumannian style, tends to reduce truth to logic and mechanism, to what an automaton can do. This is distinguishable from the other, perhaps more essentially human, thinking and reflection that grapples with meanings. Our age excels in and has extolled scientific thought, seeking to exclude all other forms of thought. Its capacity for them has atrophied and has been denigrated in our age. To extol the achievements of one's own time is easier than to attend to its deficiencies. Yet these deficiencies show themselves in our culture in a myriad of ways. Many of those struggling with meanings of activities, finding traditional scientific thought inapplicable and knowing no other style, turn to a total rejection of thought as a means for understanding the human situation. Thus the Wienerian style of thinking is a step toward relating scientific thought to thinking generally. Understanding of the human situation, of the relations among people and of people to their environment, all of which are preliminary to wisdom, is very limited if it is carried out either without thought or using only scientific thought. Other kinds of reflective thought can be efficacious, and their development is highly desirable.[39]

In particular, it appears that present intellectual and human requirements demand the placing of human beings within the total environmental context, not as conquerors but as part of an ecological pattern. Wiener's emphasis on process and his view of scientific observation itself as part of process are congenial to the new ecological perspective. His general orientation prefigures value premises that at the present moment in history are preferable to the narrow "scientific ethos" and are likely to gain prominence. The combination of holistic understanding, technical competence, and psychologically meaningful (nonalienating) work is a desideratum not only for

geniuses. We rightly suspect synthesis, lest it be rigid,
closed, or authoritarian. However, the type of synthesis pre-figured by Wiener is better characterized as a respectful ob-servation and description of patterns in the world, patterns that must include people as well. This type of synthesis is open-ended, empirical, and loose.

The traditional ethos of the scientific community, dictated by the god of scientific and technological progress, needs mod-ification so as to allow for the reemergence of suppressed human and cultural needs. I have not yet described the political, social, and economic aspects of science, but in later chapters I will argue that a shift in the ethos of the scientific community—if such a community of any size persists—is im-perative from the point of view of the viability and well-being of human society. The possibility that the character of what we now know as science may be radically transformed, such as by a new emphasis on patterns and interrelationships, should be welcomed by scientist and nonscientist alike.

We have emphasized differences between Wiener's and von Neumann's styles, but the reader will appreciate that they both belong to the species mathematician and that even the differences reveal kinship. Not only were they both blessed in abundance with the three "highly respectable" motives for mathematical research extolled by Hardy—intellectual curiosity, pride of craftsmanship, and ambition[40]—but their similar aesthetic sense, by which they identified "serious" and "beautiful" mathematics, led each to admire and ap-preciate the other's work.

Like many mathematicians, Wiener and von Neumann re-garded themselves as creative "artists" working in the special medium of mathematics.[41] Yet a sharp contrast is often drawn between art and science (which includes mathematics). Sci-ence is cumulative and progressive; each scientist makes only some small addition to the great edifice, participating in the collective growth of knowledge. In contrast, art is not pro-gressive. Each artist creates complete works from the begin-ning. Mathematics is the epitome of abstraction and generali-zation, while art and poetry tend to seek the concrete. In fact the contrast is less sharp. A scientific work by one man, espe-

160
John von
Neumann
and
Norbert
Wiener

cially one of the stature of Wiener or von Neumann, may be experienced by him as an all-but-completely-worked-out original creation, even though he has used the results of the work of many predecessors as tools. It can also make some contact with concrete instances.

Both Wiener and von Neumann were motivated to find universality or unity in the diversity of phenomena and experience, but they moved along different routes toward a unified comprehension. Both Wiener's conceptual-mathematical approach and von Neumann's formal-mathematical approach are esoteric routes to this goal.[42] But the mathematician, the scientist, and the philosopher have no monopoly on available routes toward a perception of unity in variety. Art and poetry, which can be esoteric but need not be so, are clearly other routes. Love, sex, and empathy need little erudition. Through empathy a sense of unity with a fellow human being, or even with other life and the inanimate, is possible; but it is a different route, with a different kind of unity in variety and has a different kind of impact on the world.

The concern with epistemology and the interpretation of quantum theory, the study of the bases of mathematical reasoning, mathematics itself, and the philosophy of mathematics and science are esoteric, otherworldly preoccupations. They have formed the subject of the last five chapters. But the beauty and personal enjoyment a mathematician finds in creating mathematics cannot be isolated from his effort to achieve success and advancement within a bureaucratic institutional structure. The down-to-earth events of everyday life, the social environment, political and economic circumstances also play their role.[43] Rarely does a serious mathematician follow his muse as a freelance artist might, wholly outside the establishment. Mathematics is distinctly useful to the state and to industrial enterprises, and directly or indirectly they supply the mathematician's salary. Thus aesthetic pleasure and worldly practicality dovetail all too neatly.

To focus on elegance, truth, and beauty within mathematical and scientific work, or on personal styles of creativity, serves to point away from the social and political forces. As noted earlier, in the 1920s the German scientific establishment of professors, the mandarins, favored a conservative, antidemocratic political outlook to protect their own exalted social status, while ostensibly merely being loyal to their profession, merely championing "the pursuit of truth and beauty." This was at a time when the conflict between liberal prodemocratic forces and right-wing racist forces hung in the balance. From the perspective of the internal history of science, it was a time of significant new discoveries, but the broader outlook of human history reveals an unwitting cooperation between the ideals of the "pure scientist" and the social forces leading to the Nazi regime and World War II.

The romantic vision of the mathematician-scientist in search of beauty and truth has not yet found its Cervantes to provide a corrective. Nonetheless we recognize that appreciation of the human context and the social forces underlying such work can help provide the needed realistic perspective. Thus, even if we were to examine the high peaks in mathematics and abstract science of ancient Greece, a proper historical evaluation would require that we also understand the social context. The nature of Athenian democracy, the crucial role of slavery, the debasement of women, the attendant denigration of manual work and the exaltation of contemplative thinking are inextricably part of the social and value context of the flowering of Greek science and mathematics.[44] The romanticizing of the latest in modern scientific-technological achievement has often been so frivolous as to constitute a grotesque turning away from awful realities: there is a US government film that shows a technically elegant hydrogen bomb explosion with a background of uplifting symphonic music and lingers over the lovely beaches and swaying palm trees of South Sea islands, matching the most inviting of travel commercials on television. It is as if a professional musician were extolling the beauty of the music played, according to legend, by the Roman emperor Nero on his lyre, while neglecting the context of the event, namely the city of Rome in flames from the fires Nero had ordered set. This example should be sufficient to illustrate that evaluation of scientific activity is impossible without consideration of the social context. Beauty, elegance, and truth may entrance some mathe-

162
John von
Neumann
and
Norbert
Wiener

maticians and scientists, but science and mathematics provide only partial truths, which if torn entirely out of the human and social context can become grotesque distortions. If fascination with the half-truths of "beauty and truth" is indeed an opiate for some scientists and even some science historians, a remedy lies in advancing consciousness of the human dimension, the social and political contexts, and overall ecological patterns.

In the 1930s Wiener and von Neumann were living in academic environments. This necessarily involved some of the petty politics normal among university faculties. The more significant "politics of pure science" impinged on them as well, that is, the action of scientists and their advocates to secure funding and positions for research, for themselves or their friends. Insofar as these political efforts were successful, they protected mathematicians and scientists from the general economic depression of the 1930s. But at the same time the plight of their German colleagues drew scientists in America into relation with international political events. Finally World War II would stimulate deep involvement in the external political world, outside the confines of science proper.

In 1920 the Massachusetts Institute of Technology, as its name implies, was concerned with technology, not with basic science.[1] It was an institute for training civil engineers, mechanical engineers, mining engineers, naval architects, sanitary engineers, and the like. Mathematics and physics were also taught, because they were seen as useful subjects in the training of engineers, but research in pure mathematics and in basic physics for its own sake was an anomaly in such an institution. While MIT's emphasis on the practical and the applied was consonant with the tradition of American science, such an orientation was frustrating to the young mathematics instructor Norbert Wiener, who aspired to make a major contribution to mathematics proper. It seemed an unlikely place for a reputed mathematical prodigy. Fortunately for Wiener, he was not alone. Others on the mathematics and physics faculties at MIT valued basic science and mathematics. These young Turks of the 1920s wanted to transform MIT into an institution that would take basic science seriously. Thus they made an albeit unsuccessful effort in 1926 to bring Max Born to MIT as a permanent member of the faculty, in order to build up a department in basic (as opposed to applied) physics.[2] Wiener, probably the most gifted and creative among the young Turks, was crucial to the group because of the intellectual standard he set and the esteem in which he was held internationally by the late 1920s. After Karl T. Compton, who shared the ideals of the group, became the president of MIT in 1930, the desired transformation was indeed brought about.

This change at MIT was symptomatic of the transformation American science was undergoing at that time.[3] The impetus came not only from young mathematicians and physicists. Quantum theory, developed in Europe, had served as a particular stimulus to American statesmen of science, most of whom were prominent experimental physicists (R. A. Millikan, R. W. Wood, K. T. and A. H. Compton, H. M. Randall, and G. W. Pierce among them) but some of whom were mathematicians. They had helped persuade the administrators of the Rockefeller fortune and other sources of private wealth to donate many millions of dollars to the cause of upgrading work in the neglected theoretical sciences in America, and especially to put the United States in the forefront of research in quantum theory. Rockefeller funds had provided the fellowship that brought Max Born from Göttingen to visit MIT in 1925, and a Guggenheim fellowship had brought Wiener to Göttingen in 1926. The fellowship programs were one means of transmitting European theoretical knowledge to America; another was the General Education Board, a US government agency largely financed by Rockefeller funds, which distributed nineteen million dollars during the 1920s to a few carefully chosen American universities for the specific purpose of upgrading their mathematics and science departments.

Since the United States was already strong in experimental physics, the physicists wanted primarily to build up mathematical physics. Although Princeton University emphasized the liberal arts and the humanities, its small science departments had been able to obtain a million-dollar grant from the General Education Board in 1925 and two million dollars from other sources, much of which they used to establish new faculty positions. Physicists and mathematicians, particularly Karl Compton in physics (the same man who was later to head MIT) and Oswald Veblen in mathematics, set out to bring to Princeton the most talented European mathematical physicists. Albert Einstein and Werner Heisenberg were approached but did not accept. Hermann Weyl came for a year. Then in 1929 von Neumann was invited to visit Princeton for one semester; his duties would be to deliver two or three lectures a week "on some aspect of the quantum theory."[4] Von Neumann's collaborator and friend, Eugene Wigner, was simultaneously given a similar invitation.

Coming to America, "the land of opportunity," was an adventure, and von Neumann was immediately enthusiastic,[5] but he wrote to Veblen from Berlin, "Some personal affairs to which I have to attend in the course of the next week have to be arranged before I can give a definite answer," and again five days later, this time from Budapest, "As I have now succeeded in satisfactorily settling the family matter with which I was occupied (I intend to marry in December), I am in the agreeable position of being able to accept your very kind offer."[6] Von Neumann's father had recently died of a bleeding ulcer; henceforth his mother and his younger brothers in Budapest would look to him for leadership. Meanwhile he was preparing to bring his new bride, Mariette Koevesi, the daughter of a Budapest physician and an old family friend, with him for the one-semester visit to America, which was to begin von Neumann's lifelong association with Princeton and his Americanization. For the next few years he spent alternate semesters at Princeton and at Berlin. He liked America from the start: the friendliness, the lack of hampering tradition, the absence of stuffiness.[7] Moreover, his own cheerful, buoyant optimism, his natural sense of mastery over difficulties, and his unashamed belief in competence and technique all were in the American grain, indeed almost un-European.

A few years later, in 1933, after the Nazi takeover in Germany, Jewish scientists and mathematicians were summarily dismissed from their positions at Göttingen and other German universities. Albert Einstein, who at that time was visiting friends outside Germany and who openly opposed the Nazi regime, was regarded as a public enemy who had better remain outside of Germany, and was severely censured by the Prussian Academy of Science for his lack of patriotism. Another Berlin professor already mentioned, Erwin Schrödinger, although not a Jew, emigrated from Germany that year because of his distaste for the regime. Of the Göttingen group, Max Born and Richard Courant were among those who were dismissed in the spring of 1933. Physicists and mathe-

166

John von
Neumann
and
Norbert
Wiener

maticians who had contacts on both sides of the Atlantic, von Neumann and Wiener among them, were instrumental in helping unemployed German colleagues to find positions in America. The long-range efforts to bring quantum theorists and mathematicians from Europe to America turned into an urgent humanitarian effort to somehow find positions for the flood of recently dismissed scientists from Germany in the midst of the American depression of the 1930s. During World War II many of these talented immigrants would contribute to the success of the atom bomb, while Germany's simultaneous atom bomb project under Werner Heisenberg's direction floundered.

Von Neumann terminated his position in Berlin in 1933 and became a full-time resident of Princeton, New Jersey, far away from the politics-torn German universities. Of the situation at the latter, von Neumann predicted correctly in 1933 that "if these boys continue for only two more years (which is unfortunately very probable), they will ruin German science for a generation—at least. . . ."[8] Meanwhile, in Princeton, American philanthropists had begun to model the Institute for Advanced Study (which was incorporated in 1930) after the German research institutes. On the academic side, Oswald Veblen had been particularly instrumental in guiding the establishment of the institute, which began to operate in 1933. Von Neumann, like Veblen, changed his administrative affiliation from the university to the institute, where he was the youngest permanent full-time member.[9]

The Institute at Princeton, the first of its kind in the United States, was set up to provide a place for outstanding scholars to pursue their academic interests, protected even from the distractions encountered by university faculties, such as routine work and administrative responsibilities. "The Institute," says its founder and first director, Abraham Flexner, "was conceived as a paradise for scholars and such it really is." But in his selection of faculty members he was aware that "not all men—not all gifted men—know how to live in paradise. The earth is their proper habitation, and upon the earth, such as it is, most of them do the best of which they are capable." The institute was set up to "provide the facilities, the tranquillity, and the time requisite to fundamental inquiry into the unknown." In particular it was decided to begin the institute with the appointment of mathematicians, since "mathematicians deal with intellectual concepts that they follow out for their own sake regardless of their possible usefulness, but, through this very freedom to pursue the apparently useless, they stimulate scientists, philosophers, economists, poets, and musicians, though without being at all conscious of any need or responsibility to do so. . . . It requires little—a few men, a few students, a few rooms, books, blackboards, chalk, paper and pencils."[10] It would be fifteen years before von Neumann's interest in high technology and its applications would intrude on the tranquillity of the "paradise for scholars," which also included Albert Einstein and a highly select group of mathematicians: Oswald Veblen, James Alexander, Marston Morse, and Hermann Weyl. But even then von Neumann was never the reclusive scholar. The von Neumann house, one of the grandest residences in Princeton,[11] became a center for social gatherings, and as his first wife recalled, it mattered to Johnny that they be the best parties in Princeton. Often they were occasions for sophisticated and witty conversations, sometimes the rugs would be rolled back for dancing, and there was always plenty to drink. Johnny would sometimes absent himself for a couple of hours to work in his study and then return unobtrusively. In all, the parties contributed considerably to social life in the scholarly world of Princeton.

Only in the summer months would von Neumann return to Europe, vacationing, visiting his family, and giving lectures. A letter written during one of these summer visits illustrates his style, the touch of humor and mathematics in telling of a personal mishap:

We had one of our "major" accidents: driving from Genoa to Budapest, on the highway Vienna-Budapest, we got in a heavy rain, on one of the rare concrete roads in Hungary. As our driving was somewhat faster than what is wholesome under these conditions (as was unambiguously proved by the outcome), we started skidding, and finally hit a tree with the car. Simultaneously we hit the car with our noses. All this did no good, to any of the 4 objects mentioned. The whole affair looked quite discouraging, but we were able to drive with the car 5 miles to the next doctor, who bandaged our noses, and then we came by train to Budapest (110 more miles). It seems that ultimately everything will straighten out, but it will take several weeks until all visible marks will have disappeared from all objects injured, including the car. Mariette was very brave, but the first few hours, until we did not know how serious her bruises were, where [sic] quite disagreeable. In the interest of historical completeness I must mention, although I do not like to lay too much stress on this fact, that I did the driving. . . . Besides I have an idea, how to try to reform the Dirac equation. . . .[12]

Von Neumann was notorious for driving with the bravado of an adolescent.

By contrast to von Neumann, Wiener's life style was modest. However widely he traveled, Wiener was and remained a New Englander. His food was vegetarian. For diversion from his academic activities, he tramped through the hills of Massachusetts and New Hampshire with the Appalachian Mountain Club or with his friends, as he had done earlier with his father. He and his wife Margaret lived simply in a suburb of Boston. The summers they spent in Sandwich, New Hampshire, a town in which Norbert's parents had rented a summer cottage when he was a boy, and which he commends "for the loveliness of its scenery, for the walks and climbs afforded by its mountains, and for the sober dignity, reserve, and friendliness of its country people."[13] When Norbert and Margaret Wiener looked for a permanent summer home there, this is what appealed to them:

The house was uninhabited. It had been only recently inhabited, however, and was in good condition. When Margaret and I made the circuit of its weed-grown lawns and peered through its cobweb-covered windows into its graciously proportioned rooms, we knew we had found what we wanted. . . . The region had been going downhill from the Civil War to that time, and real estate prices were at a dead bottom.[14]

This became their summer home, the place where Wiener found the kind of relaxation he needed:

In those days [1928–30] we were without a telephone, without electricity, and even without a stove. We prepared Barbara's [the baby's] formula in the fireplace and did our rudimentary cooking there until such a time as we could get a second-hand two burner oil-stove. To the present day [1956] we remain without running water, although we have found a very satisfactory substitute for this in the form of a force pump and gravity tank.[15]

Not that Wiener was deprived of intellectual contact in the summers, for other academic people also spent their summers in that part of New Hampshire, some of them induced to do so by the Wieners.

Wiener took his paternal responsibilities seriously, and in particular sought to avoid the pattern of education imposed on him by his father. In appraising him as a father, his alternating exuberance and depression and his poorly disciplined emotional life, which must have made him difficult, have to be balanced against his lively, imaginative companionship and his warm interest in the welfare of his daughters.

By the 1930s Wiener's and von Neumann's talents were generally acknowledged by the mathematical community, and they were recognized as two of the leading mathematicians in America.[16] Notwithstanding the depression and the political upheavals in central Europe, on the whole both men spent the decade living the normal life of academic mathematicians. Creating interesting mathematics, lecturing, and teaching were their primary occupations. It was also during this decade that the professional friendship between von Neumann and Wiener developed.

In 1930 von Neumann had lectured at Princeton University on quantum statistics, and in the following year he lectured on mathematical hydrodynamics. Later, at the institute, he continued to give his fellow mathematicians courses of lectures, which often contained new and original material. Mimeographed copies of notes taken at these lectures by a listener or by von Neumann's assistant made the lectures famous within the mathematics community. The notes of his 1933–34 lecture series on measure theory "were for a long

170

John von
Neumann
and
Norbert
Wiener

time one of the major sources of measure-theoretic information in the United States."[17] The subject of his lectures in the
following year was "Operator Theory," and again notes were
prepared. His own major innovations in lattice theory were
fully described by him not in published articles but in the
lecture series he gave at the institute in 1936 and 1937; they
too were issued in mimeographed form.[18] Young mathematicians everywhere vied for the opportunity to spend a year
at Princeton, which became the major center for mathematics
in the United States. Aside from Veblen and von Neumann,
there Albert Einstein, Hermann Weyl, Kurt Gödel, Marston
Morse, James Alexander, and Deane Montgomery were
among the faculty of the institute, and others were at the University. Nevertheless, von Neumann, because of his well-
known personal accessibility and his lectures, remained a
major drawing card. "The story used to be told about him in
Princeton that while he was indeed a demi-god he had made
a detailed study of humans and could imitate them perfectly.
Actually he had great social presence, a very warm, human
personality, and a wonderful sense of humor."[19] His written articles differed from his lectures in that they omitted
some of the thoughts, including his reasons for choosing one
method over another, that he shared with his listeners—
however helpful they might be for understanding the subject.[20]

Johnny's friends remember him in his characteristic poses:
standing before a blackboard or discussing problems at
home. Somehow, his gesture, smile, and the expression of the
eyes always reflected the kind of thought or the nature of the
problem under discussion. He was of middle size, quite slim
as a very young man, then increasingly corpulent; moving
about in small steps with considerable random acceleration,
but never with great speed. A smile flashed on his face
whenever a problem exhibited features of logical or mathematical paradox.[21]

Earlier I described von Neumann's work in quantum theory
and game theory and commented on his work in axiomatic set
theory. In the 1930s he enlarged the areas of his mathematical
research, working in ergodic theory and measure theory, lat-

tice theory and algebras, operator theory, rings of operators, and topological groups. This work, though not described here, has an important place in the history of mathematics. His proof of the quasi-ergodic hypothesis (1932) was seminal in the field.[22] A colleague has called von Neumann's proofs in lattice theory (1935–1937)

a truly remarkable feat of logical analysis and ingenuity. . . . Anyone wishing to get an unforgettable impression of the razor edge of von Neumann's mind, need merely try to pursue this chain of exact reasoning for himself—realizing that often five pages of it were written down before breakfast, seated at a living room writing-table in a bathrobe.[23]

It is clear, then, that during the 1930s von Neumann continued to do mathematical work of the same depth and quality as he had in the previous decade.

For Wiener the 1930s were not only a decade of more significant mathematical work but also of increasing interdisciplinary activity with his colleagues in electrical engineering and biology. At MIT he was promoted to associate professor in 1930 and then full professor in 1932. He continued to make major contributions in generalizations of harmonic analysis and in proving Tauberian theorems, and he proved new theorems pertaining to Fourier transforms in the complex plane.[24] This work was mainly in the branch of mathematics known as analysis, whereas von Neumann's work at that time tended to emphasize algebras. Still, their interests overlapped in many places, especially when in the later 1930s Wiener worked in ergodic theory, in which he and von Neumann shared a strong interest. Wiener and von Neumann would encounter each other at mathematical conferences. They sought each other out whenever the opportunity presented itself, in order to engage in mathematical discussion, and these long conversations, the interaction between two extremely quick and profound minds, were the core of their friendship.

Wiener's chief mathematical collaborators, except for students, were Europeans. The Englishman R. E. A. C. Paley, the German Eberhard Hopf, the Russian J. D. Tamarkin, and later

172

John von
Neumann
and
Norbert
Wiener

in the decade the Frenchman Szolem Mandelbrojt and the Hungarian Aurel Wintner. At the time he regarded his own work as closer to European mathematics than to American; he felt that European mathematicians had been quicker to understand and appreciate his work than the American mathematical community.[25] He also was active in bringing up a number of graduate students, some of whom would make major contributions to mathematics in their own right. One of them, Norman Levinson, recalled participating in a graduate seminar with Wiener in 1933:

At that level he was a most stimulating teacher. He would actually carry on his research at the blackboard. As soon as I displayed a slight comprehension of what he was doing, he handed me the manuscript of Paley-Wiener for revision. I found a gap in a proof and proved a lemma to set it right. Wiener thereupon sat down at his typewriter, typed my lemma, affixed my name and sent it off to a journal. A prominent professor does not often act as secretary for a young student. He convinced me to change my course from electrical engineering to mathematics. He then went to visit my parents, unschooled immigrant working people living in a run-down ghetto community, to assure them about my future in mathematics. He came to see them a number of times during the next five years to reassure them until he finally found a permanent position for me. (In those depression years positions were very scarce.)[26]

Levinson's is the genre of experience many people had with Wiener, particularly students, toward whom Wiener showed great generosity and to whom he gave extraordinary help— although Levinson also notes that on some occasions "Wiener was capable of childlike egocentric immaturity."[27]

When Frank Scimone, an inmate serving a life sentence for murder at Attica prison, wrote to Wiener (whom he had never met), alluding to his lone mathematics studies in prison and requesting that Wiener send him something on "chaos," Wiener not only complied, but, after learning the extent of the prisoner's mathematical knowledge, provided him with suitable textbooks and assurances of his personal interest in him. After three years' correspondence Wiener paid him a visit in prison. Soon thereafter, Scimone asked Wiener if he had some

problem he'd like him to solve, and Wiener sent him an interesting problem—not a textbook problem, but a problem
useful in applied mathematics.

Wiener had given up the study of biology in 1910 at his
father's insistence and because of his physical clumsiness in
the laboratory, but once he had thoroughly established himself as a mathematician, the dormant interest made itself felt
again. He struck up a friendship with British biologist J. B. S.
Haldane, and the two men enjoyed long discussions on
philosophical, political, literary, and scientific topics. But
this was only on those few occasions when Wiener could
visit Cambridge, England. In Cambridge, Massachusetts, at
the Harvard Medical School, Mexican physiologist Arturo
Rosenblueth conducted a monthly seminar on scientific
method; the participants were mostly medical school scientists, but one of Wiener's former students invited him and he
became an enthusiastic regular participant. Wiener's later
collaboration with Rosenblueth, which prepared Wiener for an
eventual expansion of his professional activity into problems
in medicine and biology, grew out of these meetings.

At the same time Wiener continued his friendship and collaborative work with electrical engineer Vannevar Bush. In
spite of his lack of manual dexterity, Wiener had a very good
understanding of apparatus and readily thought in engineering terms. Bush was developing new analogue computers,
the most advanced for their time, and in this he was aided by
Wiener's imaginative ideas and advice.[28] To work out one of
his ideas, Wiener got a Chinese student in the electrical engineering department, Yuk Wing Lee, to do a doctoral thesis
with him. An invention having to do with apparatus for
analyzing electrical networks resulted, and Lee and Wiener
patented it.[29] Wiener's explorations into biology, computers,
and electrical engineering during this time provided him with
some of the knowledge and experience that enabled him
years later to perceive parallels between living organisms
and electrical engineering systems and to create the grand
synthesis he would call cybernetics. But these developments
would not come until after World War II.

For Wiener one of the high points of the 1930s was his one-

174
John von
Neumann
and
Norbert
Wiener

year visit to Tsing Hua University in Peking, China. Lee, after having finished his degree with Wiener, soon obtained a position as professor there and was able to invite Wiener.

I am absolutely sure that you and Mrs. Wiener and your children will enjoy a sojourn in China. . . . A short distance from Tsing Hua there are the famous Western Hills which will give you all the climbing you want. And at Tsing Hua there is a most friendly group of people who will be most happy to have your friendship. . . . We have quite a few men here . . . who are producing contributions to mathematics, physics and electrical engineering. . . . Under your direction and with your inspiration and ideas, we could produce much better and a great deal more results.[30]

Shortly after arriving in China, Wiener reported enthusiastically back to Bush at MIT: "My wife and I start Chinese tomorrow. Actually, I have already thirty-odd characters, and can direct a rickshaw coolie by the points of the compass, as is the North Chinese custom."[31] The Wiener-Bush correspondence during that academic year (1935–36) contains detailed discussion and evaluation of ideas about computers, some of which Wiener managed to persuade the engineers in Peking to put into hardware form. Wiener became quite proficient at Chinese; after he returned to Massachusetts he would never miss an opportunity to exhibit his knowledge of the language with Chinese students or with Chinese waiters or grocery clerks. When Wiener was in China, the Japanese had already begun their military takeover of certain provinces. In 1937 the Chinese-Japanese war broke out, and he became busy "trying to influence American opinion to give increased aid to China."[32] He was thus engaged in at least a certain amount of political activism. Although Wiener did not like communism, he was aware of the inefficiency of the Kuomintang army and was willing to sponsor American aid specifically for the Communist soldiers in China.[33]

Curiously, Wiener did not accept an official invitation from Veblen, one of the long-time leaders of American mathematics, to spend a week as a guest lecturer at the Princeton Institute in autumn 1934. His stated reason for declining the invitation was that he did not see "sufficient financial advantage" in the modest honorarium offered.[34] But it is unlikely that he was financially hard-pressed—his $6,000 annual salary was adequate then, even for a family with two children, and it is more probable that the issue of the honorarium was a symbolic one for him. Wiener suffered from hypersensitivity about the reception of his work by the American mathematical community.[35] His anger that only in Europe had his early work been fully appreciated, although at that time he felt very much in need of recognition in his homeland and would have liked an appointment to a more distinguished intellectual center than MIT, persisted long after he had been accorded that recognition. Just before the invitation to Princeton Wiener had been elected a member of the elite National Academy of Sciences, and the Princeton invitation had made a reference to that honor. Wiener would have been sensitive to the fact that the invitation happened to have come after rather than before his election.

Some years later Wiener resigned from the academy in protest over its official power and exclusiveness and its inherent tendency to injure or stifle independent research "not pleasing to whatever group is at the moment running the scientific politics of the country." He also commented in his letter of resignation about honors given by this "body of self-appointed judges":

As to medals, prizes, and the like, the less said of them the better. The heartbreak to the unsuccessful competitors is only equalled by the injury which their receipt can wreak on a weak or vain personality, or the irony of their reception by an aging scholar long after all good which they can do is gone. I say, justly or unjustly administered, they are an abomination, and should be abolished without exception.[36]

While his response to the Princeton invitation seems merely impulsive and ungracious, perhaps merely an awkward effort to bargain for a larger honorarium, his resignation from the National Academy was clearly a deliberate step arising from the conflict between his ideals and an organization of whose functioning he disapproved in spite of its prestige. He continued throughout his life to hold the view that the "leaders in the field," whom Wiener regarded as a self-chosen elite, were no better judges of what is of value than outsiders.[37]

On April 23, 1937, Wiener was slated to give a lecture at Johns Hopkins University in Maryland, and he wrote a note to von Neumann inquiring if he might drop by to see him on the way. When Wiener accepted the invitation that followed asking him and wife "to stay several days,"[38] von Neumann replied that he was "looking forward quite particularly to have another mathematical conversation with you."[39] Wiener was at that time involved in work in ergodic theory, an esoteric branch of mathematics in which von Neumann had done pioneering work; ergodic theory, based on pure dynamics, provides the mathematical justification for Gibbs's statistical mechanics.[40] It is not surprising, then, that in his next paper on the subject Wiener had a footnote referring to some ideas derived from a conversation with von Neumann.[41]

To have a houseguest for four days involves more than mathematical conversation, however. Wiener was a strict vegetarian, and Mrs. von Neumann recalls taxing her imagination to prepare a variety of vegetarian dishes. Wiener tended to share his enthusiasms and surely told von Neumann of his work on computers with Bush. Then the Wieners told the von Neumanns of their experiences of the previous year in Peking. Von Neumann was intrigued, and Wiener sought to pave the way for a similar trip for von Neumann by a letter to Lee at Tsing Hua University:

Last week I was down at Princeton as the guest of Professor and Mrs. von Neumann . . . they seemed interested in seeing China some time in the future. Now this is a marvelous opportunity. Neumann is one of the two or three top mathematicians in the world, is totally without national or race prejudice, and has an enormously great gift for inspiring younger men and getting them to do research. . . . The Neumanns are quite wealthy. . . . Therefore if he comes to China it will not be to save any money, although he would hope that the salary would meet expenses.

The Neumanns rather like to hit the high spots socially. You know Princeton life is a bit fast and "cocktail partyish." On the other hand, Neumann is not high-hat in any way, and is most accessible to young students.[42]

Wiener's enthusiasm for von Neumann was exceeded only by his enthusiasm for China, or possibly by his enthusiasm for ergodic theory itself. The reference to "national and race prejudice" reminds us that a Chinese student in the United States in the 1930s was accustomed to experience a fair amount of it. Wiener's role, as a matter of fact, had been one of championing the victims of prejudice. But the Japanese struck in North China in July of that year, and conditions would never be right for the von Neumann plans to be realized.

The Wieners also had von Neumann as a houseguest on a number of occasions over the years. Among themselves they liked to refer to him as "Gentleman Johnny," for he was always so proper and gentlemanly in the Budapest-Vienna manner, as opposed to the by no means proper Norbert.[43] (The historical "Gentleman Johnny" was the British General John Burgoyne, who fought in the American Revolution.)

Meanwhile the clouds of war were also on the horizon in Europe. This was closer to home for both Wiener and von Neumann. Wiener's wife was German-born—they often spoke German at home—his mother too had been from a German family, and although his father had been a Russian Jew, his native tongue had also been German.[44] The persecution of the Jews in Germany evoked some of Wiener's explosive inner conflicts and emotions, and he sought help from a psychoanalyst.[45] One of Wiener's colleagues in the Harvard Medical School supper club reported an anecdote pertaining to the time, shortly before the United States entered the Second World War: Stimulated by a short story that had appeared in a literary magazine, Wiener one day urged (seriously?) that all the members of the group get those of their colleagues in Germany who were known to be zealous Nazis in trouble with their government. The device he suggested would at the least have confused the German government about which scientists it could trust and might even have caused it to put its most loyal adherents into concentration camps. It consisted of

178
John von
Neumann
and
Norbert
Wiener

sending letters to their German colleagues, which presumably would be opened by the censors, that would implicate them in plots to escape from Germany or other "disloyal" activities.[46] For Wiener, the "enemy" included fellow scientists and mathematicians, people who formerly might have been personal friends as well as professional colleagues. What anguish he must have felt!

While for Wiener the impending war in Europe generated fantasies, von Neumann felt the threat in a more direct way. In fact for him it became a kind of race against time. In 1937 his wife had taken their baby daughter and left him to marry an experimental physicist, J. B. Horner Kuper. Von Neumann took the first opportunity to visit Europe again, and specifically Hungary. It was in the autumn of 1938, and war was expected at any moment. While in Europe, von Neumann lectured in Warsaw and in Lund, Sweden; stayed for several days with Niels Bohr in Copenhagen and made arrangements for Bohr to lecture at Princeton for a semester; worked out a unitary spectral theory for non-Hermitian operators; observed astutely and in detail the state of European politics, and visited his mother and brother in Budapest. But his main concern was to help arrange for a divorce for his new bride-to-be, Klara Dan of Budapest, and obtain a visa and passport for her so they could return together to Princeton before the feared general war should break out. Klara Dan's divorce was settled October 29; she and John were married November 17, and shortly thereafter they were on the boat leaving from Le Havre to New York. World War II was still several months in the future, and von Neumann had missed only one semester at Princeton, as planned.[47] Soon the von Neumann house became again a center for social gatherings and lively parties, enriched by Johnny's own evident joie de vivre.

World War II was the watershed in the slow transformation whereby the deepest problems of science changed from the epistemological to the social and political. The war pulled the mask from the illusion that science is apolitical. The economic determinants of science were changing as well. In the 1920s and 1930s basic research in science and mathematics had been relatively aloof from and independent from the US government; research had been financed either directly by universities or through large private foundations. The war changed all that very rapidly. Scientists were quickly mobilized by the government to participate in generally costly war-related projects with practical objectives. It was not only that research with a practical objective has a very different character from pure research in the pursuit of "truth" or "beauty," scientists themselves had become workers for the state. After the war had ended and many scientists had left the military projects, the close relation between science and the government forged during wartime persisted. The primary source of research funds was henceforth US government agencies, including in particular the military services. Moreover, the funding was on a lavish scale relative to that of the prewar decades. The changed economic relation between science and the government coincided with a loss of innocence on the part of the scientific community, which had been instrumental in inventing nuclear and other weapons that had not only wrought great destruction but permanently altered the conditions of life. It is this combined change in

180
John von
Neumann
and
Norbert
Wiener

economic relations and consciousness that brought the social and political problems of science into prominence.

John von Neumann and Norbert Wiener experienced the watershed each in his own idiosyncratic way. Both men were close to events in Europe and responded with intense aversion to the Nazi regime in Germany and its military conquests, and they began to work on military projects even before the United States entered the war in December 1941. While the high technology with which each was concerned led directly to weapons development, it suggested incidentally to both men similarities in principle between machines and organisms, particularly parallels between high technology and the human nervous system. This shared insight led them at some point to make common cause. The end of the war presented them with highly consequential choices. Sartre's existentialism emphasizes that it is by fundamental choices in the face of available options that a person creates himself; he defines the kind of human being he is and is going to be. Returning to the ivory tower of pure mathematics was surely one option for Wiener and von Neumann; moving from high technology toward the fields of biology and medicine another; pursuing technology for the sake of technical innovations a third; devoting one's energy to creating ever more powerful engines of destruction a fourth, and so on. The choosing of any of these options, or some combination of them, would, whatever its personal meaning, be a political act, and for men such as von Neumann and Wiener it would be a deliberate, conscious political act. The war made it very apparent that mathematicians and physicists generally, and Wiener and von Neumann in particular, had it in their power to exert some influence on the course of history, over and above the history of mathematics, and they were conscious of this power. But this is getting ahead of the story.

Von Neumann, who had always had an equal interest in pure and applied mathematics, had continued throughout the 1930s to be close to the community of theoretical physicists; in 1938 he had been part of a group of theoretical physicists and astronomers who met to discuss the spontaneous nuclear fusion processes responsible for energy production in the sun and other stars,[1] the processes that suggested the possibility of a man-made counterpart. He had also become a scientific advisor to the Ballistics Research Laboratory of Army Ordnance before the United States had entered the war and had studied mathematical aspects of explosions and shock waves. It is only natural that J. Robert Oppenheimer, the director of the unprecedented wartime project of building an atomic bomb, would call on von Neumann—whom he had known since their Göttingen days—to become mathematical consultant for the top-priority, highly secret Manhattan Project, as it was called. Klara von Neumann described how "In 1943, soon after the Manhattan Project was started, Johnny became one of the scientists who 'disappeared into the West,' commuting back and forth between Washington, Los Alamos and many other places."[2]

Besides consulting for government laboratories, von Neumann had become very interested in methods of numerical computation.[3] Vast amounts of such computation were needed to solve some of the problems arising in connection with atomic bomb construction, as well as explosions and fluid dynamics generally. In the summer of 1944 he encountered Herman Goldstine, a mathematician and army officer, who brought him in as a consultant for the design of high-speed digital computers at the Moore School of Engineering, at the University of Pennsylvania in Philadelphia, which was under contract with Army Ordnance to build the first such machine.[4]

From 1943, then, until the war ended in August 1945, von Neumann's activity at the Princeton Institute, that "paradise for scholars," was preempted by his participation in crash programs to achieve breakthroughs in high technology. The objectives of such practical projects are of a very different kind from those of mathematical and scientific inquiries made for their own sake and so are their implications. Still, the participants were in large part mathematicians and physicists who had had little or no prior contact with such projects and were accustomed to thinking in terms of pure science and

182
John von·
Neumann
and
Norbert
Wiener

mathematics. They included many of the same group of men that a decade earlier had been debating the philosophical meaning of quantum theory.

Within a year of beginning his consultantship with the Moore School, von Neumann had laid the groundwork for enormous advances in computer design.[5] His greatest contribution arose from his characteristic ability to think in non-visual abstractions. Goldstine writes that as far as he is concerned, at the school von Neumann was the first person

who understood explicitly that a computer essentially performed logical functions, and that the electrical aspects were ancillary. . . . Prior to von Neumann people . . . concentrated primarily on the electrical engineering aspects. These aspects were of course of vital importance, but it was von Neumann who first gave a logical treatment to the subject, much as if it were a conventional branch of logics or mathematics.[6]

By focusing on the logical organization of computers, as was altogether natural for von Neumann, he separated out incidental engineering features from the essentials.

Meanwhile a new convergence of interest between von Neumann and Wiener had surfaced, very different from ergodic theory: the exploration of the possibility of analogies between engineering devices, such as computers, and nervous systems, or more generally living organisms. Von Neumann's work reflected an awareness of such analogies in his first draft of a report on the logical design of a prospective new computing machine (June 30, 1945).[7] In that draft he borrowed a formalism originally devised to describe the formal-logical organization of the human brain.[8] During the year prior to that report, Wiener and von Neumann had been discussing the subject at length with neurophysiologists and with each other.

Although Wiener had had an earlier interest in and contributed to computer development, a contribution catalyzed by his friendship with Vannevar Bush, his wartime work followed another direction. Bush had emerged as the leader in organizing American science during the war; he became director of the US Office of Scientific Research and Development, the organization that had the primary responsibility for organizing American science to support the war effort.[9] Two major technical problems required the most sophisticated scientific "personnel" (in wartime scientists turn from "individuals" into "personnel"): the long-range problem of atomic bomb construction and the more immediate problem of devising a way to attack German bomb-carrying aircraft. Bombers were the most effective weapon by which Germany was conquering and devastating much of Europe, and attempts to bring them down with antiaircraft fire were often frustrated by inaccuracy in tracking and consequent failure to hit these small, fast-moving targets. The work on the atomic bomb was concentrated primarily at Los Alamos, New Mexico, while the problem of tracking and shooting down aircraft was centered at the Radiation Laboratory established for that purpose at MIT. Wiener's work would be right at MIT, under the direction of Warren Weaver of the Rockefeller Foundation, administratively part of section D2 of the National Defense Research Committee.

The overall problem of effective antiaircraft fire required "a good gun, a good projectile, and a fire-control system that enables the gunner to know the target's position at all times, estimate its future position, apply corrections to the gun controls, and set the fuse properly, so that it will detonate the projectile at the right instant."[10] Working together with a young engineer, Julian Bigelow, Wiener dealt with the general mathematical problem of using the available information about the location and motion of the airplane to make a statistical prediction of its future course, so as to achieve the best possible control of the antiaircraft fire and improve its chances of hitting the target. What emerged was a mathematical theory of great generality, a theory for predicting the future as best one can on the basis of incomplete information about the past. It was a statistical theory throughout, relied on ergodic theorems, and made use of Wiener's earlier work on integral equations and Fourier analysis.[11] This theory helped revolutionize the whole field of communication engineering,

184
John von
Neumann
and
Norbert
Wiener
primarily by bringing statistical considerations to the fore, and forms the basis of modern statistical communication theory.

One feature of the antiaircraft problem was the cycle involving feedback: information from a radar screen is processed to calculate the adjustments on gun controls to improve aim; the effectiveness of the adjustment is observed and communicated again via radar, and then this new information is used again to readjust the aim of the gun, and so on. If the calculations are automated, one is dealing with a self-steering device; if not, the whole system including the participating human beings can be viewed as a self-steering device. Wiener and Bigelow conceived of the analogy between self-steering engineering devices involving a feedback loop of information and control and human processes, such as picking up a pencil or a glass of water, in which information from the eyes or proprioceptors is processed by the nervous system to control the hand and prevent it from overshooting or undershooting its goal. Wiener and Bigelow discussed the ideas with Wiener's physiologist friend of the medical school seminars, Arturo Rosenblueth, and by 1942 these discussions had produced a formulation in general conceptual terms of the ideas that Wiener later christened "cybernetics," subsuming under one heading a theory of "control and communication in the animal and the machine."[12]

The idea that parallels between engineering devices and living organisms, including specifically the nervous system, could be found through mathematical formulations received an unexpected boost from the work of Chicago neuropsychiatrist Warren McCulloch and a young logician, Walter Pitts. Rosenblueth had known McCulloch for years, and he knew the direction of his thinking: It was his idea to make a mathematical-logical model of the neuron, the cell that is the building block of nerves and the brain, and then to describe the functioning of the central nervous system in terms of nets composed of many loosely interconnected neurons. With the help of Pitts, McCulloch had succeeded in creating such a model at about the same time as Rosenblueth, Wiener, and Bigelow arrived at a clear formulation of their general ideas concerning cybernetics, supported by several concrete examples. On May 14 and 25, 1942, Rosenblueth and McCulloch had talked and exchanged ideas in New York.[13] McCulloch and Wiener also became friends. In 1943 young Pitts, whose mental abilities were widely regarded as on a par with von Neumann's, left McCulloch and Chicago to spend a year working with Wiener at MIT. McCulloch was hoping to bring Rosenblueth to Chicago, since Harvard was planning to dismiss him. Meanwhile the trio Wiener, Rosenblueth, and Pitts formed a working team in Cambridge.[14] Since the Pitts-McCulloch mathematical model for a neuron also happened to be an appropriate mathematical model of an electrical relay, a basic component in an electronic computer, it supported the Wiener-Bigelow-Rosenblueth thesis that efforts at unified mathematical descriptions of engineering devices and the nervous system held promise.

In the 1930s Wiener had actively pursued his secondary interests in electrical engineering and physiology, although the core of his professional work was fundamental mathematics; biology and engineering were separate interests. Given that Wiener was always seeking to harmonize and synthesize disparate intellectual interests, it is no wonder that he was elated about the ideas of cybernetics, which joined engineering and biology: the bridge was mathematics, where Wiener was most at home. He communicated his enthusiasm to John von Neumann, who in spite of his commitments at Los Alamos and elsewhere was receptive. Out of their talks came the plan for a meeting of a small group of men "to discuss questions of common interest and make plans for the future development of this field of effort, which as yet is not even named."[15]

The meeting took place on January 6 and 7, 1945, at the Princeton Institute. Wiener reported to Rosenblueth, who had departed for Mexico and could not attend the meeting, that it

was a great success. I believe you have already got von Neumann's report. . . . The first day von Neumann spoke on computing machines and I spoke on communication engineering. The second day Lorente de Nó and McCulloch

186
*John von
Neumann
and
Norbert
Wiener*

joined forces for a very convincing presentation of the present status of the problem of the organization of the brain. In the end we were all convinced that the subject embracing both the engineering and neurology aspects is essentially one, and we should go ahead with plans to embody these ideas in a permanent program of research . . . we definitely do have the intention of organizing a society and a journal after the war, and founding at Tech or elsewhere in the country a center of research in our new field. . . . When this scheme really gets going, I for one will not be content unless we can bring you and Bigelow directly into it.[16]

Wiener and von Neumann had already discussed the question of financing the prospective "center of research." Von Neumann, whom Wiener characterized as a "very slick organizer,"[17] was confident he could get financial backing; moreover they had received indications directly from the Rockefeller Foundation (W. Weaver) and the Guggenheim (H. A. Moe) that funds would be available from these sources.[18]

Thus in the midst of wartime Wiener and von Neumann found common ground in their ideas and plans for scientific work after the war. It seems that by January 1945 the coming of peace was clearly visible on the horizon.

In Wiener's vision the founding of a research center in the new field would do more than bring together disparate intellectual interests; it would bring together a group of people who had a high regard for each other and whom he himself found highly congenial. For Wiener, first and foremost among them was John von Neumann. Aside from his intellectual contribution, von Neumann would also bring his organizing ability and lend prestige to the research center, which in Wiener's eyes all but ensured the success of the venture.[19] Other members of the team in Wiener's vision included the eccentric prodigy Walter Pitts (with whom he was already working but who traveled back and forth to Chicago to work with McCulloch and to Mexico City to be in touch with Rosenblueth), McCulloch, Rosenblueth, and Julian Bigelow (the engineer with whom Wiener had worked closely during the early part of the war).

Von Neumann's enthusiasm in 1944 and 1945 had first been generated by the challenge of improving the general-purpose

computer. He had been the proponent of using the latest in
computing machines in the atomic bomb project, but he realized that for the impending hydrogen bomb project still better and faster machines were needed. On the theoretical level he was intrigued by the fact that there appeared to be organizational parallels between the brain and computers and that these parallels might lead to formal-logical theories encompassing both computers and brains; moreover, the logical theories would constitute interesting abstract logics in their own right. He was cautious in assuming similarity between a computer and the awesome functioning of the human brain, especially as in 1944 he had little preparation in physiology. Rather, he regarded the computer as a technical device functioning as an extension of its user;[20] it would lead to an aggrandizement of the human brain, and von Neumann wanted to push this aggrandizement as far and as fast as possible. In order to pursue the abstract logical problem, it would be useful for von Neumann to be in close touch with an experimental physiologist such as Rosenblueth or McCulloch so that while working on the abstract models he could also be in constant contact with experimental information concerning the nervous system. But von Neumann needed no other collaborators or "center" for this purpose.

Von Neumann was in the process of formulating an organizational scheme for computers far more efficient and convenient than that of the ENIAC, the computer at the Moore School. What he wanted to do was to get a computer built that would embody, insofar as was technically possible, all the most advanced features, especially in its overall organization. Such a computer would be far superior to and faster than any then existing, and for this project von Neumann needed funds, a "center," and collaborators.

Since the Princeton Institute, von Neumann's home base, had never ventured into developing engineering devices, he did not at first approach them. Instead he joined with Wiener to find a place to locate the center and in April advised him that "the best way to get 'something' done is to propagandize everybody who is a reasonable potential support,"[21] and to Wiener's satisfaction, MIT made von Neumann a good offer. In

188
*John von
Neumann
and
Norbert
Wiener*

July Wiener wrote to Rosenblueth, "I have had several consultations with von Neumann . . . and it really looks to me now as if the appointment and his acceptance were in the bag,"[22] and one month later, he was writing, "Johnny was down here the last two days. He is almost hooked."[23]

Wiener and von Neumann also talked to each other a good deal about their work, with the exception of von Neumann's work on the secret atom bomb project. When Wiener and Rosenblueth produced a manuscript entitled "The Mathematical Formulation of the Problem of Conduction of Impulses in a Network of Connected Excitable Elements, Specifically in Cardiac Muscle," Wiener got von Neumann to read it before they submitted it for publication, and the authors were reassured when their reviewer didn't find any loopholes in the argument.[24]

The experience and events of World War II had an enormous impact on Wiener's attitude toward his scientific work. Immediately after Hiroshima, he apparently was eager to get to work on neurophysiology. Thus he wrote to Arturo Rosenblueth on August 11, "In the present almost certain to come interval between wars (and I hope to goodness it will be a long one) I think we can do an enormous amount with our new schemes."

It seems that he anticipated an eventual war with the Soviet Union. Two months later (October 16) he reported to his close friend, Giorgio de Santillana, that he was just "recovering from an acute attack of conscience" but again alluded to the likelihood of a third world war. Contemplating this possibility, he wrote, "I have no intention of letting my services be used in such a conflict. I have seriously considered the possibility of giving up my scientific productive effort because I know no way to publish without letting my inventions go to the wrong hands. . . . I feel it most intensely personally."

Wiener always viewed the scientist as wittingly or unwittingly delivering his knowledge into the hands of whoever had the power, the money, or the inclination to use it. As the physicist Daniel Q. Posin wrote to Einstein on October 21, "Here at the Massachusetts Institute for Technology, Wiener stands aghast—as though a man in a confused dream—and wonders what we must do, and he protests at scientific meetings the 'Massacre of Nagasaki' which makes it easier, for some, to contemplate other massacres."[25]

The Wiener archives contain a signed letter dated October 18, 1945, to the president of MIT, a letter that was probably never sent, in which Wiener submits his resignation and states he intends "to leave scientific work completely and finally. I shall try to find some way of living on my farm in the country. I am not too sanguine of success, but I see no other course which accords with my conscience."[26] It is clear that the use of atomic bombs deeply disturbed Wiener, even if his first impulsive response was not his final one.

Von Neumann, with the MIT offer in hand as well as others in prospect, sought to persuade the Princeton Institute to permit him to build a prototype "all-purpose, automatic, high-speed electronic computing machine" at the institute itself.[27] There was considerable opposition to this among members of the institute faculty, who regarded the computer—hardware, engineers, technicians, large government contracts—as the unwelcome machine in the idyllic garden,[28] but von Neumann was "strong enough to override the opposition." He anticipated correctly that his friendly relations with high government officials would allow him to obtain all the funds necessary to supplement the institute's contribution.[29] The Princeton computer was funded through the Army, the Navy, and the Air Force, and later through the Atomic Energy Commission as well.[30] Von Neumann's closest coworker in the computer field was Herman Goldstine, and von Neumann arranged an appointment at the Princeton Institute for him as well. The engineers from the Moore School did not come with von Neumann and Goldstine; in fact there was considerable conflict about the relative contributions of the two mathematicians on the one hand and the engineers on the other.[31] However disappointed Wiener may have been that von Neumann did not come to MIT, they remained friends, and in particular he helped von Neumann to find a chief engineer for the Princeton computer project, namely Julian Bigelow.

190
*John von
Neumann
and
Norbert
Wiener*

Wiener regarded his former associate highly and recommended him; von Neumann interviewed Bigelow and hired him.[32]

The upshot of that year of planning and arranging was that von Neumann succeeded in obtaining what he wanted at Princeton, where he wanted it. In person he was a charming and effective persuader, maintaining cordial relations with the powers that be, but at the same time he planned his strategy and could bully and pressure effectively, where needed, to get his way. The only drawbacks of staying at Princeton were the absence of physiologists in his group and the hostility toward his project—a hostility that may have contributed to von Neumann's later choice to spend more of his time elsewhere.

Wiener's center did not materialize in the form that he had envisioned, but he was able to arrange with MIT and the Instituto Nacional de Cardiología to spend alternate semesters working with Rosenblueth, with either Rosenblueth coming to Cambridge or Wiener going to Mexico City.[33] Pitts continued to work with Wiener and keep a liaison with McCulloch, who after a few years himself moved from Chicago to MIT. Later, however, an unfortunate quarrel between Wiener on the one hand and Pitts and McCulloch on the other would make all collaboration impossible.[34]

The closing paragraph of a letter from von Neumann to Wiener in early 1945 is so typical of the rapidly moving von Neumann that it is repeated, with minor stylistic changes and changes in date, in several other letters of this period: "I am leaving on February 4 for Aberdeen, and February 6 for the West, and expect to be back in the first days of March. I hope that we shall see each other very quickly thereafter. I am sure it would be very profitable for me at least if we could have another conversation."[35] It is a reminder that in 1944 and 1945 von Neumann's first priority was his consulting work at Los Alamos, in which he spent a large portion of his time. He traveled frequently to Aberdeen, Los Alamos, and Washington, D.C., in response to wartime urgencies, while his contacts with Wiener concerned personal plans for after the war. In working on the atom bomb at Los Alamos, von Neumann had adjusted himself to the patterns of secrecy and "security" required by this top-secret military project; Wiener had already found the relatively modest security precautions imposed on his work on antiaircraft fire an at times nearly intolerable burden for his normally open personality.[36]

Von Neumann's experience and activity at Los Alamos is important not only because of his contribution to the building of the first atom bombs but also because more than any other event it signals the turning point in von Neumann's own career. Already the development of the computer in which he was participating was opening up new fields of activity for him. One possible direction was that actively espoused by Wiener: studies having to do with organisms, which might lead to apprehending something about the nature of life and the mind through formalism. Another possible direction, opened up by the atomic bomb project, was that actively espoused by the military, namely to concentrate on innovations in weapons technology. For as versatile a genius as von Neumann, one option did not necessarily preclude the other. They differed in that one tended to lead to the secluded world of scholarship while the other led away from it. What they shared was that both of them, on very different levels and in very different contexts, dealt with matters of life and death; most of the other mathematical and scientific topics available to von Neumann had no such connotation.

The experience of Los Alamos meant living in New Mexico in a community dedicated to one urgent, challenging objective, a community consisting nearly exclusively of scientists, technicians, and their families, but also of guards and some military men. Los Alamos was located on a seven-thousand-foot-high plateau rising above a vast expanse of desert and valley. Hills, pine forests, and streams were nearby, but these held little interest for the unathletic von Neumann. However, the dramatic view afforded by the site was something that not only pleased but exhilarated him, although generally he cared little about countryside and landscape:

He loved this business of looking out so that you looked practically to forever—quite literally that. He often explicitly said this to me. He liked this better than the kind of forested land around here [Yorktown Heights, N.Y.] or Princeton where the

192

*John von
Neumann
and
Norbert
Wiener*

view is quite finite, and you're well aware of the boundaries. He went out there every summer [after 1945], and whenever else he could. I'm almost sure that had he lived he would have bought himself a place out there, because it was just such a wonderful experience for him. He just loved it; it's hard to overemphasize this point. To him it was just tremendous . . . I don't know why. That's never been my thing.[37]

While others too found the landscape extraordinary, for von Neumann it may have had a highly personal meaning, since he had a special penchant for disregarding or seeking to defy supposed limitations, a penchant reflected in his working style and philosophy.

Whatever von Neumann's private experiences may have been at Los Alamos, his demeanor was good-humored and low key and his views and opinions presented judiciously. Laura Fermi, historian and wife of the associate director of the Los Alamos Laboratory, wrote that "he was one of the very few men about whom I have not heard a single critical remark. It is astonishing that so much equanimity and so much intelligence could be concentrated in a man of not extraordinary appearance."[38] Von Neumann worked closely with the leading physicists—Oppenheimer, Fermi, Teller, and Wigner—on the Manhattan Project. He was widely liked and admired by the physicists, who regarded him as a peripatetic genius,[39] an oracle,[40] and a problem solver. He was also the chief proponent of and expert in the use of computers. His most important specific contribution came at a time when the Los Alamos staff seemed to be encountering insurmountable obstacles in devising an adequate means for detonating the plutonium bomb. On the basis of very intricate calculations and his knowledge of the theory of explosions, von Neumann concluded that the previously suggested "implosion method," which had been given up as hopeless, could be made to work, and persuaded the scientific director of the laboratory, Oppenheimer, of this possibility.[41] He then helped work out the ingenious arrangements required; it was proved out in the Alamogordo test and then employed to detonate the Nagasaki bomb.[42]

Von Neumann was not only respected by the scientists but

was trusted by Army General Leslie R. Groves, a tough, decisive executive, the administrative head of the Manhattan Project. Von Neumann, on his side, had a special liking and admiration for men with political and administrative power, most
of all military men. In this respect he was unlike most of the
physicists.[43] He became part of the inner circle with whom
General Groves discussed the tactical problem of the choice
of a target for the first military use of the atom bomb.[44]

Von Neumann liked life at Los Alamos, the scenery, the
community, the project, his own peculiar role and status.
While the initial impetus for the project had been the fear that
Nazi Germany might build such a weapon first, von Neumann
saw the political purposes differently: "This was definitely a
three-way war. . . . I considered Russia as an enemy from the
beginning [of the atom-bomb project] . . . and the alliance
with Russia as a fortunate accident that two enemies had
quarreled."[45]

His hard-line view of Russia as an enemy persisted until
the end of his life. Officially Russia was an ally during
World War II, and most scientists were at that time not much
concerned about her as a potential future enemy. For example, in 1942 Wiener was making inquiries whether it
might somehow be possible for him to work together with the
Russian Kolmogoroff, who shared more closely than anyone
Wiener's mathematical interests but, Wiener feared, might be
unaware that the mathematical work had military uses.[46]
Again, it is pertinent to mention the viewpoint of Niels Bohr,
who had escaped from Nazi-occupied Denmark and was visiting at Los Alamos and Princeton. He had been concerned
with the Soviet Union and became an active proponent of
an attempt to approach the USSR very early in the hope of
heading off a nuclear arms race before it got started. As Bohr
saw it, this effort would necessitate cooperation rather than
the traditional competition between the two nations; he suggested that a means to achieving success in this effort might
consist of an offer to share America's nuclear knowhow and to
agree to have no secrets from each other in this respect.[47]
Such an agreement would require a major shift in the closed
tradition of the Soviet Union.

194
John von
Neumann
and
Norbert
Wiener

Von Neumann's view of Russia as an enemy may have had to do with his Hungarian origins. In response to a question whether his family had regarded Russia as an enemy of Hungary when he was growing up, von Neumann said, "Russia was traditionally an enemy of Hungary. . . . After the First World War everybody had reason to worry about it. . . . I think you will find, generally speaking, among Hungarians an emotional fear and dislike of Russia."[48] Von Neumann himself displayed this attitude toward Russia,[49] although as a young teenager after the First World War he would have felt threatened not so much by Russia itself as by the Russian-supported Bela Kun regime's hostility to the social and economic class to which von Neumann's family belonged. In the latter part of World War II, however, Russia had already brought Hungary within its sphere of influence. To many educated and upper-middle-class Hungarians the Germans, who murdered systematically in obedience to racist and other "principles" but who tended to be relatively educated and therefore "civilized," were much preferred to the Russian occupiers. The latter were seen as synonymous with all that is primitive, brutal, impulsive, and vulgar.[50]

A particularly noteworthy effort among scientists to prevent the military use of the atom bomb once it was created was that of physicist Leo Szilard. Szilard, five years von Neumann's senior, was also an immigrant to the United States from the privileged but insecure Budapest upper middle class.[51] After some engineering studies he had left Hungary in 1930 for Berlin, where he studied physics. Subsequently he became a privatdozent, teaching at the University of Berlin until 1933. Thus up to 1933 the careers of von Neumann and Szilard contained many parallels. They knew each other very well in Berlin, if not earlier in Budapest. Both shared, it seems to me, the energetic zeal to do the extraordinary, a near obsession to innovate lest disaster strike. For Szilard it had meant a preoccupation, as early as 1933, with the future possibility of nuclear weapons. It had been he who in 1939 had initiated the first and ultimately successful effort (through Albert Einstein) to inform the president of the United States of the prospect of developing the atom bomb, and later he worked on the bomb

at the Chicago Metallurgical Laboratory. In the spring of 1945
he arranged through various intermediaries an appointment with President Roosevelt himself to present a memorandum he had prepared urging abstention even from an experimental demonstration of the bomb. It said, in part, "Perhaps the greatest immediate danger which faces us is the probability that our 'demonstration' of atomic bombs will precipitate a race in the production of these devices between the United States and Russia and that if we continue to pursue the present course, our initial advantage may be lost very quickly in such a race." But Roosevelt died before the date of his appointment with Szilard, and the subsequent efforts by Szilard and other scientists opposing the idea of bombing Japan to reach President Truman were unsuccessful. After the war Szilard continued to exercise his imagination and engage in practical activity in the realm of new political ideas and political organizations which would serve to deter a nuclear weapons race and prevent a nuclear war. Although he never sought an official position within the political establishment, he was uninhibited about approaching men on the highest political levels to convey his concerns and ideas. Ultimately Szilard's scientific interests shifted from physics to biology.

Szilard's activities and those of von Neumann make an interesting comparative study. Both men were sensitive to and preoccupied by international political dangers. Both responded by brilliant and compulsive innovating, yet in very different ways. Both actively sought influence in the centers of political power, but by very different methods. They worked at cross-purposes, with Szilard seeking to prevent eventual nuclear war while von Neumann focused on American superiority over Russia.

One did not have to be born in Hungary to see the Soviet Union as an enemy in the early 1940s. In fact, as historians have belatedly clarified, von Neumann's formulation, according to which Russia was an enemy who only shared a temporary common objective with the United States (namely the defeat of Germany) was apparently consistent with the views and actions of dominant figures in the American government. Studies have shown that the primary consideration

196

*John von
Neumann
and
Norbert
Wiener*

in dropping the atomic bomb on Japan in August 1945 had not been speeding the end of the war—the bombs were not necessary for that purpose, and the US government knew that—but rather to flex our national muscle and be in a position to impose our terms vis-à-vis the Soviet Union in the impending peace negotiations.[52] According to political historians, the Soviet Union, whose policy is in retrospect seen as having been flexible in 1945, was defined in part by the American government's decision to regard Russia as an enemy. It follows that Niels Bohr's vision had not been unrealistic; he had simply misunderstood American objectives.

The project to build an atomic bomb, predicated on the mistaken premise that Germany would succeed in building nuclear weapons, continued unabated even after Germany had surrendered. None of the scientists, although they regarded themselves as intelligent, independent, and concerned, stopped work. Some petitioned the government not to use the bomb once it was built. Even in such blatant circumstances it was difficult to break the habit of scientists to regard their work as specialists' work and the values and political implications of their work as some other specialist's domain. And of course the momentum of the task—the interest in seeing "the gadget" (as these boys liked to call the atom bomb) completed and tried out—was also powerful.

It is quite appropriate to view the Manhattan Project as a large bureaucracy obeying political patterns characteristic of bureaucracies: To account for itself it needed tangible results. The bomb had to be produced, and preferably used in war. Moreover, the second bomb had to be used too because it worked on a different principle (the implosion method), whose effectiveness had to be justified.

To work on this fantastic weapon knowingly and to witness the "unprecedented, magnificent, beautiful, stupendous and terrifying"[53] spectacle of the first successful test explosion was a shattering experience that cannot be completely fathomed by anyone who was not there. However beautiful and however gratifying the "success," the experience also contained the prospect of death and doomsday, which those who had helped to bring about the new era might yet fall victim to. The generals' and the professional scientists' accustomed speech was temporarily undone by the vision of the nuclear ball of fire and mushroom cloud, and those who felt able to describe it drew their language from mythology and theology. I have come across no record of von Neumann's immediate reaction to witnessing the test in Alamogordo, New Mexico, on July 16, 1945, but in November of that year, a few months after Hiroshima, he made the "semi-serious" comment to Admiral Lewis Strauss that

the appearance in the heavens of supernovae—those mysterious stars which suddenly are born in great brilliance and, as quickly, become celestial cinders—could be evidence that sentient beings in other planetary systems had reached the point in their scientific knowledge where we stand now, [and] having failed to solve the problem of living together, had at least succeeded in *achieving togetherness* by cosmic suicide.[54]

According to Strauss, "von Neumann had pointed out that this piece of science fiction was not very much more fantastic than the first fission of an atomic nucleus, so minute at its birth and yet becoming the force which, six years later, pulverized Hiroshima and Nagasaki." Assuming the report of von Neumann's comment to be accurate, it indicates that he, like most of his colleagues, considered nuclear holocaust a distinct possibility in the future. There is a vision of absolute horror here, although in his gallows humor von Neumann could see the silver lining ("togetherness") in the mushroom cloud. Considering the particular quality of the logical von Neumann's emotional makeup, he might have been vulnerable to seduction by "the gadget," that is, by the promise of nuclear weapons as a means to salvation.[55] This possibility deserves elaboration.

First of all, von Neumann perceived the Soviet Union as unambiguously and unalterably his and America's dangerous enemy whom he, as an American patriot, was prepared to fight. This enemy could be reduced to harmlessness and impotence by use of "the gadget" if it were to be used soon enough. Thus, it would seem reasonable that the gadget his mathematics had helped create could remove his worst fears.[56] But his dislike of Russia was not the only remnant surviving from von Neumann's childhood. More significant is what he described as "a subconscious feeling of extreme insecurity in individuals, and the necessity of producing the unusual or facing extinction."[57] The enemy in international politics gave a particular, manageable focus to the danger, but it was heavily freighted with a less manageable enemy: death itself. Put more positively, von Neumann's interest in the bomb as an unusual tool he had helped produce reflected a desire for a long life, but focused it on the political scene in a tribal, parochial way.

Von Neumann's orientation, based on the upwardly mobile capitalist class of Hungary, had always included the desirability of a career that would give him power and influence in government. It appears that already during the war he appreciated that such a career might be open to him in the postwar period; certainly at the end of the war he was keenly aware of it.[58] His technical expertise in weapons development could be the basis for such a career, the more so if the government made innovation in this field a central pillar of its policy.

A great inhibitor to advocating the use of nuclear weapons and reliance on them in international relations lay in considerations of morality, especially if the killing and maiming of large populations was involved. For many such an advocacy of the bomb was out of bounds on account of a priori philosophical considerations of human values. However, von Neumann's whole style of thinking had always been to disregard limits imposed a priori on philosophical grounds. As he had conceived logics as limited only by superior logic itself, might he not also wish to push military technology to the limit? Within the scientific community, as well as within Los Alamos, von Neumann had always followed convention, learning and accepting the rules. That was his modus operandi. He knew from his study of history and his European experience that international relations are governed by force and power politics, usually unmitigated by moral considerations. Thus one cannot expect a priori moral considerations to have been an effective inhibitor for von Neumann. And empathy for potential victims is easily overridden by the old military masculine ideal of toughness and ruthlessness.

Though von Neumann followed conventional rules of behavior, he differed from other men in that his mental abilities allowed him to enjoy a certain mastery over the limitations of time. Wiener conceded a priori limitations owing to finite time spans in physics and mathematics, but von Neumann sought to ignore, disprove, or overcome them. According to the psychologist Robert Lifton, close association with nuclear weapons as well as witnessing the Alamogordo tests exposed scientists to an experience of the horrors of death and destruction that was not easily shaken off. One kind of response to the horror, especially for individuals who have never come to terms emotionally with their own finite time span, is to become "converted" to belief in the bomb as "good," to be fascinated with it "as a solution to our anxieties (especially our anxieties concerning the weapons themselves) and as a means of restoring a lost sense of immortality."[59] If von Neumann had been holding to the cheerful illusion of his own invulnerability, the Los Alamos and Alamogordo experience would have at least temporarily broken the spell. Could he live effectively and contentedly and yet abandon that illusion? Would he need to reconstitute it in some form? Could he avoid viewing nuclear weapons as the solution to his and the world's problems? Weapons work can have a deep psychological attraction for anyone to whom Lifton's observations apply.

Seduction by any technology is contingent on its psychological connotations, and we know that von Neumann identified strongly with the intense technological optimism that marked the period of economic growth in which he passed his youth. How, precisely, did von Neumann see his relation to technical devices? In his prewar measurement theory he had viewed the instrument as an extension of the scientist and had emphasized the arbitrariness of the dividing line between the measuring instrument and the person because the instrument expands the observer's powers of perception. During and after the war, when working with computers, von Neumann regarded these machines as exten-

200
John von
Neumann
and
Norbert
Wiener

sions, supplements to his mental powers. Thus, they too helped him overcome limitations. The continuity of this theme makes it possible to speculate that he also regarded nuclear weapons as means of self-aggrandizement; as extensions of himself.

In all, the constellation formed by von Neumann's personal background, his talents, his personality, and his experience made him more vulnerable to the appeal of nuclear weaponry than many of his colleagues at Los Alamos. He was drawn to the production of ever bigger and more effective weapons, even after the war. But he was also intrigued by the direction in which Wiener's work pointed—mathematical models that encompassed both living organisms and machines. In addition, he had numerous mathematical ideas he wanted to pursue further, and all kinds of possibilities were opened up by the computer he wanted to have built. In spite of the speed with which he could work, he would have to set priorities among the myriad possibilities confronting him.

But wait, why can an artificial hand not feel? It is easy to put
pressure gauges into the artificial fingers, and these can
communicate electric impulses to a suitable circuit. This can
in its turn activate devices acting on the living skin. . . .
Thereby we can produce a vicarious sensation of touch, and
we may learn to use this. . . . Moreover, there are still sensory
kinesthetic elements in the mutilated muscles, and these can
be turned to good account.
—Norbert Wiener

Although the center at MIT had not materialized in the form
that Wiener had first envisioned, the spark of the ideas relating mathematical engineering to living organisms which had
animated Wiener, as well as Warren McCulloch, Arturo
Rosenblueth, and Walter Pitts, had lit a steady flame. Von
Neumann was evidently interested, and so was Julian
Bigelow. A new field in science, a new paradigm for research, was in the making. It was a good moment for initiating interdisciplinary research projects, because scientists returning from wartime projects, which had usually also been
interdisciplinary, were scouting around for new ideas and
programs. Also, funds for research soon became available,
as the successful exploitation by the scientific community of
the relationship it had developed with the government during
the war ensured a shift in the flow of government funds from
military research to fundamental scientific studies.
   Immediately after the war ended, Warren McCulloch took
the initiative under the sponsorship of the Macy Foundation—a philanthropic medical foundation—of organiz-

202
John von
Neumann
and
Norbert
Wiener

ing an interdisciplinary meeting of about twenty interested scientists to help launch explorations of the emerging paradigm. A social scientist, Gregory Bateson, had heard of the new ideas from McCulloch and Rosenblueth some years earlier and insisted that they also had relevance to the social sciences; he helped McCulloch bring participants from the "soft sciences" into the meeting. McCulloch described the participants of the planned meeting in a letter to Pitts, sent care of Norbert Wiener:

In a few days I will be writing you and Wiener identical letters concerning a conference . . . on *Mechanisms Underlying Purposive Behavior.* . . . [The invitees are] Wiener, Neumann and you on mathematical engineering, and Rosenblueth, Lorente de Nó and Ralph Gerard for Neurophysiology, Kubie (first to suggest importance of feedback in psychiatric problems) and Hank Brosin (a dandy) for psychiatry. Mead, Bateson and someone else for sociology—Bateson you will remember, is the man who insists on the importance and lack of theory in sociology. Marquis, Kluever, Lewin and Harrower for Psychology and half a dozen others from various fields not well-known to me.[1]

Wiener suggested that his friend, the MIT science historian and philosopher Giorgio de Santillana, be invited,[2] and von Neumann recommended the logician Kurt Gödel, from the Princeton Institute.[3] These recommendations—although neither Gödel nor de Santillana came to the meeting—remind us that Wiener and von Neumann approached the topic differently: Wiener sought to understand it within a broad philosophical and historical context, while von Neumann foresaw the description of the brain in terms of the most sophisticated kinds of formal logics. Von Neumann had also suggested his friend, the economist Oskar Morgenstern, who also did not come. Though Bateson and Wiener would establish good rapport, Bateson was a very different kind of social scientist from Morgenstern, and their differences mirror von Neumann's and Wiener's contrasting approaches to the social sciences. Some of Wiener's ebullient mood in anticipation of the meeting comes through in his response to McCulloch's invitation. The letter was written from Princeton, where Wiener was visiting von Neumann:

This meeting is going to be a big thing for us and our cause. I am now down with von Neumann discussing plans and I can assure you that his part and mine will be well coordinated. Pitts and I are also getting busy together and so is Rosenblueth. . . . Meanwhile, we are impatient for the meeting of the Macy Foundation and when we shall see you and talk over many things of common interest. I am very much pleased with the tentative program and I am delighted that you are chairman."[4]

The meeting took place on March 8 and 9, 1946, at the Beekman Hotel in New York City—twenty-one scientists sitting around a table, talking for two days about the new ideas.[5] The discussion continued through cocktails and dinner and into a night session. Von Neumann and Wiener were the star performers. Whereas von Neumann described the organization and operation of the most advanced general-purpose computers under construction at the time, Wiener spoke about "purposive" machines and the role of information and feedback in their operation. Wiener also drew examples involving purpose from the physiology of organic mechanisms, showing how feedback and information were elements essential to implementation. Von Neumann described his theory of games, and Wiener spoke about the undecidability of some propositions (Russellian paradoxes), which would cause a computer to oscillate yes-no-yes-no-yes-no. . . .

This meeting was to be the first of a series of ten two-day conferences with the same set of participants each time, plus a few invited guests. This was one context in which Wiener and von Neumann would see each other regularly over the next five years. The group called itself the Conference for Circular Causal and Feedback Mechanisms in Biological and Social Systems, but when Wiener, after some philological research, came up with the Greek word *cybernetics* (derived from the word for "helmsman"), the group acceded to his urging and changed its title to Conference on Cybernetics.

Throughout 1946 von Neumann had harbored some doubts about the feasibility of creating fruitful formal-logical models of the brains of even simple organisms, or of the general patterns of communication and control involving nervous systems, beyond what McCulloch, Pitts, Wiener, and Rosenblueth had already devised. Although Wiener had visited von Neumann in Princeton in February and they had seen each other at the Macy conferences in March and October, von Neumann did not communicate these doubts to Wiener until later in November. The doubts, and von Neumann's letter to Wiener describing them, are of considerable interest. As von Neumann expressed it to Wiener, he felt "an intense need that we discuss the subject extensively with each other,"[6] while Wiener "was extremely eager to . . . talk over developments."[7] The two men arranged to spend the better part of December 4 together at MIT discussing the problem.

What troubled von Neumann was the state of experimental neurophysiology and its relation to theory. He compared the subtlety of available microscopic cytological techniques to that of studying the circuits of an electronic computer by means of dropping cobble stones onto them or directing a fire hose at them. On the other hand, the formal-logical theories, however interesting in themselves, were so general that they did not yield specific predictions about the outcomes of experiments. The difficulties, according to von Neumann, "reside in the exceptional complexity of the human nervous system, and indeed of any nervous system." Von Neumann's letter to Wiener was not entirely negative. He had become acquainted with the experimental work of biophysicist Max Delbrück on bacteriophage, living organisms that, although primitive, do reproduce.[8] The complexity of such an organism seemed mathematically manageable, and the observational means to study them in detail, particularly X-ray diffraction and electron microscopy, were available. Moreover, understanding these organisms "in the exacting sense in which one may want to understand a detailed drawing of a machine—i.e. finding out where every individual nut and bolt is located, etc.," might be the great step forward science needed toward understanding organisms generally. In his polemical letter, von Neumann in effect challenged Wiener to rebut the conclusion that bacteriophage is a far more promising subject for mathematical research than the nervous system.

Whatever transpired on that December 4, Wiener and von Neumann continued to attend the semiannual conferences on cybernetics. Max Delbrück was invited to become a regular participant, but after attending only one of them he had had enough: he did not find the discussion relevant to his own interests.[9] Von Neumann's old friend, coeval, and fellow Hungarian Leo Szilard was one who turned to Delbrück's kind of biophysics at that time. Young physicists who followed up Delbrück's research on bacteriophage would a few years hence reveal the genetic code and the structure of DNA and win Nobel Prizes for their discoveries.

Wiener and von Neumann cut rather different figures at the semiannual conferences on machine-organism parallels, and each had his own circle of admirers. Von Neumann was small and plump, with a large forehead and a smooth, oval face. He spoke beautiful and lucid English, with a slight middle-European accent, and he was always carefully dressed; usually a vest, coat buttoned, handkerchief in pocket, more the banker than the scholar. He was seen as urbane, cosmopolitan, witty, low-key, friendly, and accessible. He talked rapidly, and many at the Macy meetings often could not follow his careful, precise, rapid reasoning—it was the other mathematicians in the group and those inclined toward logics with whom he communicated best. Nonetheless his thoughts and scientific judgment were respected by everyone. Because of his many other commitments, he often did not spend the full two days at the meeting; especially at the later conferences of the series his "meteoric appearances" were sometimes only for lunch or a half-day. But even in his absence discussion sometimes centered on just what he had said or what the significance of his comments had been at an earlier meeting. For Gregory Bateson, the impression von Neumann gave at the conferences was softened by a later personal one:

I'd seen him only in Cybernetics meetings where he was pouring all the stuff out and punching with both fists, you know. And then I went up to Princeton some time or another and not knowing what to do, I gave him a telephone call and visited him, and Morgenstern came in. We had a lovely evening together . . . talking about a trip they'd made to the Grand

206

John von
Neumann
and
Norbert
Wiener

Canyon. It was a complete revelation that there was a sort of gay, sweet holiday-making side to this character . . . family, children (laughter), all that.[10]

Again and again one notes that others find something nearly inhuman in von Neumann's rigorous style and are pleased when they discover a comprehensible human being in him as well.

Wiener was the dominant figure at the conference series, in his role as brilliant originator of ideas and enfant terrible. Without his scientific ideas and his enthusiasm for them, the conference series would never have come into existence, nor would it have had the momentum to continue for seven years without him. A short, stout man with a paunch, usually standing splay-footed, he had coarse features and a small white goatee. He wore thick glasses and his stubby fingers usually held a fat cigar. He was robust, not the stereotype of the frail and sickly child prodigy. Wiener evidently enjoyed the meetings and his central role in them: sometimes he got up from his chair and in his ducklike fashion walked around and around the circle of tables, holding forth exuberantly, cigar in hand, apparently unstoppable. He could be quite unaware of other people, but he communicated his thoughts effectively and struck up friendships with a number of the participants. Some were intrigued as much as annoyed by Wiener's tendency to go to sleep and even snore during a discussion, but apparently hearing and digesting what was being said. Immediately upon waking he often would make penetrating comments. The psychoanalyst present, L. Kubie, was professionally interested in Wiener's "obvious ability of processing material while asleep, and then without forgetting it, carrying it right over into waking speech. My theory is that this is the key to his creativity."[11]

A similar story, but with a special twist, is told at MIT, concerning a time when Wiener was known to bring nearly every conversation around to the ergodic theorems and was preoccupied by the central importance of ergodic theory. One day, during a seminar, he fell asleep and began snoring. The speaker, annoyed but knowing Wiener's predilections, announced after having proved some lemma, "And this has nothing to do with the ergodic theorem!" Wiener immediately woke up, said, "It does too," and walking up to the blackboard proved to everyone's satisfaction that indeed the lemma had very much to do with the ergodic theorem. Apparently his snores had not prevented him from following the proceedings closely.[12]

One of the participants in the Macy seminar series, a thinker in the social sciences, L. K. Frank, happened to live close to Wiener's home in Belmont. Another, the philosopher F. S. C. Northrop, had a summer home not far from Wiener's in New Hampshire. Wiener, always in need of communicating his ideas, was accustomed to drop by his various friends' homes in Belmont on Sunday mornings to talk. After the Macy conference series was concluded, L. K. Frank's house became a regular stopover on the Belmont "Wienerweg," and the two men became fond of each other. Northrop's cottage also became an occasional stopover on Wiener's analogous New Hampshire route; Wiener always arrived unannounced, eager to talk and try out his ideas.

Alex Bavelas, a young sociologist who later became a participant in the Macy group, received his first impression of Wiener while a graduate student at MIT. Bavelas was grading papers while a number of other graduate students were in the room with him. Norbert Wiener opened the door, walked in, and introduced himself to Bavelas and the other graduate students. Spying a chessboard on one of the desks, he asked Bavelas if he would like to play a game. During the game Wiener dozed between moves. Wiener lost the game, but didn't seem to care in the least and wanted to play another. When Bavelas encountered him subsequently, Wiener often asked him what he, although he knew little mathematics, thought of Wiener's work, and what others at MIT thought of him.[13] Whereas von Neumann appeared, at least on the surface, to be lacking human frailties, Wiener appeared to have a surfeit of them.[14]

At the March 1946 meeting, Wiener's and von Neumann's contributions supplemented each other, and in the minutes their joint description of memories (whether in computers or organisms) is referred to as a "duet on recorded memory."

208
John von
Neumann
and
Norbert
Wiener

However, at subsequent meetings of the conference series, there was a noticeable coolness and even friction between the two men. When von Neumann spoke, Wiener would ostentatiously doodle or go to sleep very noisily. Wiener and von Neumann also attended, besides the cybernetics meetings, some of the same mathematical congresses.[15] On one such occasion, when Wiener was lecturing, von Neumann sat in the front row and as ostentatiously and noisily as possible read the *New York Times*, to Wiener's annoyance. The great difference in their personalities of course could easily give rise to friction. Moreover, their mutual admiration also involved rivalry in problems ranging from quantum theory to ergodic theory, from weather prediction to automata.

It is a peculiar characteristic of the mathematical and scientific communities that they attach stupendous importance to the matter of priority, of who thought what first; science and mathematics are highly competitive enterprises.[16] For example, von Neumann (among others) had appreciated the connection between entropy and information even before Wiener made it central to cybernetics and was peeved at the credit that fell to Wiener for exploring that connection and assessing its importance at full value, which von Neumann had failed to do.[17] While this kind of petty rivalry was an element in the complex relation between the two men, the more significant source of conflict arose from their differing relation to the military. In fact this conflict was so sharp that, had a joint center for research materialized at MIT, it surely would have become a place of disharmony.

By the end of 1946 Wiener had arrived at a position concerning the relation of his future work to the military, a position to which he adhered henceforth: "I do not expect to publish any future work of mine which may do damage in the hands of irresponsible militarists."[18] Von Neumann, however, occupied a very different position. He worked closely with the military, continuing his consultantships relating to weapons development and applying computers to weapons; in particular, he worked on hydrogen bomb development, in anticipation of a possible war with the Soviet Union. It was to be expected that the New Englander with a social conscience would be at odds with the hard-boiled, weapons-minded "realist." A faint, brief echo of the disagreement is heard on one occasion during the formal recorded part of one of the cybernetics conferences: the conference chairman, Warren McCulloch, had just announced that von Neumann and his chief engineer Bigelow could not be present at the meeting because "they are at present overrun by civil servants inspecting the apparatus [the IAS Computer]," whereupon Wiener impishly interposed the question, "Civil servants or civil masters?"[19]

Yet in spite of these various sources of friction between the two great mathematicians, it was in their mutual interest to maintain their professional friendship and continue to talk with each other about their new scientific activities relating machines to organisms, activities that they both loved and enjoyed. Both did brilliant work in this field, but whereas von Neumann sought to explicate primarily through formal logical structure, Wiener sought an intellectually comprehensive synthesis as a context within which to examine a few concrete topics with mathematical rigor.

The organization of the human brain fascinated von Neumann, and he envisaged a description of the brain in formal-logical terms as some kind of ultimate objective of science. He learned from sensory psychologists and neurobiologists the details of a small piece of the problem, namely those pertaining to the mechanism for the perception of visual form. He considered the problem of describing in formal terms how one recognizes a triangle as a triangle or identifies a particular friend's face. Von Neumann concluded that far too little was known about the neurophysiological details and that no existing logics were at all adequate for dealing with their complexity. Thus it was a problem for some future generation of scientists. While he seemed to believe in the possibility of some day devising a mathematical or a logical system to fit the biology of the brain, he saw clearly that such a day was not at hand.[20] However, he understood how to organize and describe formally computers that have inputs and outputs and are able to reason mathematically and logically, act on general instruction, store information so that it

210
John von
Neumann
and
Norbert
Wiener

is more or less accessible, recognize patterns with photo-receptors, and so on. Human brains do analogous things, and von Neumann proceeded, in a rough quantitative way, to compare the two, since comparison might give some preliminary ideas about the nature of the brain, or at least raise concrete questions. For example, von Neumann noted that the size of electronic components (his earliest estimate was based on vacuum tubes, his later one also on transistors) in computers is so large relative to the size of neurons that for reasons of space alone computers will of necessity have far fewer components than a brain has neurons. According to his quantitative estimate, even if the size of a computer is a hundredfold that of a human brain, the brain would still have a million times as many elements as the computer. On the other hand, a pulse (message) is transmitted or altered via an artificial component much faster than via a neuron, roughly ten thousand times as fast. The computers excel through their speed, while brains have greater complexity. The essential characteristic to compare is the number of electrical events that can take place in one second. The greater complexity of the brain more than compensates for the greater speed of computers, so that far more basic electrical events (firing of synapses) take place in the brain in one second than in a computer. (The ratio is $10^8/10^4 =$ 10,000 if brain and computer are of comparable size.) This simple bit of arithmetic already gave von Neumann a quantitative clue to the minimal degree of complexity the human brain must have. He reasoned further that because nervous systems have more but slower components than computers, while with computers it is usual to carry out operations in sequence, the characteristics of central nervous systems favor carrying out operations in parallel.[21] This kind of reasoning is striking in its simplicity and clarity and illustrates the crude yet systematic quantitative comparisons that von Neumann made between "artificial" and "natural" automata. He also began to reason in a preliminary way about the extensions of known logics and mathematics that were likely to be useful in describing the patterns of organization of the central nervous system.

The starting point in the making of logical models for the brain had been the work of Pitts and McCulloch, and von Neumann had used their model to design circuitry for a computer. Two simple examples will illustrate the basic relation of logic, neurons, and computer circuits.[22] Consider the formal statement, "If either A or B is true, then C is true." Now imagine an electrical circuit with two wires, labeled A and B respectively, coming into the circuit and one wire, C, coming out of the circuit (figure 6). The circuit is such that if an electric pulse arrives via either A or B (or both), an electric pulse will be caused to flow in wire C; otherwise there is no electricity flowing in wire C. Such a circuit is a hardware representation of "If either A or B is true, then C is true." In a nervous system the "circuit" is a single neuron; the input wires correspond to dendrites, and the output wire to an axon along which the electrical impulses in a nervous system travel. Some impulses act negatively on the neuron, preventing it from firing and passing any impulse on to the axon. For example, in figure 7, pulses coming from dendrite B have an inhibitory effect whereas those along dendrite A are excitatory. Suppose A excites the neuron at regular intervals, but an impulse from B occasionally interrupts it. This neuron-circuit corresponds to the statement, "Assuming A is always true, C is true whenever B is *not* true." With complex nets of neurons it is possible to represent very complex statements.

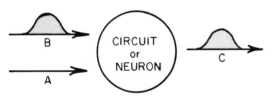

Figure 6

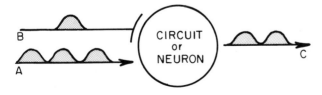

Figure 7

212
*John von
Neumann
and
Norbert
Wiener*
Von Neumann occupied himself with what he called automata theory, formulating axioms and proving theorems about assemblies of simple elements that might in an idealized way represent either possible circuits in man-made automata or patterns in organisms. Although he was stimulated by the problems of computers and organisms and continued to take account of them, he found the new abstract mathematics intrinsically interesting—regardless of whether or not it was applicable to actual organisms or actual machines. One question with which he dealt in his automata theory is, If one has a collection of connected elements computing and transmitting information, i.e., an automaton, and each element is subject to malfunction, how can one arrange and organize the elements so that the overall output of the automaton is error-free?[23] The other problem was closer to genetics and Delbrück's phage work than it was to neurobiology, that of constructing formal models of automata capable of reproducing themselves.[24] These would be models of a basic element of "life." His theory of self-reproducing automata is among von Neumann's most original work, and although he had elaborate plans for developing the theory in various directions, he became sidetracked by more practical projects, from computers to hydrogen weapons. Ultimately ill health kept him from carrying out his plans for the logical theory of automata. He did show, however, that in principle machines that can reproduce themselves can be built.

Wiener had been working in a different direction; he was concerned about the social implications of the new technology, and adopted on some occasions an attitude of moral disapproval toward colleagues engaged in weapons work. He had become a public figure, sought out by journalists for interviews. He wrote lightheartedly to von Neumann, "I am very much interested in what you have to say about the reproductive potentialities of the machines of the future. As Grey Walter in England has just made a machine with ethics of its own, it may be an opportunity for a new Kinsey report."[25] Von Neumann, so famous for his jokes and funny stories, did not think this was amusing. He was worried; he wanted no mention of the "reproductive potentialities of the machines of the future" to the mass media.[26] It may be that he regarded his own fantasies, if any, about self-reproducing machines, machines that could in effect perpetuate themselves endlessly, as a very private matter. To Wiener he wrote admonishingly, "I have been quite virtuous and had no journalistic contacts whatever." It is an apparently bizarre fact that the traditional ethos of the scientific community severely condemns dramatic statements to reporters concerning possible social consequences of research, yet accepts the exploitation of science to devise tools for mass homicide for use by the government. This attitude appears more comprehensible in the light of the single guiding value from which the ethos of the scientific community is derived, which is to implement the progress of science.[27]

Wiener's work relating organisms, mathematics, and machines proceeded on various levels: (1) The analysis and interpretation of actual experimental data in physiology, done jointly with Arturo Rosenblueth and Walter Pitts; (2) the development of sophisticated devices to aid the crippled, the blind, and the deaf; (3) the formulation and exposition of a coherent synthesis of ideas concerning "communication and control in the organism and the machine," ideas that in some contexts could be made precise and mathematical, while in others they were only philosophical and verbal. In his joint work with Rosenblueth, Wiener could be close to the experimental apparatus. In this new field he needed to have his feet on the ground, for then he could work in biology in spite of having become a mathematician. His collaboration with Rosenblueth was particularly active in the years 1945–48, but their friendship, full of warmth and charm and generosity, lasted until Wiener's death. In the 1940s Pitts was to them like a young apprentice, if not a son, but Pitts's lack of self-discipline upset Wiener, who may have been reminded of his own earlier struggles with himself.[28] The collaborative work with Rosenblueth and Pitts included the mathematical description of heart muscle during a normal heartbeat as well as during heart flutter and fibrillations;[29] the mathematical description of electrical potential variation as a single nerve impulse travels along an axon;[30] and the

214

John von
Neumann
and
Norbert
Wiener

mathematical analysis of a muscle stimulated to the point where it enters into periodic contractions ("clonus"), when it behaves analogously to certain engineering systems.[31]

Wiener's activity in the development of medical technology had its beginning in 1947 and was carried out in collaboration with Jerome Wiesner, who was trained as an electrical engineer. At that time Wiener expressed the hope that he might contribute something to the development of prosthetic devices to replace a lost sense or a lost limb, and he already had some preliminary ideas concerning these problems. He referred to the problem of sensory prosthesis as "an extremely hopeful field of work."[32] His interest in the subject seemed to stem, at least in part, from a desire to use his scientific knowledge and talent for humane ends, to turn swords into ploughshares. It reveals a gentleness that was often hidden by his awkwardness, a wish to heal, to repair the kinds of damage done by the weapons of war on which he had worked. The subject also exemplified his general ideas on cybernetics. Wiener's and Wiesner's first specific effort in this area was the translation of the sounds of speech into tactile sensations via a special glove worn by a deaf person.[33] The intelligibility of the tactile messages left much to be desired, and the project did not immediately result in a useful invention. However, the criteria for the design of sophisticated prosthetic devices, which Wiener had spelled out, have proved to be valid guiding principles, and for this reason he must be considered as one of the pioneers in the field.[34] An impetus to his work on limb prosthetics was given by the circumstance that in 1962 he had to spend several weeks in a hospital to recover from a fall. "The physicians of the hospital, aware that Wiener was around in the position of sitting duck, used to gather at his bed and talk with him."[35] To Wiener's great satisfaction these sickroom discussions led to the collaboration of a team of orthopedic surgeons, neurologists, and engineers on the limb prosthesis problem,[36] which eventually resulted in the effective prosthetic "Boston Arm."

Even though prosthetic devices constitute a technology designed to ease the hardships of individuals deprived of a limb

or of their hearing, it is clear that Wiener had moved away
from his 1945 impulse to avoid all work with technological implications. Even medical technology is often used by manufacturers holding the patents to make excessive profits; the physically handicapped person may easily be deceived and exploited rather than helped. Norbert Wiener consciously addressed this problem in his selection of coworkers and through the agreements he made with the Liberty Mutual Insurance Company, which funded the research on prosthetic limbs and ultimately manufactured them. Wiener was very pleased that he had been able to ensure that any inventions coming out of the project would be "in good hands."

Wiener's general synthesis is, to begin with, a collection of heuristic concepts that identify a new theme in science: "Control and Communication in the Animal and the Machine." This, the subtitle of his book *Cybernetics* (1948), stresses that the animal and the machine are treated by the same theory, and that the emphasis is not on physical constituents but rather on a new theme: patterns of communication and control. It is with respect to just such patterns that one finds common features in animals and machines.

One scientific concept that Wiener and his collaborators revived is that of purpose, a concept that had once been discredited because of facile teleological explanations. Consider a thermostat: in a sense it has the purpose of achieving the goal of maintaining constant temperature, which it accomplishes through a thermometer to identify the present temperature and a mechanism to turn the heater on or off, depending on the difference between the desired temperature and the actual temperature. Similarly, warm-blooded animals have mechanisms for maintaining their temperatures. Such homeostatic mechanisms in animals can be regarded as purposive or goal-directed. The driver of an automobile acts to move the car forward, to stay on the road, and to avoid collisions. Part of the mechanism of accomplishing his goal involves a complex cycle involving visual perception of other cars and the road; the translation of this perception via eye, nervous system, and muscle into the physical motion of the steering wheel or gas pedal; monitoring the consequences of

216
John von
Neumann
and
Norbert
Wiener

these actions by continued visual observation; and so on. The behavior is purposive. Anticipating criticism, Wiener and his collaborators stated that "although the definition of purposeful behavior is relatively vague, and hence operationally meaningless, the concept of purpose is useful and should, therefore, be retained.[37]

One feature typical of goal-directed behavior is a communication and control loop dependent on "negative feedback." In the example of the car and driver, when the car veers off course, the deviation is seen and corrected by means of a route whose elements are vision, brain, hand and foot, gas pedal, and steering wheel, all monitored by visual messages received from outside the car. These messages "feed back" to the driver information concerning how far and in which direction he is off course. If he responds so as to decrease the observed deviation, the overall communication and control loop is said to have "negative feedback." This negative feedback mechanism prevents him from landing in a ditch or colliding with another car. In the case of the car and driver the feedback loop is located partly in the environment and goes through a person as well as a machine. In the case of the thermostat the feedback loop traverses the environment and the thermostat-heater. Wiener contended that such feedback loops involving information and control are ubiquitous in the interaction of both organisms and machines with their environments.

The feedback loop always involves some input information (e.g., the measurement of temperature or the visual observation of the driver), and it involves some output (the heating of the room or the motion of the car); moreover—and this is the crucial feature—the input information is affected by the output (the motion of the car in relation to the road determines in part whether the driver perceives the car to be on course or the extent to which he sees it as off course; the output of the heater will determine the ensuing temperature reading, which in turn will determine whether the heater will be turned on or off). More specifically, the information transmitted to the control mechanisms is that of the difference between a desired situation and the existing situation, and the mechanism (organic or artificial) acts to reduce that difference. Engineers have long been familiar with such devices, which are known as servomechanisms. The typical servomechanism can be schematized as in figure 8, which shows the connection from the output that modifies the input.[38] Calling attention to the servomechanism principle as a heuristic principle that manifests itself in many guises and forms is one aspect of the burden of Wiener's *Cybernetics*.

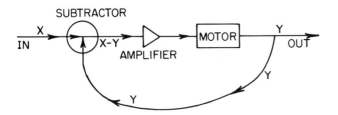

Figure 8

Why is the recognition of the ubiquity of the servomechanism principle a cause for great excitement and enthusiasm? Servomechanisms in engineering can be described mathematically and quantitatively.[39] In practice the loop in the diagram feeding back information from the output to the input may not represent precise or completely error-free communication. But Wiener's communication theory, developed earlier, was capable of describing the usual imprecise or inaccurate communications with mathematical rigor by means of statistical analysis. Thus, in spite of normal kinds of fallibility of the feedback loop, mathematical description was feasible. Previously the domain of matter and energy, namely physics, had been regarded as peculiarly amenable to exact mathematical treatment. Wiener observed that since information and communications can also be described by mathematics and are coupled to matter and energy, the precision of physics can be extended to a much larger domain than physics proper. In particular a new mathematical biology seemed in the offing. The new high-speed computers, such as the one von Neumann was designing, were an outstanding example of a mechanism that could most naturally

218
John von
Neumann
and
Norbert
Wiener
be described in terms of information, feedback loops, control, message, and so on, rather than in terms of energy and matter. Still, a description of its organization could be given strictly in terms of logic and mathematics. In that case the heuristic principles take precise mathematical form.

To notice the ubiquity of negative feedback circuits is also to raise the question of what happens if the feedback mechanism is interrupted or works incorrectly. What happens if the driver falls asleep? The car soon crashes or goes off the road. A malfunctioning thermostat will either heat a room to intolerable temperatures or will not heat it at all. The malfunctioning of a feedback system usually spells vulnerability to disaster. Feedback is necessary to achieve stability in the case of the thermostat, or, more generally, to achieve the goal. When feedback mechanisms fail, uncontrollable oscillations leading to disaster may result, be it in an organism or in a machine. Instances in engineering were well known and understood. Wiener and his medical friends quickly realized that many kinds of pathologies in organisms can be analyzed in these terms. The failure to feel pain or the failure of the blood-clotting mechanism are obvious examples. Loss of hearing or sight also interrupts normal feedback loops; prosthetic devices reconstitute them. Most examples pertaining to organisms, if analyzed in detail, involve highly complex patterns of information storage, communication of messages, and control mechanisms.

The general idea of communication involved, as a special case, language. Moreover, communication may be not only inaccurate or statistical (i.e., containing both a message and noise); it may be paradoxical. Wiener also included that interesting possibility among his heuristic observations. A prototype paradoxical message is "Everything I say is false"; if a computer is asked to analyze logically whether the statement itself is true, it will oscillate; that is, it will reply, yes-no-yes-no, and so on. Can one also have paradoxical communication in nervous systems? If so, what is the mechanism and the consequence? Social scientists at the Macy meeting called Wiener's attention to the potential usefulness of his heuristic concepts in describing human and animal societies, although

Wiener was skeptical of the possibilities of quantitative descriptions in these fields.[40]

When Wiener formulated his ideas in a broad way, he made two far-reaching assertions of a general kind, one historical and the other philosophical. Modern (1948) civilization, according to Wiener, differs from that of previous centuries:

The thought of every age is reflected in its technique. . . . If the seventeenth and early eighteenth centuries are the age of clocks, and the later eighteenth and the nineteenth centuries constitute the age of steam engines, the present time is the age of communication and control.[41]

His philosophical view is based on the idea that while organisms, life and mind, cannot be adequately described in the terms of the traditional physics of matter, the new, nonmaterial concepts of message, control, and feedback will supplement physics sufficiently so that it will in principle be possible to give a fully scientific description of an organism.[42] Concepts such as "mind," "life," and "soul" are not needed in the description. This view represents a modification of traditional materialism in biology. I do not believe that Wiener held to the "ultimate mathematical vision" mentioned earlier, which von Neumann shared with the philosopher Leibniz, because Wiener lacked enthusiasm for mathematical logic. Still, Wiener fully appreciated that some of Leibniz's ideas about mind were echoed in his own cybernetics: "If I were to choose a patron saint for cybernetics out of the history of science, I should have to choose Leibniz."[43]

The task, then, is to examine various concrete problems involving organisms in terms of an extended set of scientific concepts. Indeed, the terminology Wiener introduced—information, message, feedback, and control—has become pervasive in sensory and neurobiology, in enzyme chemistry, genetics, and other disciplines dealing with organisms. The depth and breadth of his knowledge allowed him to subsume an extraordinary range of examples and their ramifications in his generalizations. His was not so much a mathematical discovery as an insight into how to think about problems.

Von Neumann gave his appraisal of Wiener's *Cybernetics*

220
John von
Neumann
and
Norbert
Wiener

in a book review in a magazine for physicists. The review indicates how one genius looks upon another whose ideas are close to his own. It also shows how von Neumann was won over, in spite of his professional caution, by Wiener's nearly unbridled enthusiasm. Von Neumann refers admiringly to "the book's brilliancy" and "its deeply original character." In a statement more critical, but reflecting precisely their different temperaments, he remarks that "the freshness of the approach excuses some exaggeration in emphasis." Von Neumann appears to be debating with himself about Wiener's general thesis:

The author is one of the protagonists of the proposition that science as well as technology, will in the near and in the farther future increasingly turn from problems of intensity, substance, and energy, to problems of structure, organization, information, and control. Any statement of this kind and generality is risky, inviting—and not quite innocently inviting—misinterpretation, and of dubious value outside of its technical context. It may, nevertheless, be important and valuable, if its true context exists and if it has a background of fruitful ideas. The reviewer believes that this context, this background of ideas, indeed exists in the case under consideration. Wiener's book constitutes an important systematic effort at exhibiting this context.[44]

John von Neumann. Photograph
by Alan W. Richards, courtesy of
AIP Niels Bohr Library.

Norbert Wiener. Courtesy of MIT
Historical Collection.

Wiener with hearing aid, 1949. Photograph by Alfred Eisenstaedt.

Von Neumann speaking before
meeting of the National Planning
Association, December 1955.
His bone cancer had been diag-
nosed a few months earlier.
Photograph courtesy of Chase
Ltd.

Wiener playing chess at MIT
Faculty Club, 1953. Courtesy of
MIT Historical Collection.

Von Neumann at afternoon coffee, Institute for Advanced Study. Photograph by Alfred Eisenstaedt, *Life* magazine © 1957 Time Inc.

John von Neumann. Courtesy of
AIP Niels Bohr Library.

Norbert Wiener at MIT. Courtesy
of MIT Historical Collection.

11 VON NEUMANN AND THE ARMS RACE: TECHNICAL ADVISER IN THE CORRIDORS OF POWER

Now . . . the hydrogen bomb is ready to dwarf [the atom bomb] . . . and citizens and scientists stare at one another and ask, How did we blunder into this nightmare?
—J. Bronowski

It is very easy to forget the sufferer, perhaps because there is less glamor or enchantment in suffering than there is in action.
—Philip Hallie

So often it is the innocent bystander who is a victim. Destruction of bystanders en masse rightly strikes us with particular horror. During World War II the calculated, cold-blooded murder of six million Jews in Europe, ordered by Hitler and dutifully carried out by his loyal officials, is probably the most extreme instance of this kind on record. Fear of what Hitler and his fellow Nazis would do should nuclear weapons become available to them originally moved scientists in America, many of them European-born, to be the first to build atom bombs. The United States with its democratic government and long tradition of civil rights could surely be trusted with such a weapon. Only the most severe threat would cause the president of the United States to order use of atom bombs to kill large numbers of people. This is especially true if the president is a modest man, decent, normal, commonsensical, and plainspoken, as Harry Truman was. Such was the naive thought of many scientists working on the bomb.

The truth is more troubling: By mid-July 1945, the Japanese had been "utterly and completely defeated" by the United States and its allies, and President Truman had received the information that they were ready to surrender.[1] But the danger in Truman's eyes was that "Japan might surrender—*to the Russians,* or at least through Russian channels. And then where would American power be in the Far East?"[2] Truman's deliberate tactic was to prevent Japan's surrender until the Americans had a chance to drop the atomic bomb, and to drop the bomb before our "ally" Stalin could send troops and tanks to the Japanese theater and "assist" us in defeating Japan. Thus the victims of Hiroshima and Nagasaki were in effect bystanders in a power conflict between the United States and the Soviet Union. Japan was not one of the parties to the conflict, but control of Japan was part of the spoils. Retrospective analysis of the forces impelling the use of atomic bombs against defenseless population centers has shown the workings of the bureaucratic imperative. The use of the bombs, which had been recommended to the president by an advisory committee of distinguished citizens, was a decision "not to stop a bureaucratic process in which more than $2 billion and four years of incredible effort had been invested."[3]

A British journalist described what he saw one month after the Hiroshima bombings:

Thirty days after the first atomic bomb destroyed the city and shook the world, people are still dying mysteriously and horribly—people who were uninjured in the cataclysm—from an unknown something which I can only describe as the atomic plague.

Hiroshima does not look like a bombed city . . . it gives you an empty feeling in the stomach to see such man-made devastation. . . . I could see about three miles of reddish rubble. That is all the atomic bomb left. . . . The Police Chief of Hiroshima . . . took me to hospitals where the victims of the bomb are still being treated. In these hospitals I found people who, when the bomb fell, suffered absolutely no injuries, but now are dying from the uncanny after-effects. For no apparent reason their health began to fail. They lost appetite. Their hair fell out. Bluish spots appeared on their bodies. And then bleeding began from the ears, nose and mouth.

At first, the doctors told me, they thought these were the symptoms of general debility. They gave their patients Vitamin A injections. The results were horrible. The flesh started

232

*John von
Neumann
and
Norbert
Wiener*

rotting away from the hole caused by the injection of the needle. And in every case the victim dies.[4]

Uncanny aftereffects, long-term destructive consequences of exposure to the invisible nuclear radiation, in many instances manifested themselves only years after the original explosion. Two hundred thousand people died from the two bombs within the first year.[5]

The thinking behind the use of the atom bomb would probably not have been acceptable to most Americans in 1945. The public assumed that it was necessary to end the war with Japan. Subsequent policy considerations by American strategists usually took it for granted that any undisguised "preemptive" nuclear attack by the United States on another country would be unacceptable to the American people, although it became United States policy to plan the nuclear bombing of Soviet cities as a response to any attempted Russian military conquest with conventional forces in Western Europe. However, nonnuclear wars in which the victims would be primarily "colored" people (Korea, Vietnam) were thought to be relatively acceptable, even if those colored people were only bystanders caught up in a conflict between superpowers.

John von Neumann, although a patriotic American, was anything but a typical member of the American public. In his political thinking he differed with many of his scientific colleagues. Not too surprising for an upper-middle-class Hungarian immigrant, animated by ingrained hatred and fear of Russia, he had throughout World War II regarded our Soviet ally as in fact an enemy. As he put it, "We were involved in a triangular war, where two of our enemies had done . . . the nice thing of fighting each other."[6] Thus, in von Neumann's mind it had been utterly rational for President Truman to make the decision to use the atom bomb on the basis of the anticipated power struggle with the Soviet Union, and von Neumann had been one of General Leslie R. Groves's closest advisers in defining criteria for selecting particular locations in Japan to bomb. Von Neumann had strong apprehensions that a war between the US and the USSR would immediately follow World War II.[7] Although his fears did not materialize, he saw Russia

taking control of most of the Eastern European countries, including Hungary, in the process ruthlessly liquidating the anticommunist leadership.

Within a few weeks of Hiroshima, however, scientists at Los
Alamos formed an organization "to promote the attainment
and use of scientific and technological advances in the best
interest of humanity" and formulated a public statement concerning prospects and dangers.[8] Within a few months this
group joined forces with other newly formed scientists' groups
to organize the Federation of American Scientists (FAS). The
federation organizers especially hoped to avoid a nuclear
arms race as well as a nuclear war. It was among their explicit
aims "to urge the United States help initiate and perpetuate
an effective and workable system of world control [of atomic
energy] based on full cooperation among all nations" and
furthermore "to counter misinformation with scientific fact and
. . . disseminate those facts necessary for intelligent conclusions concerning the social implications of a new knowledge
in science."[9] The FAS's publication, the *Bulletin of Atomic
Scientists,* would become a broad forum for discussion and
debate concerning the social issues deriving from nuclear
weapons.

A few scientist-administrators had a particularly active role
in formulating American nuclear policy in 1945 and 1946.
Vannevar Bush, Norbert Wiener's friend from MIT, had become the director of the Office of Scientific Research and Development. He composed a document under the direction of
the American secretary of state which declared a commitment
to the international control of atomic energy by the United
States, Great Britain, and Canada and which on November 15,
1945, was signed by the three heads of state.[10] The document
espoused the ultimate objective of the elimination of nuclear
weapons from the world. J. Robert Oppenheimer, who had
been scientific director of the Los Alamos Laboratory, was
one member of the panel appointed by the State Department
to spell out American nuclear policy more fully. The report of
the panel, the Acheson-Lilienthal report, in its later version
known as the Baruch plan, was supported by President Truman and was officially proposed by the United States to

234

John von
Neumann
and
Norbert
Wiener

the United Nations. It too advocated international control of atomic energy and the supervised abolition of nuclear weapons. However, a crucial feature of the plan was unacceptable to the Soviet Union, namely an effective freezing of American nuclear superiority, in fact a monopoly, for a number of years, until the international authority was established.[11] The plan also called for extensive inspection to prevent violations.

The forum for debating America's Baruch plan was the United Nations Atomic Energy Commission, established by the recently founded UN for the express purpose of considering means to control nuclear weapons and eventually to eliminate them from national armaments.[12] The widespread desire for peace and the prevention of a nuclear armaments race prevailing while the memory of Hiroshima and Dachau was still fresh competed with the concurrent jockeying for power and the hostile mistrust between the United States and Russia. The Soviets made a counterproposal to the Baruch plan, banning the manufacture of atom bombs and requiring the destruction of existing stockpiles, but their plan, which provided no means for inspection to detect violations, was unacceptable to the Western powers. No agreement was reached, although for many years to come the United Nations provided the locus for continued efforts to find a procedure for halting the weapons race acceptable to all the major powers; some leading scientists from the United States, Britain, France, and Russia who were particularly interested in avoiding an arms race participated in them. But no agreement was reached; instead an all-out arms race, particularly between the United States and the Soviet Union, had been set into motion. As more countries proceeded to build nuclear weapons, and as the American and Russian stockpiles grew and the technology for mass homicide advanced by leaps and bounds, the chances for total nuclear disarmament became ever more slim.

In 1945 von Neumann began to define his own stand and carve out his own political and social role in relation to nuclear weapons. When the Association of Los Alamos Scientists was formed in August, von Neumann at first joined along with practically everyone else but within a few days changed his mind and withdrew, with an explanation to the effect that he saw his own future role to be in a different direction.[13] In the spring of 1946 he was asked by Norman Cousins, editor of *Saturday Review,* to sign a letter endorsing the importance of the Acheson-Lilienthal report, as well as a book of articles, *One World or None,* that contained essays by nuclear physicists, among them Hans Bethe.[14] Von Neumann declined, explaining that he did not agree with the contents of these publications in several essential respects.[15] Von Neumann was out of sympathy with the general idea of international control and with the concept of the ultimate elimination of nuclear weapons; he wanted the United States to prepare for victory in a war with the Soviet Union.

In 1948 von Neumann was formally invited to join the board of advisers of the *Bulletin of Atomic Scientists,* which provided an open forum for the airing of views concerning nuclear weapons. In his reply he spells out something of the political role he had chosen, stating that "as a matter of principle . . . I would prefer not to join the Board, since I have throughout the last years avoided all participation in public activities, which are not of a purely technical nature, and it seems to me that the objectives of the Bulletin are defined somewhat more broadly, than would be the case for a technical publication."[16]

A second invitation from the *Bulletin of Atomic Scientists* received the reply that "I do not want to appear in public in a not primarily technical context."[17] Von Neumann's militaristic views would be unpopular, and he had chosen not to express them publicly. He had apparently chosen to adopt the public image of the apolitical expert, the neutral scientist. But was von Neumann then planning to return to the ivory tower of pure mathematics at the Princeton Institute? Very much to the contrary. As a disinterested expert he was that much more valuable to the military and government establishments. The kind of power he sought was behind-the-scenes, quiet, free of bombast, and under the guise of technical advice. He did not seek popularity or recognition from the general public, but he

236
John von
Neumann
and
Norbert
Wiener
valued military and government approbation highly. He understood that leading Manhattan Project scientists would be in a position to have political power if they chose to.

The significance of von Neumann's historical role derives from the highly effective manner in which he used his technical talents in the service of strongly and emotionally held convictions. Politically he was strongly partisan. From 1945 until his death, the central issue at the interface between weapons technology and government was the question of how fast and how far to go in the escalation of destructive power. Von Neumann, with his brittle brilliance, was a consistent advocate of the "quantum jump," from the atom bomb to the hydrogen bomb and from the bomber as delivery system to the nuclear-tipped intercontinental missile, and he supported carrying out this escalation as rapidly as possible. "I don't think any weapon can be too large," he has been quoted as saying.[18]

The most famous of von Neumann's colleagues at the Princeton Institute at that time was Albert Einstein. A loner, an outsider, in his seventies, he was not taken seriously even by the scientific community, and he saw himself regarded in the United States "as a black sheep."[19] He did not engage in political arguments with von Neumann at the institute, but his views were well known, and he expressed them publicly:

The idea of achieving national security through national armament is, at the present state of military technique, a disastrous illusion. . . . The armament race between the U.S.A. and the U.S.S.R., originally supposed to be a preventive measure, assumes hysterical character. On both sides, the means of mass destruction are perfected with feverish haste—behind the respective walls of secrecy. The hydrogen bomb appears on the public horizon as a probably attainable goal. . . . In the end, there beckons more and more clearly general annihilation.[20]

Einstein consistently urged disarmament under an international authority. Niels Bohr, in this respect at the same end of the political spectrum as Einstein, urged establishment of candid interchanges with leaders of the Soviet Union on the subject of nuclear weapons and, subsequently, international control. The Baruch plan had (belatedly and in considerably flawed form) given expression to some of Bohr's ideas and indeed represented for a brief time official American policy. However, the way in which atom bombs had already been used as a political weapon against the Soviet Union was hardly conducive to an atmosphere of genuine trust.

Although during the war von Neumann had come to the Manhattan Project only as a mathematical consultant, he had also become an intimate adviser of the energetic and tough military commander of Los Alamos, General Leslie Groves. Von Neumann often sought relationships with men of power, especially those dealing with the interface between technology and military planning, with himself in the role of technical adviser. Unlike Wiener, von Neumann found leading industrialists, bankers, military men, and political figures personally highly congenial. His friend and fellow mathematician Stanislaw Ulam has observed that von Neumann

seemed to admire generals and admirals and got along well with them. Even before he became an official himself . . . he spent an increasing amount of time in consultation with the military establishment. . . . I believe [his fascination with the military] was due more generally to his admiration for people who had power. . . . I think he had a hidden admiration for people or organizations that could be tough and ruthless.[21]

The one person central to an understanding of von Neumann's political life after 1945 is the politically influential Lewis Strauss, in many respects a prototypical member of the American power elite.[22] Strauss provided for von Neumann a conduit to political, financial, and military thinking on the highest level. Strauss was, among other things, a trustee of the Princeton Institute of Advanced Studies, von Neumann's home base. He was an American-born Jew (in that respect atypical of the power elite), seven years von Neumann's senior, and he had received his first government job in World War I as private secretary to Herbert Hoover, who was then food administrator under Woodrow Wilson. Hoover's organization was not only in charge of food production, conservation, and distribution within the United States but also provided food relief to Belgium and France and master-

238
John von
Neumann
and
Norbert
Wiener

minded a food blockade of Germany. After the armistice, Hoover's activities broadened to organizing relief and reconstruction of war-ravaged Europe; then in 1921 he was named the secretary of commerce. The young Lewis Strauss had meanwhile taken a position with one of the largest American banking houses, Kuhn, Loeb and Company. In 1928 Strauss became a full partner of the firm, and his former boss, with whom he had stayed in contact and whom he had consistently backed politically, became president of the United States. Thus, at age thirty-two, Strauss was already a major figure in world banking who through his association with Hoover had also some acquaintance with prominent political figures both in Europe and America. Strauss's casual mention in his memoirs that his firm "had financed the Japanese Government during the Russo-Japanese War" reminds us that politically banking can seem as deceptively neutral as technology.[23] Strauss always took a particular interest in financing new inventions. He had a strong penchant for science and technological innovation and at one time had had ambitions to become a physicist. During World War II he obtained a desk job with the Navy in Washington, was made a rear admiral, and worked directly with the secretary, James Forrestal, himself a former banker.

John von Neumann's long-standing interest in high finance and its political face was matched by Strauss's enthusiasm for high technology and science. Each was comfortable with the other's area of special competence. While Strauss was a devout Jew,[24] whereas von Neumann showed no interest in religion, in general they found each other's views congenial, since Strauss's politics could be characterized as Republican, militaristic, and militantly anticommunist.

In the late summer of 1945 von Neumann talked with Admiral Strauss in the secretary of the navy's office about the possibility of building a general-purpose computer, a quantum jump in advance of the currently available technology of computers, at the Princeton Institute. Von Neumann was seeking funds from the navy, and emphasized possible applications to militarily important weather prediction.[25] Strauss soon became von Neumann's accomplice, helping him to get funds and making the required arrangements for building the computer at Princeton. Although it was primarily paid for out of government funds, von Neumann of course wanted to retain control of the machine's time so as to be able to use it for problems of his own choosing. In May 1946 he had to insist on this autonomy to Admiral Strauss, stating that control of the computer's time must not be given over to the government but suggesting that duplicates of the machine might be built for governmental use. In the same letter he responded to Strauss's solicitation of his views concerning the suitability of J. Robert Oppenheimer, whom Strauss, as a trustee of the Princeton Institute, was proposing as a possible new director. Von Neumann wrote simply that he had "serious misgivings. These matters are better suited for an oral discussion."[26] Von Neumann's choice was Detlev W. Bronk, a physiologist.[27]

In 1946, after long congressional debate and internal political conflict about how to deal with the newly unleashed nuclear power administratively, the Atomic Energy Act became law.[28] Under it, the president had to appoint five people as Atomic Energy commissioners to help create national policy concerning nuclear technology. They were also charged with the production, stockpiling, and testing of nuclear weapons, controlling research and development in the nuclear field, distributing radioisotopes for medical purposes, defining radiation hazards and setting standards, monitoring nuclear tests by foreign powers, and related responsibilities. The chairman of the commission was to have cabinet rank and report directly to the president. Lewis Strauss was appointed in 1946 by President Truman to be one of the first five commissioners, perhaps as a conservative counterweight to the more liberal chairman, David Lilienthal. Since being a Commissioner was a full-time position, Strauss did not then return to his prewar financial activities.

Meanwhile von Neumann was continuing with various government consulting activities at the Army Ballistic Research Laboratory and at the Navy Bureau of Ordnance; in particular he continued to spend a portion of his time at Los Alamos working on the possibility of a hydrogen bomb.[29] Increasingly he was neglecting his purely scientific and mathematical

240
John von
Neumann
and
Norbert
Wiener

interests in order to pursue the development of weapons technology, just at the time when most scientists were returning from wartime preoccupation with weapons to the backlog of scientific ideas left all but unexplored during the preceding years.

Immediately after the first atom bombs were dropped on Japan, Lewis Strauss had in his official capacity recommended that the ability of various types of navy ships to withstand the blast of atomic bombs be tested experimentally, by actually dropping an atomic bomb near some naval vessels.[30] The portion of the scientific community associated with the Federation of American Scientists objected strenuously to this military exercise in destruction, which might disturb ongoing negotiations for international control, was of no fundamental scientific interest whatsoever, might expose people to radioactive fallout, and which in any case could not answer the question of whether navies were obsolete.[31] The tests in effect constituted a message to the Soviet Union, undercutting the message of trust and nonmilitary intentions implied by the Baruch plan. Von Neumann, however, shared Strauss's point of view, and when the tests—Operation Crossroad—were finally held on July 1 and July 25, 1946, at the Bikini atoll off the Marshall Islands, von Neumann was an interested official scientific observer of the event.

In addition to his many government consultantships, in 1947 von Neumann also accepted a new consultantship with the Standard Oil Company at $6,000 per annum, for advice in connection with locating, producing, and refining oil.[32] Eventually he supplemented his income by a number of other industrial consultantships. Also in 1947, he was the recipient of two honorary awards from the government for his service, the Presidential Medal for Merit and the US Navy Distinguished Civilian Service Award, which were a source of great pride and satisfaction to him. In this he was unlike most scientists, who while they may covet symbols of recognition by their fellow scientists tend to be rather indifferent to official government honors.

As I have shown, von Neumann differed from the mainstream of the scientific community at the end of the war in his continued enthusiasm for weapons development, in the apparent parochialism of his loyalty to his adopted country, in his opposition to international control, and in his personal preference for associating with the worldly and powerful. By contrast, Wiener was moving in just the opposite direction. He would have absolutely nothing more to do with helping the military or with weapons development. He communicated his views to the general public, speaking out against militaristic and elitist thinking. And he was much disturbed to see his erstwhile colleague and friend Vannevar Bush become occupied with government and high-level administration and urging the cooperation of scientists with the military even in peacetime. This fact was sufficient to sorely strain Wiener's friendship, even with the relatively liberal Bush.

During the years 1945–47, while the abortive attempt at internationalizing the control of atomic energy was being made, President Truman launched the policy of "atomic diplomacy," which involved using the implied threat of nuclear weapons as a primary tool of diplomacy. The early use of nuclear weapons both in Japan and in the Bikini tests served to demonstrate to the Soviets that it would be unwise trustingly to accept American dominance in the field of nuclear weapons, as the Baruch plan required. In response to the threat, the Soviet Union, although profoundly war-weary but "having no direct means of deterring an atomic attack on its territory," increased the strength of its army in Eastern Europe "to a point where it could march across Europe in the event of an American attack on Soviet territory. In a very real sense, Western Europe had become a Soviet hostage in order to deter an American attack on Soviet territory until the Soviet Union could develop its own atomic arsenal and delivery system."[33] Such was the primitive fear and rivalry that gave a strong early momentum to the nuclear arms race, which from July 1945 on consistently overwhelmed all attempts to inhibit it.

The conceptual framework of Truman's foreign policy in relation to the Soviet Union had been provided by the idea of "containment" of Soviet expansion, which consisted of an acceptance of the territorial status quo but a readiness to combat any attempts at military expansion.[34] This concept was

242

John von
Neumann
and
Norbert
Wiener

coupled with the peculiar balance between the threat of the use of American nuclear weapons on Russian cities and the counterthreat of Russia's use of ground forces to invade Western Europe. Whether either threat constituted a bluff was never fully determined.

Surveying the trend signaled by von Neumann's activities in the 1940s and amplified in the 1950s, one is struck by the emergence of personal characteristics and interests that his mathematical colleagues in the 1920s and 1930s had not much noticed. It is as if his wartime activity had touched off an atavistic impulse, a reversion to the aspirations of his father's generation in Central Europe, an eager identification with the conservative and militaristic power elite. In von Neumann's unmitigated enthusiasm for the rapid development of nuclear weapons in response to the political conditions after 1945, one recognizes a naive scientific and technological optimism that belies his intellectual sophistication. In his remarkable readiness to serve military and governmental officialdom, even in the very quality of his patriotism, he seemed no longer the intellectually independent scientist but a captive of the function he had come to fill. He seemed more like a talented nineteenth-century banker to some Central European court, aspiring to aristocracy, than a mid-twentieth century American scientist accustomed to egalitarian and democratic values. Perhaps this was the mark that centuries of persecution of European Jewry had left on John von Neumann.

A variety of elements enter into the complex process by which the character of human life is altered by the introduction of a major new technology. The nature of the process will depend on whether the technology is primarily of a military or a civilian kind, the nature of the prevailing economic system, and other circumstances. In any case, however, development of a major new technology requires energetic entrepreneurship and advocacy by people with the will to overcome the obstacles in its path. It requires money; it requires organization. It also often requires technical and scientific talent. If the technology is of a military kind and the pressure for a "quantum jump" in advancing the technology derives from an armaments race, considerable scientific and technical talent may be required. The decision to build the atomic bomb was made in secret by Franklin Roosevelt, who under wartime conditions found ways to finance the project, which were then implemented by a large group of scientists and technicians working indefatigably under great pressure and with great devotion to their extraordinary mission. In the postwar decade a few scientists found a regular official channel through which to influence weapons policies, the civilian Atomic Energy Commission and other government weapons committees. Unlike purely political figures, the scientist might have a double function: expert adviser and committeeman within the administration and working scientist or engineer directly contributing to the design of new weapons systems. To appreciate the full contribution of scientists to the armaments race, one needs to consider both their political role and their technical role.

It is the peculiar nature of weapons work, be it during war or peacetime, that research and development and the major decisions concerning them are to a considerable extent known only to a small elite within a country, even in the relatively democratic United States. The subject matter of weapons innovation is complex, arcane, and often distasteful to the general public. Yet billions of dollars of its tax money will be devoted to weapons, its freedoms may be severely circumscribed by the very existence of secret weapons systems, its physical health injured by radioactive fallout from weapons tests, and its outlook on life affected by a sense that doomsday is inevitable. The legitimacy of this decision-making elite is open to question: most are consultants or bureaucrats within the executive branch and consequently not elected by the public. Although legally they may be accountable for their actions, in some instances such laws as exist have not been applied. Politically, curbing their power would require a yet-to-be-developed mechanism for protecting the people against the military-technological excesses of the insiders.

During the height of the cold war, this basic inability to reconcile the growth of weapons technology with popular democracy, while discussed in the writings of Wiener and a few other intellectuals outside the mainstream, was overwhelmed

244

John von
Neumann
and
Norbert
Wiener

in the public arena by militant cold war rhetoric. It was not until January 17, 1961, that President Eisenhower, in his farewell address, alerted the public belatedly to these issues, which he had been unable to resolve during his own administration:

In the councils of government, we must guard against the acquisition of unwarranted influence, whether sought or unsought, by the military-industrial complex. The potential for the disastrous rise of misplaced power exists and will persist. . . . In holding scientific research and discovery in respect, as we should, we must also be alert to the . . . danger that public policy could itself become the captive of a scientific-technological elite.[35]

Eisenhower's policies had probably been captive ones to a greater extent than he ever openly acknowledged or perhaps even consciously understood. Although his warnings are often quoted, his successors only exacerbated the problem by streamlining the organization of the technological-industrial-military power center.[36] Since then, large military expenditures by numerous nations have tended to use up the earth's nonrenewable resources for weapons production at an increasing rate, with other human needs more and more neglected, and the likelihood of a world nuclear war has steadily grown. What has not occurred is the restructuring of national and international political and economic organizations along lines conducive to the reversal of the trend toward general annihilation. Worldwide military-industrial interests are undeniably one of the major obstacles to radical change. The United Nations, which constitutes the most ambitious effort at initiating a political restructuring so far, has been bypassed, overruled, and at times become itself captive to special interests. Could restructuring halt this malignant process, or are we ready to concede to technology itself an autonomy that no economic or political institutions can control?[37]

From 1945 to 1949 only a few of the luminaries among the Manhattan Project scientists continued to work on the superbomb. The hope of achieving international control of nuclear weapons and avoiding a disastrous arms race prevailed in the American scientific community. It was a brief period, this interval between World War II and the widespread acceptance of a state of cold war with the USSR, a period of respite from wartime militarism. However, for some (von Neumann among them), the view of the Soviet Union as the enemy had begun even before the Japanese empire had capitulated to the United States. The idea of the hydrogen superbomb, which would have an explosive power of about a thousand Hiroshima-type atom bombs (one of which had killed about a hundred thousand people), had inspired some physicists, particularly the Hungarian-born Edward Teller, well before the first atom bomb had been completed. Teller and von Neumann, even after the events of August 1945, would spend two or three months every year at the relatively deserted Los Alamos site helping to design the superbomb. Von Neumann made the indispensable contribution of bringing the very latest in high-speed computer technology to bear on the problem. In the spring of 1946 they were the two leading lights at a small secret conference on the status of the bomb which issued a technologically optimistic report considerably underestimating the technical difficulties entailed in actually designing a bomb that would work.[38] Notwithstanding the Los Alamos technical report, the superbomb project at that time had a relatively low priority for the US government, and it progressed at a slow pace.

The foreign policy of containing expansion of the Russian sphere of influence was for a time adequately buttressed by the American monopoly on atomic bombs. But in 1948 and 1949 some failures in achieving containment combined with the ending of the American atomic bomb monopoly to arouse the US government to renewed emphasis on weapons and the military. In 1948 the Czech Communists took over the Czech government in a coup under the threat of support from the Russian army, which was not far off. When in the same year Russia blockaded land access to West Berlin, the US government responded with an airlift. Some American political leaders feared that Communist forces might also stage successful coups in Italy and France.[39] In 1949 indigenous Communists took over China. The crucial event pressing toward the American decision for weapons escalation, however,

246

John von
Neumann
and
Norbert
Wiener

was the Soviet explosion of an atomic bomb in Siberia on August 29, 1949. In the lockstep logic of an armaments race, the USSR's signaling the end of the American nuclear monopoly was sufficient by itself to generate a reflex among some of the American leadership to regain "security" by a quantum jump. In any case, the American threat of the use of nuclear weapons was seen as having lost its effectiveness in deterring the presumed and feared Soviet "expansion."

Various alternative responses to the new developments of 1949 were discussed within the administration, and among atomic scientists. One important school of thought championed the buildup of American ground forces,[40] so as to be able to fight conventional wars, not only nuclear wars. Presumably Soviet expansion would be by means of the use of conventional ground forces, and atom bombs, acting as threat and counterthreat, would presumably cancel each other out if both countries possessed them. Only after the Korean War had begun in the summer of 1950 did the United States again increase its ground forces. Generally this was seen as an expensive way to increase "military strength," as compared to, say, developing a superbomb. Others emphasized the increased urgency of resuming negotiations between the United States and the USSR through the United Nations in order to achieve international control of nuclear weapons. This group was primarily concerned with preventing a nuclear arms race while there was still time. However, hardly any of these people were influential government officials. In fact, few had the needed security clearances. Their voices were largely ignored by officials and advisers to the government.

The more aggressive and militaristic options were discussed in secret debate among government officials and their military and scientific advisers. Within these circles, "the Soviet A-bomb test produced a small flurry of preventive war discussions, and the Korean War considerably more."[41] The idea of a preventive war was that the United States should initiate a direct military attack, laying waste to as many Soviet cities as possible and to the Soviet Union's still very meager nuclear weapons resources in a nuclear "first strike" by American bombers. Such a first strike could not only kill millions of people within a few weeks, it could leave the survivors and their offspring as victims of cancer a decade or two later and inflict genetic defects on future generations. The argument in favor of such a devastating action was that in 1950 the United States still had a great advantage in nuclear stockpiles and in long-range bombers over the Soviet Union and, as the aggressor, could achieve a decisive victory. Of course it was to be expected that otherwise Russia would eventually develop an effective retaliatory capacity, and a first strike would not be decisive if it were delayed for many years. Von Neumann was a strong and outspoken advocate of a preventive war: "a hard-boiled strategist, he was one of the few scientists to advocate preventive war, and in 1950 he was remarking, 'If you say why not bomb them tomorrow, I say why not today? If you say today at 5 o'clock, I say why not one o'clock?' "[42] American presidents, however, refused to consider this option very seriously, because they expected it to be utterly unacceptable to the American public.

The most intense controversy among scientific advisers to the United States government in the months following the first Soviet test of an atomic bomb revolved about the question of whether the design, invention, and development of the super-bomb might not be an appropriate response, to be pursued with the highest priority. As their friend Ulam has reported, von Neumann and Teller met immediately after they learned the news of the Soviet bomb and discussed not *whether* but *how* to get political backing for an accelerated superbomb program.[43] Atomic Energy Commissioner Lewis Strauss, von Neumann's influential ally, within a few weeks initiated the political process by a formal memorandum to his fellow commissioners, proposing a highly accelerated superbomb program so as to remain ahead of the Russians. Moreover, Strauss also went outside the AEC and argued the case for the program with other members of the Truman administration.[44] Scientific advice on how to proceed was sought by the AEC from its General Advisory Committee, chaired by Robert Oppenheimer. Oppenheimer, who had directed the wartime atomic bomb project, was still in October 1949 regarded as the most influential scientist in relation to all applications of atomic energy. The committee unanimously advised against a crash program: "We all hope that by one means or another the development of these weapons can be avoided. We are all reluctant to see the United States take the initiative in precipitating this development. We are all agreed that it would be wrong at the present moment to commit ourselves to an all-out effort towards its development."[45] Instead the committee recommended that a greater effort be put into development of smaller, "tactical" nuclear weapons. It also urged that the subject of the nuclear arms race be discussed publicly by the president so an informed public could participate in the debates. Strauss's intentions were apparently frustrated by the unanimous opinion of the advisory committee of scientists when the majority of the Atomic Energy commissioners concurred with the committee and recommended to President Truman against proceeding with the superbomb. Still, he conveyed his contrary views and recommended in a letter and memorandum to President Truman that the "highest priority" be given to development of the superbomb.[46]

The decision that President Truman finally reached and acted upon early in 1950 essentially supported the viewpoint of Strauss, Teller, and von Neumann. It gave the go-ahead to a highly accelerated superbomb program. Though it represented a minority view among atomic scientists, some of that minority had worked very actively to persuade others in the government establishment. The secretary of state, Dean Acheson, and the secretary of defense, Louis Johnson, had come to share Strauss's opinion. The most powerful legislator concerned with atomic energy, Senator Brian McMahon, chairman of the Joint Committee on Atomic Energy, also agreed with Strauss and the Truman decision. The perpetually competing military services were divided: the air force had favored the large superbombs, while the army had urged tactical nuclear weapons for battlefield conditions.

In all this, although von Neumann spent numerous hours trying to urge his own views on Oppenheimer (who reciprocated in kind) and talked political tactics at length with Edward Teller and Lewis Strauss, it appears that he did not actively lobby for his position on the superbomb with political figures to the extent that Strauss and Teller did. He was influential because his judgment and reasoning were widely respected. This picture of von Neumann's role in the controversy is consistent with his desire to appear as a rational, objective expert rather than as someone with strong political motivations. However, to his close friends and colleagues it was easy to see, as Ulam says, that "all along Johnny was emotionally involved in favor of the construction of the H-bomb."[47] Aside from talking politics, he was of course devoting several months of every year to working on the substantive technical and mathematical problems that development of the superbomb presented.

The sequence of events beginning with the building of the first atom bomb, followed by the 1945 decision to use the atom bomb on Japanese urban centers, and then by the 1950 decision to proceed at full speed with the superbomb, along with the corresponding decisions in the Soviet Union, are so deep in their implications that it is not sufficient to look at them solely in terms of politics or strategy. Since they touch the marrow of many people's outlook on life they deserve consideration from a more existential perspective.

Viewed rationally, the sequence of decisions and actions has placed the vision of a man-made, worldwide doomsday on the horizon of realistic probability. Insofar as people permit such a vision to enter their consciousness, they are obliged to confront a far more comprehensive essential phenomenon than death itself. Just as we reconcile ourselves to our own individual death, we are now obliged to reconcile ourselves to the likelihood of thermonuclear war and the destruction, within a few days' time, of all community and civilization as we know it, with the survivors contaminated with cancer, condemned to a life of suffering and the horror of giving birth to genetic mutants in many cases. This situation is unprecedented, although one might find a pale, unsatisfactory parallel in the kind of destruction that did indeed take place in some earlier civilizations, such as the massacre of some native North American tribes. We might appropriately say that through these actions and decisions, a few individuals have

250
John von
Neumann
and
Norbert
Wiener

altered the conditions of human existence, but it must be added that it would seem to be entirely possible for the human community again to alter conditions, to reduce radically the likelihood of a man-made doomsday, conceivably even to abolish it. We have no reason to regard thermonuclear war as inevitable. We have only historical analogies to other weapons innovations. It is true that in nearly every case the weapons, once invented and built, were eventually used in wars. But from historical studies we have some insight into how social and political instabilities have propelled men to use them; men made choices, and those choices do not have the quality of inevitability that we must assign to physical death.

Regarding the impact of the 1945 and 1950 American and Soviet political-technological decisions, one is struck by the bizarre incongruity between the relative ordinariness of the political decision-making process itself, similar to the process by which other major political decisions are made, and the enormity of the consequences. It is both frivolous and Panglossian to regard this kind of incongruity as inextricable from the human condition. Rather one must consider it as a reflection of the political structure of the governments involved, and the pattern of international relations prevailing at that time. Political institutions, then and now, are not geared to far-reaching technological decisions of this kind.

One element contributing to the dismissive attitude toward the consequences of the superbomb decision in the United States is the nature of the elite group that was involved in debating the alternatives. Herbert York, who was one of the group, wrote, "The participants in the secret debate were very few: the members of the GAC, the members of the AEC and a few of their staff, the members of the JCAE and a few of their staff, a very few top officials in the Defense Department, and a very small group of very concerned scientists, mostly from two of the AEC's laboratories. Altogether, there were less than one hundred people."[48] The structure of the committees and the security requirements had automatically excluded men with such antimilitary views as Linus Pauling, Albert Einstein, and

Norbert Wiener. The group probably included no women, and perhaps no one who felt more strongly identified with the victims of society than with the lords of the earth. All moved in the world of military-technological thinking, which provided the language of discourse for the debates. It is an unfortunate language for consideration of human issues. Oppenheimer, who saw himself as nearly tormented by the existential issues, and the GAC had advocated open public discussion of the superbomb issue, which would have led to a very different decision process. But it is notable that Oppenheimer and the GAC, while cool to the superbomb, had urged an "intensification of effort to make atomic weapons available for tactical purposes." Thus they too recommended an escalation of the armaments race, but in the direction of smaller, more mobile, and less expensive bombs rather than bigger, more powerful ones. The secret debate over the superbomb had ranged over a narrow set of options, all of which required increased innovation in nuclear weaponry of one kind or another.

In 1950, after Truman's decision on the superbomb had become definite, von Neumann became a member of some additional committees in Washington, D.C., dealing with nuclear weapons. He attended meetings of the Weapons Systems Evaluation Group, and made an effort to bring other scientists and scholars into it.[49] The Weapons Systems Evaluation Group was seen by von Neumann as, among other things, a vehicle for achieving an orderly relationship between the military services and the Atomic Energy Commission.[50] He also participated in the Armed Forces Special Weapons Project committee meetings, and on June 21, 1951, he formally became a member of the Scientific Advisory Board of the US Air Force.[51] In 1951 the air force was the only branch of the military with the means of "delivering" atom bombs. Von Neumann was increasingly recognized as "one of the best weapons men in the world" as the new AEC chairman, Gordon Dean, who succeeded Lilienthal, expressed it.[52] The number of von Neumann's activities in the post-Hiroshima decade is simply astounding. The fast-moving, fast-thinking, fast-talking von Neumann seemed to be in many places at once, in each

252
John von
Neumann
and
Norbert
Wiener

place functioning with unflagging energy and full mental capacity, as if stimulated by the very multiplicity of his responsibilities.

Meanwhile Lewis Strauss, having victoriously concluded the battle for the superbomb, resigned from the AEC and returned to high finance; in particular he was made consultant and financial adviser to the Rockefeller brothers.[53] AEC Chairman David Lilienthal also retired. The new chairman, Gordon Dean, was the one commissioner who had supported Strauss's view, and Strauss, according to his own memoirs, had personally spoken to President Truman to convey the recommendation of Senator Brian McMahon to him that Gordon Dean be appointed.[54]

President Truman's decision to go full steam ahead did not automatically make the superbomb a reality, but now the project had a high level of support. Work at Los Alamos continued, and by March 1951 the needed theoretical innovations had been achieved.[55] In June, a meeting was held at the Princeton Institute to which the members of the AEC and the General Advisory Committee were invited, in addition to some Los Alamos scientists and consultants, including von Neumann. The scientists at the meeting agreed that now indeed no major technical obstacle stood in the way of designing, constructing, and testing the superbomb. Even Oppenheimer got caught up in the technological enthusiasm. Von Neumann was continuing to push forward the construction of the JOHNNIAC computer at Princeton, since computing capability was badly needed for the calculations the bomb project required. One offshoot of the Los Alamos group working on the superbomb was located at Princeton, primarily because the computer would presumably be available there, although as it turned out, computer construction took somewhat longer than expected. As von Neumann later reported, he managed the needed calculations for the superbomb mostly without the JOHNNIAC:

I would say about two-thirds of the development of the superbomb took place under conditions like this: That heavy use of computers was made, that they were not generally available, and that it was necessary to scrounge around and find a

computer here and find a computer there which was running
half the time and try to use it, and this was the operation I was
considerably interested in. . . . As far as the Institute is concerned, and the people who were there are concerned, this
computer came into operation in 1952, after which this first
large problem that was done on it, and which was quite large
and took even under these conditions half a year, was for the
thermonuclear program. Previous to that I had spent a lot of
time on calculations on other computers for the thermonuclear
program.[56]

The first experimental test of a type of hydrogen bomb that
qualified as a superbomb—it yielded an explosion as powerful as about a thousand Hiroshima bombs—took place in
November 1952.[57] With the new scheme thus verified in practice, it was now clear that it was technically feasible to build a
superbomb of arbitrarily large explosive power. The destructive power one could build into a single bomb was henceforth
unlimited. Limitations on the size of bombs would be determined not by technical considerations but only by question of
military strategy, economic considerations, and the issue of
the viability of earth, especially the United States, as a habitat
for the human species.

With the prosuperbomb faction in the saddle, von Neumann
became increasingly active and eligible for participation in
important committees advising the government on weapons
policies. Whatever one's views, von Neumann was a good
man to have on a committee involving complex technical issues. His ability to keep in mind many considerations and
technical factors simultaneously and to analyze and interre-
late them clearly and logically were second to none. He was
low-key, unemotional, cheerful, charming, and politic. As a
committee chairman his style was to "press strongly his technical views, but defer rather easily on personal or organizational matters."[58]

With the General Advisory Committee recommendation
overruled, one of its members, Cyril Smith, disappointed with
the Truman decision on the superbomb, resigned his position,[59] and von Neumann took his place. Lewis Strauss, in
his congratulatory message to his friend Johnny, expressed
pleasure "for the sake of the country" that von Neumann had

254

John von
Neumann
and

Norbert
Wiener

been appointed.[60] When Robert Oppenheimer's term on the GAC expired in mid-1952, he was not reappointed, and his official advisory roles and influence rapidly waned.

Nineteen fifty-two was a presidential election year, a time of politics. Already Senator Joseph McCarthy of Wisconsin had with considerable success engaged in a wild and paranoid crusade against people within the United States whom he suspected of communist sympathies. The leaders of the American Communist party had been given prison terms; accused atomic spies were also making headlines, and the proven Soviet atom bomb capability was popularly credited to information passed on by spies. Members of the State Department of the Truman administration were accused by McCarthy of communist affiliations, and Truman himself was no longer a viable presidential candidate. Lifelong Republican Lewis Strauss had backed for his party's candidate the conservative isolationist Robert Taft, who in his preconvention campaign had capitalized on and echoed the anticommunist hysteria of McCarthy. But the convention finally chose the moderate Eisenhower, and subsequently Strauss supported the nominee. Dwight Eisenhower was known to the public as the general who had directed the successful Allied European Operation in World War II. The Democratic candidate was the literate and liberal Illinois governor Adlai Stevenson, who expressed some hope of realizing the possibility of banning the superbomb. Yet on November 1, shortly before election day, the bulky, sixty-five-ton superbomb was tested by American scientists on a small island in the South Pacific, Elugelab in the Marshall Island group, after the islanders in the area had been evacuated. The bulkiness of the bomb derived from the need to use refrigerated liquid deuterium as the nuclear fuel, rather than the more compact but less available lithium-6.[61] But the test, dubbed the "Mike Shot," was successful, and the island was completely removed from the face of the earth. President Truman delayed the announcement of the Mike Shot until his final State of the Union message, two months after the Republicans had won the election.

Soon after his election, President Eisenhower created the position of Special Presidential Advisor on Atomic Affairs and appointed Lewis Strauss to the position. Von Neumann wrote to Strauss that he was "glad and reassured" by this appointment but was also now eager to get on with the actual job.[62] A few months later Gordon Dean was eased out of his position as chairman of the AEC and replaced by Strauss.[63] Holding these two positions simultaneously, Strauss had a controlling influence, both advising on and executing the president's decisions on American nuclear weapons policy. He had the president's ear, and his trust. The relationship that developed between Strauss and Eisenhower is reflected in the ringing eulogy Eisenhower gave him in his memoirs, speaking of his "respect and admiration" as well as of his "personal feelings of sympathy and affection" for Strauss.[64] In commenting on Strauss and his wife, Eisenhower described them as "two of the ablest, most personable and dedicated persons who have ever graced the Washington scene."[65] Like Secretary of State John Foster Dulles, Strauss participated with Eisenhower at conferences with other heads of state whenever nuclear weapons policies were to be a major subject of discussion. In political conflict with Oppenheimer and Lilienthal or the Federation of American Scientists, groups and individuals opposed to a crash program on the superbomb, Strauss had emerged the complete victor. Strauss's political good fortune also gave von Neumann a direct link to the center of nuclear weapons policymaking; affording him knowledge and influence over and above his own participation on government committees.

In his work as consultant to Los Alamos and as a member of the GAC, von Neumann was primarily involved in the research and development of nuclear and thermonuclear bombs, as well as the possible peaceful uses of nuclear energy. Von Neumann also participated in committees advising the secretary of the air force and the secretary of defense. Here the primary focus of his work was not the atomic bombs themselves but the consideration and development of means to hurl the bombs accurately at the targets, either by dropping them from bombers or with missiles. A bomb, no matter how powerful, is not a complete weapon until it is combined with a means of hurling it at the target and detonating it. The ultimate

256
*John von
Neumann
and
Norbert
Wiener*

achievement von Neumann was pursuing is what he himself has called "nuclear weapons in their expected most vicious form of long-range missile delivery."[66] The journalist Robert Jungk, referring to von Neumann, said that with his eventual success in achieving the objective, "He united two monsters, hitherto considered impracticable for war purposes, in unholy matrimony and presented them to his grateful employers as the 'absolute weapon.'"[67] This description is not inappropriate, but it is also misleading, in that it personalizes a task that von Neumann performed as part of the function expected of him in his designated role and that he carried out in concert with other members of the government, the military, and the associated industrial firms. The superbomb, combined with the intercontinental missile, could realize devastation on a huge scale, a "push-button war" initiated thousands of miles away from the target.

Even though President Eisenhower sought frugality in military spending, the development of the "absolute weapon" was compatible with the weapons policy of his administration. That policy, known as the "New Look," was based on the belief that Soviet territorial expansion would be adequately deterred by the threat of massive attack with nuclear weapons. Consequently, conventional military armament could be cut, but the development of the efficacious use of superbombs could not be neglected. The foreign policy devised by Secretary of State John Foster Dulles and the general nuclear policies advocated by atomic adviser Lewis Strauss were of a piece. Regardless of what the Soviet Union might do, it seemed that the von Neumann-Strauss-Dulles cooperation would inevitably force an escalation of the arms race. However, since Eisenhower had rejected the idea of a preventive nuclear attack as "contrary to all American principles and traditions," he had made it clear that only if provoked by Russia's possible "violent attempts to alter the territorial status quo" might American atomic bombs or superbombs be unleashed.[68] In other words, an action involving only the use of conventional weapons by Russia could lead to the use of thermonuclear weapons by the United States. This was the threat.

It is quite likely, given von Neumann's general approach to strategic thinking, that he regarded the escalation in nuclear armament and its effectiveness and destructive power as the most promising route for deterring a nuclear world war. With no faith in any disarmament efforts whatsoever, he placed his confidence in superior military power of genocidal proportions, even if both sides in a conflict should have similar military capabilities. Could the mutual threat of annihilation, mutual terror, become the world's way of preventing nuclear wars? Could such threats and counterthreats of world disaster become the normal condition of life for very long without actually precipitating the catastrophe to end all catastrophes?

A more concrete idea of von Neumann's views and activities in the early 1950s within the practical context of his membership on the General Advisory Committee can be drawn from the comments and opinions he expressed at the committee meetings.[69] The committee held about a half-dozen meetings per year, each with a duration of two or three days. At the meeting of April 27–29, 1952, two members of the GAC (Conant and Rabi) had recommended that the conclusions reached in recent studies regarding "the upper limit on the number of bombs which can be exploded without serious global hazard" be somehow directly communicated to the president. Von Neumann, among others, agreed that these studies deserved a high priority and that the president should be aware of the results, which should be continually reassessed. This concern shows that all those present were conscious of at least some of the limits the finite size of the earth imposes on the scale of nuclear war, lest the earth itself become unsuitable for human habitation.

Von Neumann became chairman of the weapons subcommittee of the GAC, and at the meeting of February 5–7, 1953, he reported on its recommendations. Since many crucial phrases are deleted from the unclassified excerpts available, all that is clearly conveyed is the general impression that the committee thought work on both fusion and fission weapons should be pursued, including work on small atomic bombs. Von Neumann personally argued in favor of support for ul-

258

John von
Neumann
and
Norbert
Wiener

trahigh-energy accelerators, machines that were then at the very forefront of physics. "Next, Dr. von Neumann expressed his strong conviction that there is needed a long-range study of possibilities in atomic weapons and their use. . . . He said that such a study would not be based on what is certain, but on what is possible." This note was struck by von Neumann repeatedly at the meetings, and it is clear that he was taking practical steps to set such studies in motion. Again and again he emphasized the need to look at long-term weapons possibilities, which allowed determination of the very worst the enemy might do, and identified the direction one's own technological quantum jump might take.[70]

Von Neumann also reported at this meeting on the needs of the superbomb program for tritium and lithium-6, whose unavailability had forced the first superbomb tested to be so inordinately bulky and heavy as to be impractical as a weapon. It was apparent at this and subsequent meetings that von Neumann had made himself the GAC's expert on the requirements, availability, costs, and alternate methods of production of lithium-6 and tritium, crucial to superbomb weapons development and production. His special effort was to ensure that sufficient lithium-6 would be available for the scheduled Operation Castle tests of superbombs that, unlike the first model tested, were compact enough to serve as "deliverable" weapons.

At the same meeting, one member, Henry Smyth, raised the question whether "one could see any signs of diminishing returns" in the testing of nuclear weapons. Von Neumann "felt the situation was very much to the contrary." This discussion was repeated with minor variations at successive meetings, with von Neumann invariably commenting "vigorously" that the tests were absolutely necessary for national defense and that testing might even need to be increased.

At the meeting of November 4–6, 1953, von Neumann brought up the subject of "larger thermonuclear weapons." "He argues that the Strategic Air Command is confident it can make deliveries with its large planes and wants the largest possible bang. . . . He felt that a weapon should be developed

with the largest possible yield in the 50,000 lb. weight class."
He mentioned that the "weapons which look good right now are in the 20,000 lb. and less range." This comment again illustrates von Neumann's interest in developing bombs of the greatest possible destructive power. It also reflects his concurrent participation on the Air Force's weapons committees. (Note that the term "delivery"—as in delivering a baby or delivering a letter—was already established as the euphemism for the explosion of nuclear weapons in enemy population centers, which might be bluntly and accurately described as "mass slaughter." The euphemistic idiom characteristic of most discussions of nuclear weapons helps us to be emotionally unaware of the destruction implied, but since it is not my purpose to perpetuate the unawareness, these euphemisms will be avoided in this chapter.)

The background to the GAC meetings was the anticommunist witch hunts led by Senator Joseph McCarthy and the first stirrings of new efforts within the United Nations to put an end to the arms race. Eisenhower hoped to make progress in this direction during his tenure as president, yet the very advisers he chose, the vested interests of the "military-industrial complex" and the "scientific-technological elite" against which he belatedly warned the country, and the political situation and attitudes he confronted when he assumed office all seemed to conspire to interfere with any real steps toward demilitarization and disarmament. In sum, the arms race continued during the Eisenhower presidency with the president participating actively, although apparently in spite of himself. Secretary of State John Foster Dulles formulated and implemented a foreign policy based on the threat of massive retaliation against the Soviet Union. The Baruch plan for international control of nuclear weapons was dead and forgotten, and discussions at the United Nations degenerated into propagandistic speeches on both the American and the Soviet side. After Stalin died in the spring of 1953, the Soviet Union was under a new, more flexible leadership. The capacity and the desire to build superbombs on both sides was a ubiquitous element in United States-Soviet relations, and a

260
John von
Neumann
and
Norbert
Wiener

new step in arms escalation—the switch from bombers to missiles as the means for implementing terror and mass homicide—was coming up for discussion.

Von Neumann was involved in a number of the notable technology-related political developments of the early years (1953–1955) of Eisenhower's first administration. They include (1) McCarthyite attacks on US scientists, especially the political attempt to destroy J. Robert Oppenheimer; (2) the public's awakening to the danger of radioactive fallout from nuclear tests and proposals for a nuclear test ban; (3) new proposals in the United Nations for nuclear disarmament; and (4) escalation of the arms race through new advances in missilery. (The apparent incompatibility of items 3 and 4 reflects policy ambiguities and conflicts; the right hand of policy—the arms race—and the left hand—disarmament proposals—opposed each other.)

Von Neumann intensely disliked the political harassment of scientists, which he had witnessed in Hungary and in Nazi Germany, and he especially disliked the harassment of exceptionally gifted ones. The Oppenheimer case hit particularly close to home. Not only had Lewis Strauss taken the initiative in seeking unnecessarily to vilify Oppenheimer, with whom he had disagreed about the superbomb crash program, but Oppenheimer was von Neumann's boss at the Princeton Institute, and when he had become director of the Institute in 1947 he had given full administrative backing to von Neumann's controversial computer project. The scientific community, with hardly any exceptions, rallied to the support of Oppenheimer. By arranging inquisitional proceedings against the director of the institute, Strauss, a trustee, helped to destroy it as an enclave of peaceful scholarship; the Institute became painfully politicized. At the hearings, conducted under the auspices of the Atomic Energy Commission, von Neumann testified strongly in favor of Oppenheimer, whose loyalty had been questioned because he had urged a slower development of hydrogen bombs, and expressed wholehearted confidence in his integrity and loyalty,[71] even though he and Oppenheimer disagreed about political issues. Had he done otherwise, von Neumann might well have

become anathema to the scientific community, as not only
Strauss but even Edward Teller in fact did—although eventually his colleagues forgave Teller for what they perceived as his betrayal of Oppenheimer and the scientific community. The official purpose of the hearings was to determine whether Oppenheimer, who had directed the Manhattan Project, was a "security risk" who should be deprived of access to secret and top secret information. As Philip Stern points out, in 1954 Oppenheimer had hardly any official connection with the government anyway, and the AEC had the "easy and graceful 'out'" of letting his remaining connection lapse; the connection would have ended automatically in June of that year.[72] The hearings were held in April and May. Lewis Strauss wrote the majority opinion of the AEC, which denied clearance to Oppenheimer—a document that does far more discredit to Strauss and his cosigners than to Oppenheimer.

Throughout these proceedings the Strauss–von Neumann friendship remained intact, and von Neumann's politically astute behavior helped pave the way for his own later ascendancy to the office of commissioner; he agreed with Strauss on weapons policy but nevertheless was respected by and viewed as loyal to the scientific community. On June 14, in the aftermath of the Oppenheimer case, von Neumann, testifying before a congressional committee, articulated his point of view concerning security hearings:

Public attention is now focused on this problem, and the public frequently views security proceedings as minor trials for treason. Yet we are still living with a system where this is an administrative matter and handled administratively and not judicially. This is a very bad situation. . . . Consequently, I think you ought to consider very seriously how to get this from the administrative into the judicial area, presumably by legislation . . . it is very dangerous that large numbers of people live in constant fear because some associations they had in the past, some indiscretions they have committed in the past may come back and damage them and nobody who is not absolutely unblemished knows for sure where he stands. . . . To have once been dropped for security reasons, is for the average person, and especially for small people, a professional catastrophe . . . to lose your clearance in some

262
John von
Neumann
and
Norbert
Wiener

industries is about as pleasant as for a doctor to be expelled from the AMA.[73]

Von Neumann recommended to the subcommittee that in security clearance cases the civil rights of the scientist be protected, just as they would normally be in judicial proceedings. Although he was strongly anti-Soviet, he had no sympathy for the anticommunist campaigns of the McCarthy era in America.[74]

I have already noted that von Neumann repeatedly insisted at the GAC meetings that atomic bomb and superbomb testing must in no case be hampered because it was imperative for national defense; that without testing, the arms race would be slowed. And von Neumann consistently favored the acceleration of the arms race because he believed the US could stay ahead in that way. In March 1954, the US government carried out Operation Castle, in which a number of hydrogen bombs were exploded in the Marshall Islands region. The total explosive power of the bombs tested was equivalent to forty-seven million tons of TNT. In one of these bomb tests, the "Bravo" shot, a shifting wind caused a Japanese fishing trawler, the Lucky Dragon, more than a thousand miles to the west, to be showered with some radioactive fallout. The crew of the fishing vessel experienced radiation sickness; the radio operator, Kuboyama, died, and the other crew members were hospitalized upon returning to Japan. Their radioactive catch immediately raised in Japan and elsewhere the fear that nuclear testing would lead to concentration of radioactivity in fish on a wide scale and do great damage to this important source of food. The unexpected wind also took fallout from the Bravo shot toward an inhabited island, Rongelap, where radiation led to severe skin burns and other immediate effects; eventually it was reported that "many of the adults and almost all of the children have contracted cancer."[75]

The Lucky Dragon incident, the first of a number of occasions on which the public learned that nuclear testing may involve human sacrifice, is historically important in that it stimulated a growing worldwide movement to ban such tests. Both the Lucky Dragon incident and the Oppenheimer hearings were in the public eye during the spring of 1954. Lewis Strauss and the AEC tried to reassure the public by minimizing the biological danger of fallout from nuclear tests,[76] while physicist Ralph Lapp, chemist Linus Pauling, and geneticist H. J. Muller called attention to their harmfulness, both immediate and long-term.[77] The fallout controversy reinforced among scientists the impression from the Oppenheimer case that the AEC, and Lewis Strauss in particular, lacked honesty and human decency. They could not be trusted with nuclear technology because they had no respect for human life. After all, it was well understood that the worldwide fallout from nuclear tests was likely considerably to increase the rate of death by cancer and produce thousands of birth defects from genetic mutations. The AEC sponsored the "Sunshine" studies (the name serves to put a cheerful face on gruesome phenomena) concerning the anticipated long-range impact of nuclear tests on the population. Von Neumann, at the GAC meeting of May 27–29, 1954, reported that the Sunshine studies indicated that the general human population could tolerate up to about 10,000 megatons, but he suggested that the study's concept of "tolerance" may have been too strict. In effect, he was saying, there is no need whatever for inhibiting nuclear tests. In connection with a proposal that the United Nations undertake a worldwide study of the effects of nuclear tests, von Neumann offered the following analysis:

The present vague fear and vague talk regarding the adverse worldwide effects of general radioactive contamination are all geared to the concept that any general damage to life must be excluded, and at least with certainty. . . . Every worthwhile activity has a price, both in terms of certain damage and of potential damage—of risks—and the only relevant question is, whether the price is worth paying. . . . Is the price worth paying? For the U.S. it is. For another country, with no nuclear industry and a neutralistic attitude in world politics it may not be.[78]

Of course measuring the effects of fallout, such as a worldwide increase in the cancer rate, against American military advantage has no objective basis whatsoever. In spite of the apparently rational assessment of costs and benefits, the

264

John von
Neumann
and
Norbert
Wiener

question is one of values, and to von Neumann the cold warrior, the primary values were American political, economic, and military power. In 1954 and 1955, as weapons testing continued to enhance both American and Soviet destructive capability, it was von Neumann's view that no international restrictions should be permitted to inhibit the arms race. Inasmuch as he was skeptical that the Soviets would abide by such restrictions, he did not wish to handicap the United States in the insane race to develop more efficacious versions of these weapons.

At the thirty-eighth meeting of the GAC on January 6–8, 1954, there was a more general discussion dealing with Eisenhower's arms policies, and von Neumann clearly indicated where he stood in relation to them. One side of the Eisenhower policies was the "new look" in weapons: emphasis on the large superbombs, with a lower priority for the small atomic bombs and an actual cut in conventional military weapons appropriations. At the meeting, physicist I. Rabi had criticized the new look; in particular, he feared that plutonium production for the smaller atom bombs would be neglected and some obstacle might develop to limit "deliverability" of the superbombs, leaving the United States with limited striking power in both the small- and large-bomb categories. Von Neumann responded reassuringly that "the proposal was clearly preliminary in character and would certainly be reviewed after tests had determined the suitability of the superbomb for military use. Von Neumann "went on to give several arguments in favor of the 'new look' in weapons," but also reassured Rabi that improvements in plutonium production methods would obviate the necessity of a cutback. He agreed with Rabi, however, in that "we will need a very large number of small bombs" and suggested the possibility "that a large fraction of the fissionable material produced should be used in small bombs" rather than for other purposes. Of course, the new look was coupled to the policy of massive retaliation in case of provocation by the Soviet Union, and von Neumann agreed with that policy. As the GAC minutes state, he "felt that our main defense against Russian strategic bombing is their belief that we have the capacity to inflict massive retaliation." The relatively "small" (for example, Hiroshima-size) atom bombs, Secretary of State John Foster Dulles threatened, might be used directly against any country making an attack on America's allies with mere conventional weapons. While the Korean conflict had been fought entirely with conventional weapons on both sides, should a similar situation occur again, the United States, according to Dulles, might well use nuclear air strikes.[79] This policy allowed the United States to neglect its conventional armaments. It was a risky policy. Any war with Hiroshima-type bombs could easily escalate to a superbomb holocaust. And if Dulles was bluffing, what if somebody were to call his bluff?

Even as the nuclear arms race was rushing ahead, the opinion was widely held that eventually it would lead to impossible decision problems, and that the only long-term mechanism for avoiding nuclear holocaust was disarmament or international control. Thus, even though the Baruch plan and the Soviet counterproposals had failed, by 1953, with the Americans and Russians in possession of the superbomb and new leadership in both countries, there were some new efforts at disarmament. These efforts reached a high point characterized by the British statesman Philip Noel-Baker as "The Moment of Hope: May 10, 1955,"[80] but the hope was to be dashed to the ground by a reversal of American policy to which Lewis Strauss contributed in his position as presidential adviser on atomic affairs and as AEC chairman.

The new disarmament efforts had their beginning with a report to President Eisenhower made early in 1953 by a State Department advisory panel on disarmament, a panel headed by J. Robert Oppenheimer, a year before Oppenheimer would become the victim of security hearings. The report urged that the president speak candidly to the American people about the nuclear weapons race and the hydrogen bomb, its nature, its implications, and the urgency of dealing with it. Eisenhower, much troubled, indeed put his speech writers to work on "Operation Candor," but he frequently discussed the issues and problems with Lewis Strauss, and with the advice of Strauss, Operation Candor, which would speak truthfully of possible horrors, was scrapped and replaced by "Operation

266

John von
Neumann
and
Norbert
Wiener

Wheaties," which involved emphasizing in an optimistic way the possible uses of "atoms for peace." In a sense, this was characteristic of Strauss's approach. His failure to deal candidly with the realities of nuclear weapons is illustrated by his later absurd hailing of a form of superbomb with somewhat less fallout than the Bravo test bomb as a "humanized" superbomb.[81] And earlier, when the *Lucky Dragon* incident had caught public attention, Strauss at a White House press conference had simply given the press a false report about the injured fishermen, playing down the harm done.[82]

While Strauss explicitly recommended against any sweeping disarmament plans, when President Eisenhower proposed for study to his own staff the idea that both the United States and the Soviet Union contribute a certain amount of their stockpiles of fissionable materials to the United Nations for peaceful uses, Strauss favored it, seeing that it neither depended on Russian "good faith" nor required "inspection" and was more a symbolic act of co-operation that did not seem to involve serious risk. Thus on December 8, 1953, Eisenhower made his famous "atoms for peace" speech, which was partly written by Strauss. It contained the proposal for sharing nuclear materials and generally urged use of nuclear materials for peaceful purposes. He made the much-publicized speech at the United Nations in front of 3,500 people, and even the Russian delegation participated in the enthusiastic applause.[83] So it was that at the January 6–8, 1954, meeting of the GAC, the president's UN proposals and the methods for implementing them were a major topic of discussion. One of the GAC members, James B. Fisk, the director of research at Bell Laboratories, stimulated by the possibilities of working toward disarmament, proposed that the Atomic Energy Commission "explore the possibility of negotiating with the Russians to shut down all present plutonium producing facilities." He pointed out that the suggestion was consistent with the objectives expressed by Eisenhower and with the expressed Russian attitude toward disarmament, and would test the Soviet's "good faith." Committee members Rabi and Egel Murphree, president of the Standard Oil Company, liked Fisk's suggestion, and Murphree in fact wanted to go much further. Oliver Buckley, the former chairman of Bell Laboratories, and von Neumann disagreed.

To begin with, von Neumann stated that "he was not hopeful about the objectives of armament reduction."

He pointed out that we believe the Russians are better equipped for clandestine operations than we are and that we would be competing, at a disadvantage, with them in a clandestine armament race. Disarmament agreements would shift the area of conflict to territory less favorable to us than the present. . . . Such agreement probably could not be reached anyway, and that if it were, it would have the very unfortunate effect that the U. S. would not be willing to spend as much on air defense measures. Also, it would increase the AEC's difficulty in getting money from Congress.[84]

Von Neumann opposed disarmament agreements; it was more armaments rather than disarmament that he favored. The threat of massive retaliation was America's protective shield, he believed, not any disarmament agreement. The "atoms for peace" proposal itself entailed no disarmament. It primarily generated propaganda for nuclear reactors, which would help financially in the development of an American nuclear industry for export and would help to justify funds for research in applied nuclear physics.[85] As the reference to congressional funding indicates, one major argument against disarmament had to do, then as now, with special economic interests. The "atoms for peace" proposal served as a benign veneer over the public fear of the deaths and miseries resulting from use of nuclear weapons. It was intended to reduce international tension, even though it was not a direct move toward disarmament. It was also the second major instance in which Strauss's views had prevailed over the recommendation of a committee chaired by Oppenheimer.

Since January 1952, however, a disarmament commission set up by the United Nations had been at work. The work of the commission appeared stymied by Russia's persistent opposition to disarmament proposals made by other countries, while at the same time the Russian disarmament proposals were always of a form unacceptable to the West. Six months

268

John von
Neumann
and
Norbert
Wiener

after Eisenhower's UN speech, on June 11, 1954, in an effort to reconcile conflicting viewpoints, a proposal was put forth jointly by the French and British delegates for a treaty that would lead to complete nuclear disarmament under international control. In spite of Strauss's earlier recommendation to Eisenhower against any broad disarmament proposal, in spite of the increasing confidence placed on the policy of using the threat of massive retaliation as if it were an alternative to disarmament, the United States nevertheless endorsed the British-French proposal, leaving again only Russia in opposition. In the spring of 1955, the Western delegates were appealing, urging, and challenging the Soviet delegates to be more open-minded in considering the proposal; the Soviets assured the West they were doing so. Then, unexpectedly, on May 10, 1955, the Russians presented their own plan, which itself contained all the main features of the British-French proposal. After two days of consultation with the administration, the American delegate gave the official US response: "We have been gratified to find that the concepts which we have put forward over a considerable length of time, and which we have repeated many times during this past two months, have been accepted in a large measure by the Soviet Union."[86] This was the "moment of hope," as Philip Noel-Baker termed it. Finally, all the nuclear powers had agreed on a concrete approach to disarmament. The arms race would be called off. It was too good to be true: on September 6, the new US delegate, Harold Stassen, announced in a sudden and dramatic reversal of American policy that the United States would "put a reservation" on all of the substantive positions it had previously taken in the disarmament commission or at the UN on questions relating to levels of armament.[87]

The US betrayal of the disarmament effort again reflected the Dulles-Strauss line. In the official jargon, the United States had "reappraised its position." In fact, Eisenhower himself was clearly chagrined and disturbed when the Russians had essentially agreed to nuclear disarmament on the terms urged by the Western powers, and he began to look for a graceful way to extricate the United States from the agreement. This led him to issue, with much fanfare, the Open

Skies proposal, of which he later said, "We knew the Soviets wouldn't accept it. We were sure of that."[88] It appears that not only had Strauss and Dulles been opposed to engaging in significant nuclear disarmament but that President Eisenhower accepted their view, even though "Ike" had presented a convincing public image of a president eager to reduce the threat of war.

A potential force favoring nuclear disarmament was the many poor African and Asian countries, whose national interests were not served by continuing tension between the superpowers. Already in 1953 Dulles had given Prime Minister Nehru of India to understand that America was prepared to use nuclear weapons in Asia against communist China.[89] Presumably the Soviet Union would be equally ruthless about its political interests. India, like many other poor African and Asian nations, understandably preferred to be uninvolved in the conflict between America and Russia, lest it become a victim. The nuclear weapons race is a form of terrorism continually threatening even innocent bystanders. When in April 1955 the leaders of the Afro-Asian nations had their first international conference in Bandung, Indonesia, they initiated the movement of "nonaligned" nations, neutral in the Russian-American power struggle. They favored a halt to the nuclear arms race and amelioration of the conflict between the giants, and they especially wanted to prevent nuclear war. They were the potential "third force," the force for mediation and conciliation, and the natural allies of any political group within the superpowers that genuinely sought nuclear disarmament. But the position of neutrality was seen by the US government as a betrayal of the American side in the cold war.[90] Eisenhower, after discussing the matter with his advisers, would not even send greetings to the Bandung Conference. The Manichean Dulles, commenting on the nonaligned countries, states that "except under very special circumstances, neutrality is an immoral and short-sighted conception."[91] Zealous cold warriors on both sides persistently tried to enlist Asian and African nations in their own cause, although the very nonalignment of these countries was the seed that might have enabled them to become an inter-

270
*John von
Neumann
and
Norbert
Wiener*

national force for disarmament and avert the danger of nuclear war.

The issue of nuclear disarmament is, of course, a profound and persistent one. Von Neumann had defended practically unlimited testing of nuclear weapons in spite of radioactive fallout on the grounds that "every worthwhile activity has a price, both in terms of certain damage and of potential damage—of risks—and the only relevant question is, whether the price is worth paying." Is it worth taking the risks of worldwide nuclear disarmament (possible violation) and paying the price (some loss of national sovereignty) to obtain freedom from the threat of nuclear holocaust? The events of 1954–55 accentuate Richard Barnet's comment that "the primary question is not how to disarm but whether to disarm. It is this question that the United States has never really faced."[92]

Von Neumann, in his time, had unambiguously answered the question in the negative. This clear choice of values gives his statements and actions a particular clarity and simplicity. Having for all practical purposes settled this basic question in his mind in a relatively unsophisticated way, von Neumann could devote himself to the merely technical problem of implementing superiority in arms. This whole mode of procedure was altogether in tune with von Neumann's characteristic style. Eisenhower and other public figures appeared more ambiguous, either because that was their cultivated image or because they were genuinely unsure of what policy to adopt. One reason that Robert Oppenheimer so deeply disturbed and fascinated people was that he was preoccupied and torn by the very question that Barnet raises.[93] The American chemist Linus Pauling was in the 1950s an articulate proponent of nuclear disarmament, as was the British statesman Philip Noel-Baker, and both men worked energetically in their own ways to implement their convictions. They each received a Nobel Peace Prize, but they were swimming against the stream. To determine where the US government actually stood during the Eisenhower administration, one might measure the size (in dollars or employees) of the groups working on disarmament against those involved with armaments. In such terms it is evident the disarmament effort was minuscule.[94]

Aside from the design of bombs, von Neumann took an early interest in the development of missiles. Since 1951 he had been a member of the Scientific Advisory Board of the US Air Force. After Eisenhower was elected, von Neumann worked closely with the two men given the task of making a full review of all weapons research and development programs, Donald Quarles of the Department of Defense and Trevor Gardner of the Air Force. Eisenhower was examining existing weapons programs in order to establish new priorities, and especially to identify projects that could be eliminated from the budget.[95] Until then bombers were the only extant means to propel nuclear weapons toward whoever or whatever is to be annihilated, but the possibility of building missiles to carry nuclear bombs would naturally have to be explored under Eisenhower's new weapons policy. A review of their feasibility, their potential capabilities, and their cost was of primary concern. If they were feasible, they might have technical advantages over bombers: greater speed, more automatic and dehumanized, and less vulnerable. Herbert York, an insider, described the organization of the weapons review procedure:

The reviews themselves were carried out at first by a number of committees working separately in each of the military services and reporting to different levels of program management. The committees had a strongly interlocking membership, a feature of the advisory apparatus which . . . allowed information to travel both up and down within agencies and laterally between agencies and thus to pass over and around the various barriers of secrecy, propriety, and bureaucracy which would otherwise cripple technical progress. . . . After about a year, in the interest of making more rapid progress and more profound changes, the originally large number of committees were regrouped into a much smaller number and John von Neumann became the chairman of the most important of them.[96]

The streamlining of the organization made the von Neumann committee, as it was usually called,[97] the decisive influence on major weapons decisions. York adds that von Neumann's "combination of scientific ability and practicality gave him a credibility with military officers, engineers, industrialists, and

272

John von
Neumann
and
Norbert
Wiener

scientists that no one else could match. He was the clearly dominant advisory figure in nuclear missilery at the time, and everyone took his statements about what could and should be done very seriously."

His simultaneous membership in the GAC and the von Neumann committee gave him a particular authority on the means for transporting destructive power, such as when he sought the GAC's endorsement for ascribing "great urgency" to the development of nuclear-powered bombers;[98] in turn, his up-to-date and quantitative familiarity with the characteristics of atom bombs and superbombs was useful in specifying criteria for missile designs. Specifically, information about the anticipated performance as well as the weights of the superbombs tested in the Castle series of explosions and the information gathered about the extent of damage done were essential to the von Neumann committee for making policy recommendations. Should nuclear-tipped missiles be built at all? And if so, is it preferable to build a missile to travel moderate distances, thus permitting more accurate aim and lighter weight, or should we try to build a missile that could carry a superbomb all the way from the United States to the Soviet Union? A missile of intermediate range might travel from a US base in Western Europe into Russia. Should we start by building missiles tipped with relatively light and compact atom bombs or should we try right away to build a missile to carry the heavier superbomb? These were the questions confronting the von Neumann committee.

The members of the von Neumann committee concluded early in 1954 that since the radius of destruction of the superbomb was so large considerable inaccuracy in aiming the missile could be tolerated. Nor did the weight of the most recently developed superbombs seem prohibitive. The committee concluded that a militarily effective intercontinental ballistic missile (ICBM) carrying a superbomb was technically feasible, and recommended this quantum jump in weapons technology. It urged that the United States initiate the development program, giving it "the highest national priority." However, the committee went still further by offering suggestions about how the program might be implemented: it

should not adhere to the traditional framework of weapons development; rather, "a management-scientific-technical team
of extraordinary competence had to be swiftly assembled; and
. . . normal procedures would have to be circumvented, eliminated, or temporarily set aside."[99] The urgency was so great
that the development, production, and deployment phases of
the new weapon should all be carried out concurrently.[100]

The proponents of the ICBM had to overcome the prevalent
view, favored especially by the Army, that it made more sense
to develop missiles of more modest range as a stepping-
stone to the ICBM. The von Neumann committee not only
was in close touch with Quarles and Gardner of the executive branch but also worked closely with Air Force generals (McCormack, Schriever) and found additional support
from influential senators (Henry Jackson, Clinton Anderson).
Working through the Secretary of the Air Force and the Defense Department, the committee began to implement the innovative organizational structure it had recommended, [101] but
it also continued to seek stronger political backing. In the
summer of 1955 Trevor Gardner, General Bernard Schriever,
and von Neumann jointly gave a personal briefing to President
Eisenhower,[102] and indeed Eisenhower did then assign the
ICBM (Atlas) project the highest priority status. In his memoirs, Eisenhower recalls how the findings of the von Neumann committee influenced the level of funding of the ICBM:
"By May [1954] the Castle bomb tests in the Pacific had
substantiated the von Neumann findings. In accordance with
this report, the Air Force reshaped its program and began
to accelerate work on an ICBM. By early 1955 its Atlas project
was mushrooming: from the fiscal year 1953 figure of $3 mil-
lion, in 1954 it went to $14 million and in 1955 to $161 million."[103] The government's zeal to develop nuclear-tipped
missiles as tools for "massive retaliation," heightened by interservice rivalries, led a few years thereafter to a multitude of
simultaneous crash programs carried out at vast cost to the
taxpayers.[104]

The cast of characters in this scenario that led to an acceleration of the arms race by means of the frenzied ICBM program provides a glimpse into the process: Trevor Gardner,

274

John von
Neumann
and
Norbert
Wiener

the energetic promoter and organizer within the executive branch; von Neumann, the authority on weapons, equally effective and enthusiastic; the famous and glamorous aviator Charles Lindbergh, a member of von Neumann's committee; some younger scientists on the committee, upwardly mobile within the government, like von Neumann himself;[105] General Schriever, the innovative, competent military administrator of the program, who continued to work in liaison with the von Neumann committee; and the industrial component of the military-industrial complex, Simon Ramo and Dean Wooldridge, both members of the committee, who resigned to form the private Ramo-Wooldridge Corporation, which then received the large contract for the technical direction and systems engineering of the Atlas project. The men at the top, such as Dulles and Strauss, were indispensable in generating a general policy framework within which the Atlas missile made sense; the president himself and a few influential senators and congressmen complete the cast. All received an impetus from intelligence reports to the effect that the Russians were working hard on missiles.

What was at first in effect a political coalition of a few influential men, spearheaded by Gardner and von Neumann, gained momentum and soon became a large establishment with its own vested interests. It effectively countered the traditional view that bombers were the best means for carrying nuclear weapons. It prevailed over the technologically more conservative proponents of intermediate-range missiles. It overcame the interservice rivalries between Army, Navy, and Air Force and the Republican Eisenhower administration's commitment to frugality. But this major political and technical achievement was from a broader perspective a hollow one, for all it accomplished was to initiate a crash program for building "nuclear weapons in their expected most vicious form," as von Neumann himself had accurately described them.[106]

On their own terms, the Atlas missile and its successors were greatly successful, like the hydrogen bomb before them. But when one recognizes, with Einstein, that the implication of the nuclear armaments race is that "in the end, there beckons more and more clearly general annihilation,"[107] this achievement on a deeper level represents horrendous failure of Western civilization in its use of science and technology, reinforced by the scuttling of the international disarmament agreements proposed by Britain, France, and the Soviet Union in 1954–55. The Oppenheimer hearings point up another aspect of that failure, one whose political effects are still being felt. The shallow successes are easy to describe and understand. However, in the crucible of history, it is the failures that matter most. Technological civilization's continued pattern of "deep failure," of which the aforementioned events of 1954–55 are representative, demands to be understood, and not only by the philosophers or the historians, but also by technologists, politicians, and managers in government and industry.

In May 1953 von Neumann assumed the chairmanship of the nuclear weapons panel of the Scientific Advisory Board of the US Air Force (under the chairmanship of Hungarian-born Theodore von Karman),[108] and in February 1954 he was appointed by Assistant Secretary of Defense Quarles to the advisory panel on atomic Energy of the Department of Defense.[109] In May 1954 von Neumann had occasion to make a list of all the organizations to which he was a consultant, twenty-one in all, mostly government organizations but also some private companies. The latter included the Standard Oil Company, IBM, Ramo-Wooldridge, the Rand Corporation, which carried out strategic analysis, and Van Nostrand, the publishing house. The government organizations included, besides the Atomic Energy Commission, the Air Force, and the Defense Department, the Sandia Laboratory, the Los Alamos Laboratory, the University of California Radiation Laboratory, and the Central Intelligence Agency.[110] He was also at that time in correspondence with a friend in South America concerning a possible joint business venture.[111]

In 1955 the highest official position available to scientists in the US government was that of Atomic Energy Commissioner. Usually only one of the five commissioners was a professional scientist. Sometime in 1954, if not earlier, von Neumann had confided to Strauss that he had made up his mind to resign

276
John von
Neumann
and
Norbert
Wiener
his professorship at the Princeton Institute.[112] The hostile feeling of his colleagues at Princeton toward von Neumann's computer project may have contributed to his desire to go elsewhere. Both his computer work and his participation in the weapons establishment put him in a position to make use of some sound and promising opportunities for business ventures, which intrigued him as a possible arena for his activity. But when Strauss in 1954 offered Johnny the position of Atomic Energy Commissioner, to replace one of the other commissioners who was about to resign, Johnny hesitated. Strauss's actions in connection with the Oppenheimer hearings had made him detested by most scientists. Von Neumann did not want to become a pariah in the scientific community, but he very much wanted the position of commissioner. There may be overtones here of the kind of dilemma which must have confronted the Jewish bankers of his father's generation in Budapest when a government promoting anti-Semitic policies offered them honors and status because it needed their services. Ulam, Johnny's personal friend, comments on the decision:

Just after Johnny was offered the post of AEC Commissioner . . . we had a long conversation. He had profound reservations about his acceptance because of the ramification of the Oppenheimer Affair. He knew the majority of scientists did not like Admiral Strauss' actions and did not share the extreme views of Teller. Some of the more liberal members of the scientific community did not like Johnny's pragmatic and rather pro-military views. . . . He recognized this feeling even among some of his Princeton associates. . . . The decision to join the AEC had caused Johnny many sleepless nights, he said. . . . But he was flattered and proud that although foreign born he would be entrusted with a high governmental position of great potential influence in directing large areas of technology and science. He knew this could be an activity of great national importance.[113]

Ulam adds the political observation that "as a friend of his and having pressed him to accept the offer, Strauss would be obligated to support his views and ideas." Moreover, unlike the Jewish bankers, who were always outsiders no matter how patriotic and effective their service to the Hungarian government, von Neumann's official status would reflect his power and influence. With a fixed five-year term, a commissioner was relatively invulnerable to the president's or Congress's whim, and his actions were largely secret from the electorate.

John von Neumann's name was submitted to the Senate for confirmation as AEC commissioner on January 10, 1955, but von Neumann had already been preparing for several months for the confirmation hearings. In November 1954 he attended the confirmation hearings of Willard Libbey, another new appointee to the commission, consciously educating himself on the political process in a situation where he understood all of the issues involved. "What does Senator Kefauver want?" he asked Strauss, his political mentor. Was he trying to get a "public power" proponent on the commission? "What are your views on tactics and chances of success?", he inquired.[114] Strauss had become deeply involved in the issue of whether electric power should continue to be supplied to the AEC's Oak Ridge Laboratory by the publicly owned Tennessee Valley Authority or by a private corporation to be formed for the purpose. Eisenhower, Strauss, and von Neumann were all "private power" proponents. Eisenhower feared the "encroachment of socialistic tendencies."[115] Conservative Republican Strauss had been adamant. "Estimates showing that the AEC would be forced actually to pay more for its power from the Dixon-Yates Company than it had been paying to TVA, even taking into consideration the government's ability to collect taxes from private organization, did not deter . . . Lewis Strauss . . . [and other conservatives] normally preoccupied with reducing the cost of government."[116] Not only would von Neumann be interrogated about his views concerning the political hot potato, the Dixon-Yates contract, he would also be asked about the Oppenheimer affair, as well as a letter of support he had written for a former student, the mathematician I. Halperin, who had at a much earlier time been a member of the Communist party. It was perhaps symptomatic of the times, and not merely a personal statement, that von Neumann described himself at the confirmation hearings as "violently anti-Communist, and a good deal more militaristic than most," and again, as "violently opposed to Marxism ever since I can

278

*John von
Neumann
and
Norbert
Wiener*

remember."[117] He also stated that his closest relatives still in Hungary were "only cousins." On March 22, he could write his friend in South America about the congressional confirmation that "at long last it is victoriously settled." Their joint business venture would have to be deferred until his tenure with the AEC had expired. The news media described von Neumann's appointment as a promising means to heal the split between the Atomic Energy Commission and the scientific community.

Who were the men who sat on this powerful commission with John von Neumann and Lewis Strauss? Only Thomas Murray was a holdover from the Truman administration, and on many points he was out of sympathy with the Strauss–Dulles–von Neumann philosophy emphasizing superbombs and massive retaliation. Murray had been a wealthy industrialist before his 1950 appointment to the AEC, president of the Metropolitan Engineering Company, a diversified company founded by his multimillionaire father that was mainly engaged in the design and building of electrical power plants. He had organized the Murray Manufacturing Company, which built switches and electric circuit breakers. He was on the board of trustees of some leading banks and on the board of directors of the Chrysler Corporation. He was also an engineer, holding some two hundred patents, and in 1940 had reorganized the New York subway system when it had gone bankrupt. In his early years on the AEC he had been instrumental in arranging for the United States to obtain uranium ore from South African mines as well as from Spanish and Portuguese colonies. He was a zealous promoter of the building of nuclear reactors, and was determined that the United States should maintain the lead in the field, regardless of whether the reactors were controlled by the private or public sector. In 1954, in contrast to Strauss and the rest of the commission, Murray favored a moratorium on the testing of (megaton) superbombs but favored continued development and stockpiling of the smaller (kiloton) atomic bombs.[118]

Harold S. Vance, the president since 1948 of the Studebaker Corporation (at that time a major American automobile company) and an old friend of Lewis Strauss, was another 1955 Strauss-Eisenhower appointment to the commission.[119]

He was a Republican, an Episcopalian, and an industrialist who had publicly opposed excess-profits taxes on corporations, lest they "curtail business expansion."[120] He also had been a special government consultant early in the Eisenhower administration on the orderly lifting of the voluntary wage and price controls that had been imposed in connection with the Korean War by the Truman administration.

Willard F. Libby, a nuclear chemist from the University of Chicago, had been from the very beginning a persistent proponent of the superbomb program, like von Neumann, Strauss, and Teller, and saw it as one of his major functions on the commission "to keep the weapons program really stepping." His appointment had preceded von Neumann's by only a few months. He was also a "leading proponent of the private development of power."[121] His most significant scientific work had been the development of a method of using radioactive carbon to determine the age of remains from ancient times. His technical competence was beyond question; he was respected by the scientific community. A conservative Republican, he was the kind of man who would be congenial to Lewis Strauss. He was replacing Henry DeWolf Smyth, the physicist commission member, who had dissented from Strauss's opinion in the Oppenheimer case and generally did not see eye to eye with Strauss. Smyth had resigned.

With von Neumann's appointment, Strauss had in effect packed the commission to suit him, except for the one remaining holdover, Murray, whose term would expire in 1957. It was Strauss's innovation to have two scientists on the commission rather than just one, perhaps to help him regain favor with the scientific community.

These five men carried an enormous responsibility; their actions would be highly consequential on a worldwide scale for later generations. While all five commissioners had distinguished themselves through their prior achievements, and all had survived the scrutiny of a congressional committee in their confirmation hearings as well as the selection process of the executive branch of the government, it is noteworthy that the criteria of selection were narrow and parochial, such as a preference for private over public power, enthusiasm for the

280

John von
Neumann
and
Norbert
Wiener

arms race, membership in the wealthy and powerful strata of the population, anticommunism, and possession of a political orientation and personal qualities congenial to Lewis Strauss. Those who felt a broader loyalty to the worldwide human community and subscribed to the objective of abrogating the nuclear arms race and eliminating the possibility of super-bomb wars were effectively excluded; such ideas appear all but absent in Strauss and his cohorts. The commissioners were all active promoters of nuclear technology and sought to have the United States retain hegemony over this field. This is no accident, because from its inception the mission of the AEC included these goals.[122] The promotional and the narrowly nationalistic aspects of the AEC overwhelmed any responsibility to human welfare generally, as well as respect for human life and concerns about biological injury to future generations. The AEC promulgated and implemented policies that because of the progressive stockpiling of increasingly destructive weapons made nuclear war both more potentially disastrous and more likely. The Atomic Energy Commission of 1955, as an arm of the US government, reflected the political orientation dominant in its time, just as today's counterpart reflects the political biases of our time. Insofar as biases are parochial and short-term, the dangers of nuclear disaster build up cumulatively, and the difficulties of averting such a disaster mount with every passing decade.

John von Neumann moved his family from Princeton to Washington in the spring of 1955. His active role as a commissioner lasted not much over a year; illness began to make it difficult for him to participate. Increasingly his assistant, Paul C. Fine, would be called upon for liaison. Von Neumann's confinement to a wheelchair made his usual style of travel impossible; even going to his office (at 19th and Constitution Ave. NW) and attending commission meetings, typically three long afternoons every week, was a chore.

Although the primary concern of the AEC, and of von Neumann as commissioner, was the development and stockpiling of nuclear weapons, almost every page of the AEC's semiannual report to Congress dealt with civilian applications. The weapons work, except perhaps for a list of new

AEC laboratories being built, is usually described only in a sentence or two, such as "work continues to increase and improve the US arsenal of weapons. Work continued on designs for defensive use and on methods of reducing radioactive contamination from explosions."[123] Secrecy had become a tool for the AEC to protect itself from Congressional criticism, and even more from journalistic scrutiny. Since only a small portion of the AEC files (mostly not dealing with weapons) were available to me, some of what follows may overemphasize von Neumann's role in regard to nonmilitary applications. His public statements during that period, insofar as they pertain to atomic energy policy at all, must be viewed as reflecting his self-perception as always representing his official role: "In the AEC, we have learned never actually to 'go off the record.' . . . Even the appearance of an 'off the record' position must be avoided to prevent any possible public impression we are not properly careful of classification and security."[124] According to the official AEC historian, von Neumann usually said little at the meetings; occasionally he made an incisive comment, but he was invariably polite in his style and civil when he disagreed with a colleague.[125] It seemed that von Neumann was more active behind the scenes than at the formal meetings.

At the time of von Neumann's tenure, all the commissioners except Murray, the one pre-Strauss holdover, shared Strauss's disposition in favor of escalating the nuclear weapons race and apparently harbored similar fears and obsessions. It was a corollary of their collective outlook to disregard injuries inflicted on individuals in the process of promoting nuclear weapons development and stockpiling:

The AEC's . . . great failures have been in not foreseeing sufficiently, even when it was pointed out to them for years, firstly, the dangers associated with radioactive fall-out from atmospheric testing; secondly the hazards of breathing uranium-mine atmospheres where no effort was made to reduce the airborne radioactivity therein (despite the fact that the AEC knew *before* it began its massive uranium procurement that a thousand European uranium miners had aleady died under conditions identical to those the AEC was to create on the Colorado Plateau in the 1950's).[126]

282

John von
Neumann
and
Norbert
Wiener

The AEC had been a glamor agency, administering the atom. With the Oppenheimer hearings and the publicity about the *Lucky Dragon*, it had lost some of its luster. Yet, it continued in its high-handed manner to belittle or ignore expected lethal radiation effects, such as deaths from cancer of uranium miners and victims of fallout. Whatever the details of the cost-benefit analysis which underlay the killing of Colorado miners to keep the cost of uranium low, individual government functionaries participating in the decision were not held accountable.

During most of the late spring and summer of 1955, von Neumann was acting chairman of the commission, as Strauss was away. Strauss had other responsibilities, such as attending meetings of the National Security Council, negotiating with foreign ambassadors, and traveling to international conferences in his role as the president's atomic adviser. One major topic with which the AEC had to deal at that time was the development and production of nuclear reactors. The Strauss-Eisenhower "atoms for peace" proposal, which put a benign facade on nuclear technology, had become foreign policy. Export of reactors to NATO allies would help to tie those nations to the United States, providing a "bulwark against Soviet expansion westward" and at the same time spurring development of the nuclear reactor industry. Congress, on the other hand, was eager to see domestic nuclear power development get started.[127] In a memorandum he wrote in his capacity as acting AEC chairman, von Neumann succinctly summarized the policy: "Our domestic interest is the development of low-cost atomic power in view of the low cost of power in the United States, rather than any kind of atomic power for its own sake. From the international point of view, on the other hand, we will be doing more than the Soviets, since we will push the construction of power reactors abroad."[128] He also wrote a letter to the secretary of defense, reassuring him that the policy of promoting power reactors abroad would not "place atomic weapons within the custody of foreign nations."[129]

In response to congressional urgings to speed up nuclear reactor development in the US, the AEC had invited companies to submit proposals for building them, and in the summer of 1955 it was considering the first group of such proposals.[130] The AEC was also concerned with finding mechanisms to make nuclear power plants attractive investments for private corporations. Since ordinarily the electric power from the reactors would be far more expensive than power from conventional sources, it became a question of what form the government subsidy should take. Von Neumann sent a letter to Senator Clinton Anderson of the Joint Committee on Atomic Energy, reporting on the first four promising reactor proposals from private companies and at the same time suggesting new legislation: "Each of the four proposals is contingent in some way on a satisfactory solution of the problems associated with providing indemnity against possible liability resulting from nuclear catastrophe. The Commission is not authorized to grant indemnity in connection with these proposals, and it is studying the possible need for legislation."[131] Characteristically, the concern was not to protect people from the risk of nuclear catastrophe but merely to protect private corporate investors. Indeed, Senator Anderson would become coauthor of the well-known Price-Anderson Act, according to which the government would furnish free insurance to the industry in the amount of half a billion dollars per nuclear accident.[132] The awareness that the safe storage of nuclear reactor waste might pose a major problem for the future had apparently not yet arisen. The tendency of scientists and technologists grossly to underestimate the health and safety problems associated with nuclear reactors is remarkable, since the factual information available in the 1950s should have sufficed to show the dangers. The reason for their collective misjudgment may be psychological: Manhattan Project alumni, especially, had counted strongly on the benefits of nuclear reactors to ease their consciences.[133]

One of von Neumann's interests was in weather modification, and he participated in a panel on "possible effects of atomic and thermonuclear explosions in modifying weather."[134] Von Neumann's most interesting conclusion was that the most likely way to affect the weather and climate is the possible modification of the albedo of the earth. Thinking had moved toward the question of how might we change the weather at will. Von Neumann thought that the evidence so far was that nuclear explosions had only negligible effects on the weather, but that more theoretical and computer studies were needed, like the ones he and Jules Charney had initiated at Princeton.

The AEC was also concerned with ensuring an ample uranium supply for both the weapons and the reactor programs. Uranium had first been obtained from the Belgian Congo, from the Union of South Africa, and from Canada, after which the United States began a crash program to obtain uranium from domestic mines in Colorado. It was in the Colorado mines that the AEC, by refusing to entertain the additional expense of a proper ventilating system, generated many needless deaths from cancer among the miners.[135] The AEC had committed itself to paying a relatively low but guaranteed price to the mine owners for uranium, so low that the government was in no position to ask them to install safety features. Von Neumann objected to the fixed price for uranium for a reason entirely unrelated to the miners' health but nevertheless of interest in indicating his views on price fixing and the market economy. In a memorandum to his fellow commissioners he wrote,

A fixed-price is a powerful means to stimulate domestic prospecting and production—which are very important. . . . Nevertheless I am worried about the present rule because it takes the uranium business entirely out of the market economy. I would suggest some system whereby the government offers to buy domestic uranium at a sliding price scale; in other words, where the purchase price decreases as the quantities increase.[136]

Von Neumann and Strauss, and Senator Clinton Anderson, were glad to encourage American companies to engage in business related to international nuclear technology. Von Neumann was much less concerned about secrecy or rigorous US control over materials sent to other (NATO) countries than some other people. He was quite agreeable to letting private industry handle it. When a question arose whether the

United States should supply unclassified materials like zir-
conium to foreign atomic energy programs, he waved aside
anxieties about "security" and emphasized economic considerations, as for example that US industry, rather than British industry, should dominate the market, or that a particular
material might become available to the AEC at reduced
prices.[137]

Von Neumann's remarks on a large variety of issues at the
AEC distinguish themselves by his characteristic style. Von
Neumann's practice of at least taking note of all the varied
considerations that enter into an issue gave an impression of
objectivity that disarmed the reader or listener; when he then
proceeded to articulate his own position, he would implicitly
reveal his value premises by selecting some of the considerations as important and dismissing others. One could disagree radically, if one's values or one's expectation of the
likely consequences of particular actions differed from von
Neumann's. One could also perceive the issue in fundamentally different terms and disagree on that basis. The style
permits rational discussion and largely avoids hypocrisy,
cant, and personal attacks. The premises of some of the zealous cold warriors may well have been derived from private
fears or fantasies, but von Neumann's reasoning, whatever his
premises, was usually lucid. Thus, for example, on the subject
of the necessity of a high rate of nuclear testing and the
physical injuries resulting from fallout, he might concede the
likelihood that many thousands of cases of cancer might result from weapons tests, but he would balance against this
injury the risk of America's losing some of its advantage over
Russia in the weapons race. To his mind, the latter "risk" increased the probability of a Soviet nuclear attack on the
United States. Consequently he would argue in favor of the
human cost exacted by fallout as a justifiable means of reducing the likelihood of attack.

Commissioner Willard Libby became the chief AEC
spokesman on the subject of the biological effects of fallout
from weapons tests, especially those that had taken place
in Nevada in the years 1953–1955. Von Neumann, independently going over the official AEC report on fallout from

286

*John von
Neumann
and
Norbert
Wiener*

the Nevada tests,[138] commended Libby for his exposition yet suggested that he may in one instance have very much *over-estimated* the exposure to radiation of the US population from nuclear tests.[139] Von Neumann thought the estimate should be revised downward by a factor of at least ten. The pronouncements on fallout by the AEC, with its deeply vested interest in weapons tests, had begun to be challenged by independent scientists, particularly Ralph Lapp, Linus Pauling, and Barry Commoner, who asserted that Libby *underestimated* the biological impact of fallout, or at least had presented the information in a highly misleading way.[140] By late 1955, the AEC had begun to lose its authority within the scientific community on the subject of fallout and was coming to be recognized more and more as a promoter of weapons tests whose primary interest was to assuage public fears rather than to prevent damage to public health. President Eisenhower, receiving his technical information via Lewis Strauss, indicated his ignorance (or complicity) when he asserted that "the continuance of the present rate of H-bomb testing, by the most sober and responsible scientific judgment . . . does not imperil the health of humanity."[141] A further threat to the AEC's authority and credibility on the subject of fallout came from the United Nations, which was planning a study of the effects of radioactivity on human health and safety. In his official capacity as commissioner, von Neumann expressed great concern about this prospect, writing, "I wish to be on record for my own part that I always opposed, and I am now opposed to any measure which would directly or indirectly endorse the setting up of a UN study group in this area. It is my conviction that such a step would be contrary to the interests of the United States."[142]

Does the facile, controversial claim of "national interest" provide a cogent argument exonerating the Atomic Energy commissioners, who understood the likely risks and injuries, from complicity in radiation-induced deaths and injuries? The failure to provide adequate measures for uranium miners surely constituted a form of criminal negligence, not to mention the public health menace of generating large amounts of cancer-producing fallout. But even had the effort been made, the practical achievement of legal accountability in connection with AEC policies would have been impeded in a number of ways over and above the protection the Establishment provides for its own. Like many government decisions, these instances involved the active collusion and passive acquiescence of a number of individuals, and it is difficult to define individual responsibility clearly. More specific obstacles to legal process in AEC decisions were provided by the secrecy hiding much of the AEC's activity, the highly technical nature of the assessment of risks and possible injuries, the long gestation period before the consequences of radioactive contamination become clearly manifest, and the difficulty of ascertaining the "cause" of cancer in individual instance.

Von Neumann was an ardent nationalist, and his general attitude toward international bodies such as the UN, international control of nuclear weapons, and disarmament talks was one of "hard-boiled" contempt, expressed in its most polite form as skepticism.[143] Von Neumann's consistent opposition to disarmament can be documented for the whole period of his active promotion of the superbomb, the ICBM, and throughout the period when he was active as a commissioner, although theoretically he acknowledged the consequences of the armament race in a letter to Strauss in November 1951:

The preliminaries of war are to some extent a mutually self-excitatory process, where the actions of either side stimulate the actions of the other side. These then react back on the first side and cause him to go further than he did "one round earlier," etc. . . . each one must systematically interpret the other's reactions to his aggression, and this, after several rounds of amplification, finally leads to "total" conflict . . . I think, in particular, that the USA-USSR conflict will probably lead to an armed "total" collision, and that a maximum rate of armament is therefore imperative.[144]

Throughout his tenure on the GAC, von Neumann consistently poured cold water on any other member's outburst of hope or enthusiasm for disarmament or international control. Although by the mid-1950s he was no longer speaking about the upcoming "total collision" with the Soviet Union, he never modified his opinion that "a maximum rate of armament is im-

288
John von
Neumann
and
Norbert
Wiener

perative." Nor did he ever put any stock whatsoever in disarmament efforts.

Disarmament negotiations, with the exception of the 1946–47 period of the Baruch plan, had been treated until 1955 as an operation irrelevant to the mainstream of government action. "The negotiators themselves never had any access to the White House nor even to the Secretary of State. . . . The United States negotiators had never received adequate support either in terms of finances or personnel to carry out their duties effectively."[145] The dominant policy advisers, especially Dulles but also Strauss, banking on the threat of massive retaliation, continued to stockpile nuclear weapons. President Eisenhower's impulses toward a durable peace based on disarmament had, with the aid of his advisers, been channeled into creating a public-relations image of an Eisenhower who favored peace, which meant tolerating a group of international disarmament negotiators who were not seriously expected to achieve the agreements they were supposedly working toward. The unexpected success of the negotiators and Soviet acquiescence to the Western nations' proposals generated a crisis, since Dulles's whole foreign policy was to rely on terror as a guarantor of peace.

In spring 1955 Eisenhower began a reorganization to upgrade the importance of the disarmament issue.[146] In particular he created a new position of cabinet rank—that is, ostensibly the same as Strauss's and Dulles's—with the title of special assistant to the president on disarmament and appointed Harold Stassen, a liberal Republican and former governor of Minnesota, to the post. Stassen and Dulles had a long-standing difference of view concerning nuclear armament: "Where Dulles thought in terms of the deterrent use of power, Stassen looked toward breaking the nuclear deadlock and reducing the arms race."[147] Stassen, who had been a delegate to the charter session of the United Nations in 1945, was concerned with making the UN a functional locus for achieving nuclear disarmament and resolving international disputes. However, in 1955 the need was for resolution of the differences between Dulles and Stassen so as to achieve a consistent foreign policy for the United States.

Stassen's position was soon undercut when on August 5, 1955, Eisenhower set up the President's Special Committee on Disarmament, composed of senior representatives (assistant secretaries or the equivalent) of the departments of State, Defense, and Justice, the CIA, the AEC, the Joint Chiefs of Staff, and the US Information Agency.[148] As Barnet points out, "Since many of these departments were also represented on the National Security Council, which was to have the final word on disarmament policy, Mr. Stassen found himself more in the position of a coordinator than an authoritative originator of policy."[149] Lewis Strauss, who himself often attended National Security Council meetings, "after having discussed this with Governor Stassen at considerable length at dinner last night," on August 26 urged von Neumann to represent the AEC formally on the disarmament committee:

My thought is . . . that you might take on the representation with the idea that you would only attend the first meeting or two and thereupon assign your deputy who would be either Paul Fine or . . . someone else you might select. This will have an advantage, from Stassen's point of view, of enhancing his position by the caliber and standing of his group. It would also improve our status with him, and would require one, or at most two appearances by you at the United Nations, unless you felt that the affair showed enough promise of success to warrant more of your own time.[150]

Von Neumann consented and (ironically, in view of his lack of support for disarmament) became the offical AEC member of the President's Special Committee on Disarmament. Paul Fine became his alternate.[151]

Before the President's Disarmament Committee had had an opportunity to work on the issues, the UN Subcommittee on Disarmament reconvened, now with Stassen himself representing the United States, and on September 6 it fell to him to inform the subcommittee that the "moment of hope" had ended, and that the United States had now "placed a reservation" upon all "earlier substantive positions" it had taken on questions pertaining to levels of armament.[152] While withdrawing from his earlier position Stassen was not yet able to spell out what the new American position would be, since it

290

*John von
Neumann
and
Norbert
Wiener*

would depend on the outcome of the conflict between Stassen on the one hand and Dulles, Strauss, and their cohorts on the other. In a letter of February 10, 1956, Strauss reported to John von Neumann, who was at that time not well, "Had a crowded day—with our boss [Eisenhower] in the morning, and with Dulles and Stassen this afternoon. . . . The Dulles and Stassen talks went well (from your and my viewpoint). I won't say more—keep the letter unclassified."

In March, Stassen began to articulate to the UN subcommittee the American viewpoint to the effect that the terror induced by the threat of massive retaliation "constituted an atomic shield against aggression" and that "it is an important safeguard of peace . . . a powerful deterrent of war."[153] Soon thereafter Stassen announced an unambiguous US opposition to major reductions in armaments, because "these reductions would increase the danger of the outbreak of war."[154] Clearly the Dulles-Strauss forces had overwhelmed the voices favoring disarmament within the American government, and Stassen, as official US delegate, had to present a policy with which he strongly disagreed to the United Nations Disarmament Commission.

For his loyal and extremely competent service to the US government and its unfortunate nuclear weapons policies, von Neumann received official recognition. Lewis Strauss proposed to Eisenhower the creation of the $50,000 "Fermi Award" and the "Fermi Gold Medal," recommending von Neumann as the first recipient.[155] The president himself presented the medal with the award to the proud John von Neumann.

12 POLITICAL POWER, HUMAN NATURE, AND SOCIETY

There is no reason, abstractly speaking, why a human society cannot be imagined which would be less concerned with increasing the potential of its machines or of its productive energies than with assuring to everyone the minimum requirements for a decent existence. This was the teaching of Rousseau. But economic history has been quite different. The will to power and wealth has carried the day. . . .
—Raymond Aron

Neither Wiener nor von Neumann was a hypocrite. Each man's political actions and theories of society formed a coherent unit. Both escaped the political ambivalence that afflicted some of their contemporaries, especially J. Robert Oppenheimer.[1] Although von Neumann was more a political animal, more the worldly man of action than Wiener, both were first and foremost thinkers. For both of them political activity was joined to an intellectual structure within which they perceived society. Although the world of politics seems far removed from mathematics, both Wiener and von Neumann based their conceptual frameworks for describing society on some mathematical ideas. For von Neumann game theory was the cornerstone; for Wiener it was cybernetics. They came to disagree about the choice of models for evaluating political action, but that disagreement was only the tip of the iceberg. They differed in their ideas about the nature of human nature, about what constitutes wisdom, and in their attitudes toward political power and its uses.

The theory of games von Neumann worked out in 1928 has been described, a theory connecting various branches of

292

John von
Neumann
and
Norbert
Wiener

mathematics and involving the difficult-to-prove minimax theorem. In itself it was apolitical, an academic exercise whose
execution most mathematicians would characterize in
terms of its elegance. But game theory is two-faced, as science generally is. Its other face—its application to economic,
political, and military objectives, first carried out in the
1940s—is highly value-laden. Applied first only to economics, game theory was eventually used by US government
strategists as a foreign-policy instrument, and a technique for
choosing parlor-game strategies proved all too easily adapt-
able to choosing strategies for international thermonuclear
conflict.

Competent gamblers have found game theory of little interest, since the complexity and subtlety even of a game like
poker exceed its capacity for guidance. Serious application
of game-theoretic models to strategic thinking about thermonuclear conflict is a political phenomenon that deserves
some attention. (Again we call attention to language: Is the
word *game* not just another one of the euphemisms that
obscure our perception of the horrors involved?)

At Princeton in 1940, von Neumann had begun a collaboration with the Austrian economist Oskar Morgenstern whose
purpose was to adapt game theory to problems in economics.
Morgenstern had had considerable contact with the Vienna
Circle of philosophers and had brought with him to America a
logical-positivistic approach to the social sciences. Long
after 1940 he recalled the experience of working together with
von Neumann virtually every day for several months as "a
wonderful time, but terribly strenuous."[2] Speaking English
with an Austrian accent, Morgenstern told me of their meeting
early for breakfast at Princeton's Nassau Inn while Klara von
Neumann was still sleeping, then spending the morning sitting side by side, taking turns writing. He reminisced about
their long walks together along the seashore, talking, and
their joint study, for pleasure, of ancient Greek history.

In their book, which appeared in 1944, von Neumann and
Morgenstern stated their belief that economics would develop
into a rigorous mathematical science, just as physics had—
except that economics was still in an early stage of devel-

opment—perhaps, they suggested, analogous to that of physics in the sixteenth century. The theory of games was the first serious step toward a comprehensive mathematical economics, which would stimulate empirical research to test and possibly modify its theories. The bold optimism of their first chapter assumes an unending and unidirectional scientific progress toward increased mechanization and a faith that careful collection of empirical data and new mathematical discoveries will "produce decisive success" in the effort to create a rigorous science of social phenomena.[3]

Game theory purports not only to describe social phenomena but also to provide a technique for making "good" decisions from the point of view of one's own self-interest when faced with other "players" whose interests may conflict with one's own. Consequently, if the von Neumann–Morgenstern belief were indeed justified, the elements of competitiveness and aggressiveness in modern society would be contained within the formalism of a strictly axiomatic mathematical theory, and the problem of "wise" choice, of "rational" action, would be reduced to a matter of calculation. It was pointed out earlier how von Neumann apparently reduced to a mathematical formalism the epistemological enigma surrounding quantum theory; he sought—and came as close as anyone—to prove the completeness and self-consistency of mathematics itself by formal means. Now economic and political decisions were to be imbedded in the formal structure of axiomatic set theory.

In the late 1940s von Neumann and Morgenstern were not alone in their optimism about prospects for mathematizing (that is, mechanizing) the social sciences; many members of the scholarly community enthusiastically embraced game theory as a tool for the social sciences.[4] Within a few years, however, disillusionment with the difficulties and limitations inherent in game theory set in among social scientists,[5] although mathematicians continued to find it of considerable interest.

Besides the stated belief in the possibility of mathematizing the social sciences, the von Neumann–Morgenstern theory of economic behavior, like every theory in economics, was prem-

294
*John von
Neumann
and
Norbert
Wiener*

ised on certain values. Thorstein Veblen and Gunnar Myrdal have taught us that the significance of an economic theory cannot be correctly understood without attention to its value premises, which are rarely contained in its explicit assumptions but may be implicit in the very form of the questions it poses. They usually reflect "tradition for one thing, but even more decisively the interests and prejudices dominating their [the theorists'] environment," as well as the individual theorist's personality. These implicit values are not only highly consequential for the theory itself; they may affect the well-being of a whole society if a particular economic theory gains enough political favor to be applied on a wide scale.

By comparison, the values and individual attitudes that underlay the creation of quantum theory conditioned the form of that theory far less, as illustrated by diverging philosophical interpretations, all consistent with the same theory. Quantum theory's most immediate impact on the broader society was to remind the reading public of some profound and fascinating issues of ontology and epistemology. However, the content of the theory eventually did make possible some technological innovations whose avid development reflected the anticipation of material rewards on the part of inventors and manufacturers, if not physicists. Thus, in the end the uses to which quantum theory was put also reflected the traditions and interests dominating the larger society.

The von Neumann–Morgenstern model of economic behavior starts with the explicit postulate that each person (or coalition) lists the alternative events that might occur and makes an unambiguous list of his relative preferences for these various outcomes of his and the other players' alternative strategies. The objective of applying game theory is then to prescribe or describe the course of action that most favors (in the precise minimax sense) the preferred outcomes—that is, the course of action most likely to bring about desired results insofar as compatible with a minimum of risk. Game theory differs from the related mathematical decision theory in that other players and their options are explicitly taken into account. In competitive situations one's own "good" strategies will in particular be guided by the conservative norm of ensuring minimum damage to one's own interests in the anticipation that competing players, so as to secure their own advantage, will make full use of their capacity to harm. Thus the question on which game theory focuses is, What is the best-calculated "strategy" or policy to follow when confronted with other, equally cunning players, so as to achieve the largest gain or one's interests that caution permits? The conceptual framework containing this question is supposed to be that of operationally measurable subjective probabilities and utility functions and the strictly mechanical process of game-theoretic formalism.

295
Political
Power,
Human
Nature,
and
Society

The ideal (least ambiguous) situation from the point of view of the theory is that of two parties with diametrically opposed interests. Von Neumann and Morgenstern, in their theory of economic behavior, assume "that the aim of all participants in the economic system, consumers as well as entrepreneurs, is money, or equivalently a single monetary commodity. This is supposed to be . . . identical, even in the quantitative sense, with whatever 'satisfaction' or 'utility' is desired by each participant."[6] The von Neumann–Morgenstern theory "is thoroughly static" in the sense that the objective of each player remains rigidly fixed throughout the economic conflict.[7] Also implicit is the fixed nature of the interest each player pursues throughout the period when a policy is in effect, although in practice the identity of a player can shift from one group, individual, or department within an organization to another.[8] Moreover, the available options remain fixed throughout the course of the game, as well as the rules of economic activity and the preferences of each player. This rigidity is built into the theory primarily so as to make it mathematically tractable. The real world has more fluidity and variability. Von Neumann and Morgenstern envisaged the possibility that their static theory might be a first step toward a more complete, more dynamic theory, but their intent was nevertheless "to establish satisfactorily . . . that the typical problems of economic behavior become strictly identical with the mathematical notions of suitable games of strategy."[9]

Game theory portrays a world of people relentlessly and ruthlessly but with intelligence and calculation pursuing what

296

*John von
Neumann
and
Norbert
Wiener*

each perceives to be his own interest. Other "players" are seen as enemies, competitors, or collaborators, depending on the degree of mutual compatibility of their objectives. This is the norm in game theory. The harshness of this Hobbesian picture of human behavior is repugnant to many, but von Neumann would much rather err on the side of mistrust and suspicion than be caught in wishful thinking about the nature of people and society. His temperament was conditioned by the harsh political realities of his Hungarian experience. The recommended style of "playing the economic game," the emphasis on caution, on calculation of expected consequences, the whole utilitarian emphasis aptly expresses the characteristic ideals of the middle class in capitalist societies. It is nearly the antithesis of the values of, for example, creative subcultures ("Bohemians"),[10] and it is also at variance with the attitudes predominant among aristocratic classes and the poor.

I will examine three facets of the assumptions underlying the application of game theory to social science: the insistence of formal-logical structure, the relation of means to ends, and the implicit scarcity assumption.

The abstract logical structure of the theory—which must be free of formal contradictions—was of course its essential characteristic from von Neumann's point of view. However, the price paid by insisting on such a structure, as Myrdal pointed out,[11] is a simplistic and incomplete relation to reality. For example, the assumption of "economic man," with his utility function, as well as the aforementioned rigidity, is inconsistent with psychological and sociological knowledge about human beings and their circumstances. According to Myrdal, the systematic exclusion of social realities which in fact come into play from abstract theories of economics such as game theory give the lie to the notion that such theories lead to a "scientific" economics, although a naive social scientist might be impressed by the mathematics. The indeterminacies characterizing the solution to those game-theoretic problems in which a measure of common interest among participants plays a role might well be taken as a signal that the language of set theory, utility functions, and mathematics generally is not the best language for discussing good courses of action.[12]

But suppose for a moment that the technical limitations of the theory could be overcome and the summa bonum of a perfect "game theory machine" could be constructed. One would only need to find the representative individual who could spell out the utility function and expectations of the particular "player-entity" (be it a government, a corporation, or a particular department). Technicians could elaborate all the strategic options, translate them into suitable quantitative terms, and feed all the information into the machine, which would then make the decision according to instructions. What pattern of decision making could be more congenial to a bureaucratic organization? The decision-making process would become almost completely depersonalized. Moreover, if one takes the language of the theorists literally, the machine will arrive at the "optimum" decision. No organization could ask for more. As Max Weber has pointed out, institutional structures enhance their stability by becoming independent of particular individuals, and consequently formal, mechanical processes for decision making are an asset for an institution that seeks a long life.[13] In this way personal responsibility for consequential decisions is also neatly diffused, if not avoided altogether. As in the case of those elixirs of immortality that killed many a Chinese emperor up to the eleventh century,[14] the efficacious instrument may be highly desirable, but to mistake pseudoscience for science is highly dangerous.

In game theory ends and means are neatly separated: the "outcomes" are judged by their desirability, while the "strategies" are judged solely by their efficaciousness in bringing about the desired ends. Myrdal rightly calls attention to the artificiality of that separation in economic theories and the consequent tendency to neglect the evaluation of strategies from any other point of view and to fail to take seriously the multitude of side effects that following a particular strategy entails.[15] In fact, outside the framework of game theory intrinsic merit attaches to certain kinds of behavior independent of the outcome. One would want to ask whether the behavior as-

298
*John von
Neumann
and
Norbert
Wiener*

sumed by game theory, including its single-minded purposiveness, accurately portrays existing norms, but such an inquiry into behavioral norms, descriptive or prescriptive, is beyond the scope of the von Neumann–Morgenstern treatise. To appraise existing norms would require examination of prevailing traditions and the nature of social institutions as well as individual psychology. And whether game-theoretic recommendations constitute "wise" choices is a philosophical inquiry.

However, the separation of ends and means, of strategies and outcomes, does permit us to characterize the sense in which game theory is related to human rationality and irrationality. The ends of each participant's actions are defined as the outcomes he most desires, or, more precisely, those he initially believes he would prefer. Of course, he might not like the actual outcomes once they were achieved, especially if unanticipated side effects had spoiled them. The game-theoretic formulation in economics seems to lend itself most naturally to the ideal of oligopolistic capitalism. However, for an economics primarily concerned with social justice, with a distribution of wealth to permit satisfaction of minimal human needs, and with humanitarian values generally, game theory is at best an extremely awkward formulation. Rationality enters into game theory through the method, not the purpose—it is only a narrow, technical, instrumental kind of rationality, which can be made to serve the most irrational ends.

The von Neumann–Morgenstern theory of economic behavior falls within the framework of the tradition of classical economics, even though it contains new formulations and techniques. In particular, the criteria for "optimizing" one's gain and risk rest on the traditional assumption of scarcity of resources and shares with other theories in classical economics the psychological underpinnings of such an assumption. One scholar who has examined the assumption of scarcity finds that it is not an empirical datum about resources and physical human needs. Instead it is an attitude characteristic of modern industrial society and a particular response to the human condition:

In the human situation means of production and need satisfaction are always scarce in relation to needs and ends which are unlimited and can never fully be satisfied. Therefore there is a continuous gap between means and ends . . . this idea as applied in economics is historically relative and culture-bound and represents the special orientation of industrial society toward economic activity and material need satisfaction. There is, however, a sense in which the scarcity principle is universally valid because it is rooted in the conditions under which human beings exist. Existential scarcity is caused by human finitude on the one hand, and on the human ability to transcend this finitude and the given existential condition through consciousness and thought on the other hand. . . . Human life is confronted with an allocation problem not only in respect to material means of production. The resources which are ultimately scarce are *Life, Time* and *Energy* because of human finitude, aging and mortality. . . . There is, however, a question whether allocation and economizing and decisions about preferences may not have to be made even with immortality and eternal youth as long as we are subject to the limitations of time and space; whatever our situations, we can actualize only limited desires here and now.[16]

The existential allocation problem obliges us to choose among alternative options. Translating game theory and utility-based decision theory to an approach to everyday life would make it a systematic, logical technique of selecting the options with the greatest likelihood of fulfilling as many of one's wishes and goals as possible within the framework of a finite and limited existence, of making the most of one's opportunities.

Von Neumann, whose combination of talents and opportunities multiplied his personal options, evolved a personal style of dealing with limitations. He liked to use game-theoretic concepts to describe practical situations in everyday life and at times was given to nearly instant, seemingly effortless calculations and evaluations of tactical situations based on a game-theoretic kind of thinking. Even some elements in his style of life as a middle-aged man suggest that it often contained a kind of natural maximization in the allocation of the scarce existential resources of life, time, and energy: a man who slept only four or five hours a night; who gave large

299
Political
Power,
Human
Nature,
and
Society

300
John von
Neumann
and
Norbert
Wiener

parties but did not let his role as host keep him from an hour or two of mathematics in his study; who traveled so extensively that he seemed to be several places at once, appearing at conferences for a few brief hours before moving on to something else; who always had numerous plans for the future.

Indeed a belief in a theory such as game theory is as much visceral as it is intellectual; it is a personality trait. L. J. Savage, an admirer of von Neumann and of his game theory, whom von Neumann at one point wanted as his assistant, used the von Neumann–Morgenstern utility-function approach to formulate a theory of "wise" choices in the face of uncertainty.[17] Perhaps more than any other mathematician in the 1940s and 1950s, Savage probed the philosophical underpinnings of the utility function approach. He told me of the reassuring experience of finding someone in these early years of the theory whose habitual thinking involved utility functions and the other concepts of decision theory Savage found congenial:

We are far apart geographically and far apart in our other scientific interests, but when we talk we know that we agree about almost everything. There is a system there that works. It's as though you had been taught in childhood what was supposed to be Albanian, but you never knew an Albanian and you weren't quite sure whether you'd been tricked. But if you got off the dock in Albania you would know. . . . It's like that. I know that it's cogent because it's in such absolute detailed agreement with Schlaifer.[18]

It appears, however, that while von Neumann may have at times been caught by his affinity for overly formalized game-theoretic thinking in very practical matters, especially in the late 1940s, at other times he clearly appreciated that the formal game-theoretic approach should not be taken too literally, for as he said in a luncheon speech on December 12, 1955, "The indications are that . . . the best that mechanization will do for a long time is to supply mechanical aids for decision-making while the process itself must remain human. The human intellect has many qualities for which no automatic approximation exists. The kind of logic involved, usually described by the word 'intuitive,' is such that we do not even have a decent description of it."[19] He thus tied the limitations of decision theory to those of automata theory, and he described the limits of devising models of the human brain in precisely similar terms. Other practitioners of applied game theory, however, did not share von Neumann's caution in this regard.

301
Political
Power,
Human
Nature,
and
Society

Von Neumann's economic orientation seems clear. He identifies with the entrepreneur and the capitalist. He consistently favors economic expansion and industrial innovation. He seems not to object to the concentration of power in monopolies and oligopolies, so long as these organizations do not block technical innovation itself. Von Neumann showed no evidence of sympathy with economic theories that emphasize the redistribution of wealth and social justice. He fulfilled the Marxist prediction that the bias of most people in matters of economics will express their own class interest. In von Neumann's life that class interest was imprinted during his years in Hungary, and he apparently never outgrew it.

Unlike von Neumann, Norbert Wiener described his ideas and many of his views on society in books written for the general public. He ranged in his writings over all the social "sciences" from politics and economics to psychology, but from a unified point of view. Von Neumann had his Morgenstern; Wiener talked in particular with political theorist Karl Deutsch, on occasion collaborating with him, as well as with social theorists Gregory Bateson and Lawrence K. Frank, each of whom elaborated Wiener's ideas further within his particular discipline. Wiener also held discussions with linguists, psychologists, and psychiatrists when his own tentative ideas impinged on their special fields. Social scientists have subsequently adapted the cybernetic mode of analysis Wiener developed to various fields, in particular psychiatry, linguistics, social anthropology, political science, and industrial management.[20]

Wiener's synthesis was based on his primary focus on the "communications" aspect of all social phenomena and on the aspect of "control."[21] He believed this to be a particularly appropriate focus for social science in the second half of the twentieth century because it appeared to him—partly on the

302
John von
Neumann
and
Norbert
Wiener

basis of the state of technology in the late 1940s—that communication and control were likely to be central concerns of society. He could give the concepts describing communication and control precise (albeit statistical) mathematical meanings in electronic systems, but in describing social phenomena Wiener introduced them merely as useful concepts, each of which could be further qualified by descriptive adjectives, such as "two-way" communication or "top-down" communication. From physiology he took the concept of homeostasis (self-regulation), which refers to the biological mechanisms giving stability and viability to an organism, and applied it to social and political systems. He also introduced the word "feedback" from control engineering into the general language, a concept more general than homeostasis, and again provided it with qualifying adjectives. From conventional descriptions of human behavior he selected a few concepts, namely, learning, memory, flexibility, and purpose, and applied them more generally to "systems." From Bertrand Russell's theory of logical types he derived the notion that in human decisions and communication several levels may be operative at once and may lead to essential paradoxes which define the limits of logics and machines.

These concepts were supplemented by an articulate set of values. Wiener's ideas were implicitly "organismic," in that elements interacting through mechanisms of communication and control (generally not in a deterministic way, but rather in one that leaves room for various contingencies) would lead to the emergence of patterns of organization. Since Wiener tended to include, besides people, other organisms and the environment, man-made or not, in his purview, his outlook can be characterized as ecological. The social scientist, Wiener emphasized, is part of the overall system of control and communication he is studying and cannot in principle be isolated from it. The paradigm for a decision maker is not the Neumannian game player but the "steersman" (recall the derivation of the word "cybernetics") who relies not only on skill and experience to achieve his goal but especially on constant feedback relating his present situation to it. Using this original conceptual framework, Wiener found himself able to articulate his views and his idea of wise choices in clear language, and also to appraise other social theories, such as psycho-analysis, Marxist economics, capitalist economics, and game theory.

The cybernetic mode of analyzing elements of society traces the patterns of communication and control, the self-regulating and destabilizing elements, the sources of learning and flexibility, the freedom to realize human possibilities, and the interrelation among different purposes. It is particularly suited to exposing the inhuman use of human beings. It reveals inadequate communication links, such as isolation, deception, or manipulative control of communication; the limitations on available mechanisms for influencing one's own destiny in a class society; and inadequate feedback and regulatory mechanisms that result in explosive changes or periodic instabilities in a society. It is a normative mode of analysis, because it implicitly recommends corrections of these defects and inadequacies in favor of humane modes of social interaction.

One implication of Wiener's model is that intact homeo-static mechanisms are highly valuable. The steersman of a ship should receive correct information about storms, rocks, unexpected leaks in the hull, or other dangers, and should (using the ship's equipment, pattern of social organization, and resources) function effectively to keep the ship afloat and intact and maintain the well-being of its crew. Homeostasis is a conservative subsidiary goal to the ship's reaching its destination; it is contrary to a suicidal goal. Thus, good visibility, accurate information, knowledge, resourcefulness, experience, cooperation, and responsiveness among crew members all come into play in heading off danger. It is desirable that the ship reach its destination, and a deviation from the course—either because the steersman is systematically misinformed, or because he receives true information but responds to it in a systematically perverse way, or even because the steering mechanism is not working properly—would be seen as a malignant process ("positive" feedback) by Wiener, unless from a more inclusive perspective this misdirection served a desired purpose.

304
John von
Neumann
and
Norbert
Wiener

The title of Wiener's book, *The Human Use of Human Beings,* indicates his underlying values. He characterizes himself "as a participant in a liberal outlook" with a long tradition and describes

> what I myself and those about me consider necessary for the existence of justice. The best words to express these requirements are the words of the French Revolution: Liberté, Egalité, Fraternité. These mean: the liberty of each human being to develop in his freedom the full measure of the human possibilities embodied in him; the equality by which what is just for A and B remains just when the positions of A and B are interchanged; and a good will between man and man that knows no limits short of humanity. These great principles of justice mean and demand that no person, by virtue of the personal strength of his position, shall enforce a sharp bargain by duress.[22]

Humanitarian, democratic, and libertarian values inform Wiener's social theory throughout, even though it is often expressed in an engineer's metaphors. The emotional experiences of Wiener's youth had helped to condition his outlook. Not only had he imbibed some of his father's Tolstoyan and utopian-socialist (but non-Marxist) values, but he had himself been bruised and had suffered in his early years, the memory of which encouraged him to be generous rather than narrow. And the political realities of his youth had been relatively benign, in an America far more liberal than von Neumann's Hungary.

Wiener discussed with Gregory Bateson the compatibility of psychoanalytic practice with cybernetics and communication theory.[23] Wiener suggested that in communication systems the crucial concept is information rather than energy, and that consequently the emphasis on libido in Freudian theory was misplaced. In 1947 Wiener conjectured that, just as in the malfunctioning of a computer, the physical basis of the so-called "functional disorders" of psychiatry may have to do with "instructions," "messages," "programs," and "memory," but he concluded that the techniques of the psychoanalyst were perfectly consistent with the point of view of cybernetics.[24] Bateson, aware that in psychoanalytic practice the concern was in fact often with communications, in spite of Freud's libido theory, carried the matter further and began to develop a description of clinical practice using the language of cybernetics.[25] This, with the help of some more discussion with Wiener, especially about Russellian paradox, led to Bateson's theory of the double bind in schizophrenia, and later to a cybernetic theory of alcoholism based on an examination of the successful therapeutic practices of Alcoholics Anonymous.[26] The focus on communications also contributed to making Bateson and his coworkers pioneers in advocating family therapy for some types of cases, rather than the traditional individual therapy. Thus cybernetics and communication theory gave an impetus to new directions in psychiatry.

305
Political
Power,
Human
Nature,
and
Society

Wiener did not expect social theory in general to be mathematical or to have the kind of solid empirical base one finds in the natural sciences. While he spoke of Bateson's "valiant work in attempting to bring psychoanalytic processes under the heading of cybernetics," he also noted that this work "is and must be sketchy" because in psychology the elementary processes themselves are only incompletely known.[27] It was Wiener's philosophical standpoint that a fully materialistic explanation of psychiatric disorders would in principle be possible if concepts like "message" were included in the explanation. He was skeptical of sociological and economic predictions because statistical runs in the social sciences are usually too short and because he thought the observations on which they were based were strongly conditioned by the unavoidable interaction between the social scientist and his subject.[28] Von Neumann sharply disagreed with Wiener and did not accept the argument concerning statistical runs.[29] According to Wiener, however, "There is much we must leave, whether we like it or not, to the 'un-scientific,' narrative method of the professional historian." Wiener found himself obliged to dispel the expectation of some of those engaged in the study of social phenomena that cybernetics and communication theory could somehow render their field of study

306
John von
Neumann
and
Norbert
Wiener

"scientific" or even provide mathematical exactness. He did not regard Marx's sociology and economics as scientific, either.

One element of psychology is choosing one action over another. Conceptualizing this process can lead to defining an idea of "good" versus "bad" choices—in effect, wisdom. Wiener, although his information theory had accustomed him to thinking of a mathematical space of alternative options and of selecting among them, and although he had been an innovator in the mathematical theory of maximization or minimization of functions, approached the matter of human choice and wisdom in terms quite different from the von Neumann–Morgenstern utility function or game theory. He emphasized personal integrity and decency more than utilitarian values, and the need for any theory to recognize the "vast range of probability that characterizes the human situation."[30] It is clear from Wiener's autobiographical writings that he saw the process of imaginatively identifying alternatives and choosing among them (at least in personally important matters) as an assertion of one's identity, and the act of selecting among the vast range of possibilities as giving character and meaning to individual human life. Wiener's notion of choice is most clear in his later writings: the range of possibilities in a particular situation can be viewed, if one insists on being formal about it, from a variety of "levels" analogous to the hierarchy of Russell's theory of logical types.[31] Valuations will differ from one level to the next, often flatly contradicting each other. Making a wise choice necessitates being aware of as many of these levels of reality as possible and requires the essentially human process of adjudicating among them, in spite of paradoxical and contradictory elements. However, such a perception is tantamount to the admission that the language of logic or mathematics is poorly suited to the description of very significant decisions.

Although philosophically more a bold romantic than a calculating utilitarian, Wiener admired game theory as a brilliant mathematical model capable of describing many of the patterns of behavior governing high-level politics and business. He himself lacked the talents of a good Neumannian game player, but at the same time he disliked these patterns intensely and viewed them as both destructive and immoral:

307
Political
Power,
Human
Nature,
and
Society

In many cases, where there are three players, and in the overwhelming majority of cases, when the number of players is large, the result is one of extreme indeterminacy and instability. The individual players are compelled by their own cupidity to form coalitions; but these coalitions do not generally establish themselves in any single, determinate way, and usually terminate in a welter of betrayal, turncoatism, and deception, which is only too true a picture of the higher business life, or the closely related lives of politics, diplomacy, and war. In the long run, even the most brilliant and unprincipled huckster must expect ruin; but let the hucksters become tired of this, and agree to live in peace with one another, and the great rewards are reserved for the one who watches for an opportune time to break his agreement and betray his companions. There is no homeostasis whatever. We are involved in the business cycles of boom and failure, in the successions of dictatorship and revolution, in the wars which everyone loses, which are so real a feature of modern times.[32]

That was in 1947. Two years later, Wiener, alarmed by the possibility that game theory might be used by governments in an entirely mechanical way, expressed concern that the software of game theory might become part of a comprehensive government apparatus for military and political domination. He urged that "the anthropologist and the philosopher" be consulted, lest we fail to evaluate properly the human purposes involved.[33] Wiener talked with political theorist Karl Deutsch and corresponded with Gregory Bateson. Bateson provided substance to Wiener's concerns. He wrote,

What applications of the theory of games do, is to reinforce the players' acceptance of the rules and competitive premises, and therefore make it more and more difficult for the players to conceive that there might be other ways of meeting and dealing with each other. . . . The theory may be "static" within itself, but its use propagates changes, and I suspect that the long term changes so propagated are in a paranoidal direction and odious. I am thinking not only of the propagation of the premises of distrust which are built into the von Neumann model *ex hypothesi*, but also of the more abstract premise that human nature is unchangeable. This premise . . .

308
John von
Neumann
and
Norbert
Wiener

is the reflection or corollary of the fact that the original theory was set up only to describe the games in which the rules are unchanging and the psychological characters of the players are fixed *ex hypothesi.* I know as an anthropologist that the "rules" of the cultural game are not constant; that the psychology of the players is not fixed; and even that the psychology can be at times out of step with the rules.[34]

He adds, incidentally, that "Von Neumann's 'players' differ profoundly from people and mammals in that those robots totally lack humor and are totally unable to 'play' (in the sense in which the word is applied to kittens and puppies)." Bateson urged Wiener to take an active interest in a critique of game theory, as only one versed in mathematics could do authoritatively.

Referring to the arms race, and in particular the escalation from atomic to hydrogen weapons, Wiener had written at an earlier time, "This military version of 'Another little drink won't hurt us' shares much of the unconvincingness of the alcoholic assertion on which it is patterned." His logic and his recommendation at that time were straightforward: "Whether we like it or not, the Russians and ourselves are living on the same planet. . . . Fundamentally, they can have no more desire than we to celebrate a nominal victory by a universal funeral pyre of both sides. They share with us a healthy hope for a longer span of life on earth." This common ground, Wiener thought in 1950, is a realistic basis for negotiating with Russia so as to make a nuclear holocaust an impossibility. But, Wiener noted—at this time the anticommunist witch hunts in the United States were at their height—"even in peacetime, after the browbeating they have received" atomic scientists are loath to "give effective but unpopular advice to the generals." Among the obstacles Wiener saw was that "we are dominated by a rigid propaganda which makes the destruction of Russia appear more important than our own survival" and that "the very idea of temporizing with Russia in any matter has been made so hateful to the people of the United States that for some time now a statesman who ventures to suggest it has become likely to find his authority cut off at home." Wiener, having no authority to lose and seeking no favors from the government, could say these things.[35]

Whereas von Neumann had assumed an instrumental rationality on the part of the Soviet government for the relatively short-term objective of competing militarily with the United States, Wiener assumed that the USSR under Stalin would be rational about the long-term objective it shared with the United States of avoiding catastrophe as well as the wasteful production of weapons one could never use, if the United States were to adopt a "humane but realistic attitude."

However, the arms race moved relentlessly forward. By 1954 Wiener had realized that the Neumannian game player is not so clever after all, because he fails completely to take advantage of the psychological characteristics of his opponent.[36] He argued that policies based on game theory do not provide a good strategy for winning in a conflict.

It was not until December 27, 1959, when von Neumann was no longer alive and the general anticommunist hysteria had eased, that Wiener in his lecture to the American Association for the Advancement of Science on "Some Moral and Technical Consequences of Automation" gave a full-fledged critique of mechanical means for making decisions and of a mechanized Neumannian game player in particular. The general topic of the lecture was the relation between humans and automated mechanical devices, but its particular and occasionally explicit application to the cold war could not have escaped any listener among Wiener's audience.

Wiener begins his analysis by pointing to the inapplicability, on several counts, of the Neumannian theory of games to military decisions. Illustrating his point with examples from military history and from the patterns of fighting between animals of different species, Wiener argued that the winning tactics and strategy in war are not the minimax solutions of a game-theoretic model but more often the ones that correctly appraise the rigidities, mental limitations, and psychological characteristics of the opposing party on the basis of previous encounters or information about his habits. By contrast, game-theoretic strategies are based on the conservative as-

309
Political
Power,
Human
Nature,
and
Society

310
John von
Neumann
and
Norbert
Wiener

sumption that the opponent is a totally ruthless and intelligent strategist, an assumption that tends to exaggerate his power. Wiener's second reason for the inapplicability of game theory to military decisions was that one must deal with several levels—tactics, strategy, and general considerations—that occupy different time spans. Generally, recommendations based on different levels will contradict each other, so that—in a manner analogous to the paradoxes in the theory of Russell's logical types—a game-theoretic "search for the best policy under all levels of sophistication is a futile one and must lead to nothing but confusion."

In a policy decision produced by a computer, one also encounters various levels: the program determining the moves, the program used to determine which program to use, and so on. Applying these ideas to nuclear policy, Wiener notes that there is no actual nuclear war experience to act as a guide, but only the necessarily less realistic war games. Moreover, objectives may easily be stated on too low a level of generality and then automated, so that "if the rules for victory in a war game do not correspond to what we actually wish for our country, it is more than likely that such a machine may produce a policy which would win a nominal victory on points at the cost of every interest we have at heart, even that of national survival." Do we know how to spell out for the computer ahead of time what we actually wish for our country, our species, and our planet, so that in every contingency the computer's decision reflects these wishes?

Wiener points out that if we examine the patterns of communication and control, it is clear that the machines on which we depend are themselves a source of communication and control and in practice are in some respects *not* subject to human interference. Computers digesting various complex items of information may recommend to the president that nuclear-tipped missiles be launched immediately in self-defense. There is no time to examine and rehash the computer's logic in detail, to understand fully the computer's meaning, to check carefully for possible miscommunication between the computer and the source of its input or between it and the recipient of its output. In the end the president is confronted with an insoluble decision problem.

311
Political
Power,
Human
Nature,
and
Society

To Wiener the primary enemy was never Russia but rather the inhuman use of human beings, and for him this included exploitation, imposition of rigidity, and absence of feedback and honest two-way communication in the social organization. Among the hazards he included the possibility of highly centralized, technocratic governments, in which "political leaders may attempt to control their populations . . . through political techniques as narrow and indifferent to human possibility as if they had, in fact, been conceived mechanically."[37] According to Wiener's analysis, any society in which the competitive drives for money or power are given full sway or assigned the highest value is one in which "there is no homeostasis whatsoever," that is, in objective terms it is disruptive of community and stymies individual human possibilities. He tended to be skeptical concerning the likelihood of genuinely favorable developments: "Who is to assure us that ruthless power will not find its way back into the hands of those most avid for it?" he asked.[38]

The totalitarian and centralized government of the Soviet Union was unsatisfactory from a cybernetic perspective.[39] Yet "a society like ours, avowedly based on buying and selling," is antihomeostatic.[40] Dirk Struik described Wiener's attitude fairly when he wrote that Wiener's "individualism, caught in the world struggle between capitalism and socialism . . . led him to take a 'plague on both your houses' position. . . ."[41] As to American society, Wiener thought "the average man is quite reasonably intelligent concerning subjects which come to his direct attention and quite reasonably altruistic in matters of public benefit or private suffering which are brought before his own eyes."[42] Wiener was democratic in outlook, opposed to the concentrated power of economic or political elites. Since, however, in American society the means of communication are severely constricted or corrupted by the game of profit and power, they fail to provide the social homeostasis their role demands, thus becoming one factor disrupting these communities and largely vitiating the

312
John von
Neumann
and
Norbert
Wiener

possibility of a self-regulating, large-scale, genuinely demo-cratic capitalistic society.

While Wiener found the large-scale communist or capitalist societies to be antihomeostatic, he considered small communities in which people have direct contact with each other a different matter: "Small, closely knit communities have a very considerable measure of homeostasis," Wiener argued, and he presented the small New England towns he was familiar with as an example.[43] Individual self-respect and integrity are consonant with the viability of such a community. With his romantic individualism and his preference for small communities, he was close to the anarchist socialists. In his personal life he sought and found such communities, or helped to create them.

But Wiener was too deeply imbued with a sense of process and change to regard even such communities as stable utopias, as a solution to the world's political problems. Ultimately they too are vulnerable to power hunger and acquisitiveness, and other pressures coming from outside and within. Thus, Wiener's tragic sense of life had its reflection in the sphere of social and political organization. His enthusiasms were for the temporary achievements. He did not share the faith in the future held by many Marxists, any more than he shared the Enlightenment belief in man's progress toward perfection. Unlike Marx, he had no systematic program for social and economic change, nor did he envisage a political revolution. His concern for homeostasis did not lead him to map out a political process or struggle to bring about favorable changes, except on a small scale for enclaves within the larger society. If the organizational pattern of society as a whole could be reformed, the pursuit of money and power would be curbed and subordinated to more humane social values. For Wiener, widespread adoption of wiser and more humane social patterns was derived from an order of consciousness incommensurable with the actual nature of the world, and he posed the issue more as a moral struggle, an affirmation of personal values. In this he reflected a New England tradition. The only special political force he recommended, and that primarily by personal example (his own re-

313
Political
Power,
Human
Nature,
and
Society

fusal to be helpful in weapons-related research) was conscientious noncooperation to an extent that—in the United States—would entail no violation of the law, although it would result in a personal loss of income and exclusion from some circles.

Wiener's overall social theory appears to form a coherent whole that places primary value on human life and development, and not on technological "progress" or political power.

Wiener's warnings against mechanical thinking in government and his opposition to continued militarization were heard by a general public and by some members of the American Association for the Advancement of Science, but they did not penetrate into the conference rooms where government policy was devised. Of course Wiener himself had no truck with the military elite or the planners of nuclear weapons policies. He was the kind of noncooperator whose existence government officials wanted to ignore. Wiener's warnings of insoluble problems and his principled opposition to increased militarization were anathema to these men.

What kind of help did military-political policymakers hope to obtain from mathematicians after World War II? Nuclear weapons were a new phenomenon of which the planners had practically no knowledge or experience. The global range of policy decisions was also a new circumstance. And the United States was unquestionably the dominant world power. It was an awesome situation for those who were seeking some advance in the technology for policymaking commensurate with the advance in weapons technology, and they looked to the mathematicians and scientists for aid. This was the era of the rise of a new style of thinker in military and world affairs, the "strategic analyst"; in particular, von Neumann's mathematical game theory became part of the arsenal of conceptual tools of American strategic thinking. At a time when social scientists were becoming increasingly disillusioned with the usefulness of game theory, the military strategists were becoming more and more enthusiastic about it. Von Neumann especially and also his coauthor on game theory, Oskar Morgenstern, were in demand for strategic advice. The latter has described a situation in which

314
John von
Neumann
and
Norbert
Wiener

military matters have become so complex and so involved that the ordinary experience and training of the generals and admirals are no longer sufficient to master the problems. . . . The initiative to seek contact with science and scientists comes normally from the military men themselves. . . . More often than not their attitude is: "Here is a big problem. Can you help us?" And this is *not* restricted to the making of new bombs, better fuel, a new guidance system or what have you. If often comprises *tactical and strategic use* of the things on hand and the things only planned.[44]

Project Rand, later known as the Rand Corporation, was set up in 1946 by the Army Air Force for study and research "on the broad subject of intercontinental warfare other than surface," and it

made many contributions to the development of the theory [of games] and has applied it to various tactical problems—such as to radar search and prediction, to allocation of defense to targets of unequal value, to missile penetration aids, to the scheduling of missile fire under enemy pin-down, to antisubmarine warfare, and to inspection for arms control. On the other hand, in studies of policy analysis, it is not the theorems that are useful but rather the spirit of game theory and the way it focuses attention on conflict with a live dynamic, intelligent, and reacting opponent.[45]

The Rand Corporation became the world center for studies in and promotion of game theory, and retained von Neumann as a consultant. Rand researchers authored not only an enormous number of monographs and reports on game theory but also several books on the subject, ranging from popularizations to advanced textbooks.[46] Rand offered mathematicians a good salary for interesting mathematical work. One Air Force military theorist reported to the Rand Corporation the result of his study showing that the traditional doctrines of military decision making have many features in common with game theory, which might even be used to improve them.[47] A Rand mathematics researcher, George Dantzig, recalled that after the war it "became clear to members of this organization [US Air Force Comptroller] that efficiently coordinating the energies of whole nations in the event of a total war would require scientific programming techniques."[48] The development of large-scale computers made this feasible, and intensive work began in June 1947. Dantzig recalls a first meeting with von Neumann on October 1, 1947, in which von Neumann conjectured the equivalence of some problems of game theory and the new field of linear programming, especially as it had been developed by Dantzig himself, and on that basis provided Dantzig with a suitable mathematical foundation for the theory of linear programming.[49]

315
Political
Power,
Human
Nature,
and
Society

Von Neumann was not only a consultant to the Rand Corporation but an active and respected participant in the making of government weapons policies. His knowledge, virtuosity, and brilliance gave the aura of sophistication to discussions in which he participated—and especially when his emotional biases dovetailed neatly with those of the policymakers—the latter listened to him eagerly. Stanislaw Ulam commented in a memorial volume to von Neumann that "[Von Neumann's] knowledge of ancient history was unbelievably detailed. . . . On a trip South . . . passing near the battlefields of the Civil War he amazed us by his familiarity with the minutest features of the battles," but to help explain von Neumann's misjudgment of Soviet intentions after the war, Ulam cautiously blames von Neumann's "inclination to take a too exclusively rational point of view about the cases of historical events. This tendency was possibly due to an overformalized game theory approach." Alternatively, the overformalized game-theoretic approach may merely have disguised the hold that hate-fear of Russia, and especially of its Communist government, had on von Neumann. To many educated, westward-looking Budapesters, Russia had traditionally symbolized all that was primitive, barbaric, and threatening.[50]

Von Neumann's Princeton colleague and collaborator Valentine Bargman sometimes had the feeling that whereas Niels Bohr was conscious of the fact that schemes and theorems catch only part of reality, not all of it, von Neumann tended to forget this and rely fully on his theoretical scheme, especially in game theory: "It was probably in 1944. The von Neumanns were at our house one evening and I think the Paulis were there too. It was already clear that Hitler had lost the War.

316
John von
Neumann
and
Norbert
Wiener

The question came up in our conversation, what will Hitler do? Johnny said, 'There is no question. The plane to South America stands ready.'" The prediction was refuted a few months later, when Hitler committed suicide, but it reflected "Johnny's notion that one can reduce actions to moves in a game."[51]

I have described Wiener's analysis of the danger of nuclear war, as he saw it from outside the military establishment. In 1955, while he was a member of the Atomic Energy Commission but after he had given up the idea of a preemptive nuclear attack, von Neumann also offered a general analysis of nuclear war, which makes it possible to compare von Neumann's and Wiener's respective concerns:

In the past . . . if the enemy came out with a particularly brilliant new trick, then you just had to take your losses until you had developed the countermeasures, which may have taken weeks or months. The period of one month is probably reasonable for a very brilliantly performed counter-counter-move. This duration is now much too long, and the losses you may have to take during this period may be quite decisive. . . .

The difficulty with atomic weapons, and especially with missile-carried atomic weapons, will be that they can decide a war, and do a good deal more in terms of destruction, in less than a month or two weeks. Consequently, the nature of technical surprise will be different from what it was before.

It will not be sufficient to know the enemy has only fifty possible tricks and that you can counter every one of them, but you must also invent some system of being able to counter them practically at the instant they occur.[52]

Von Neumann's concern was clearly "winning the arms race" rather than mitigating it. This particular essay deserves comment on several counts. First, it expresses the injunction to innovate or expect to be destroyed, which had something of the quality of an ingrained obsession and was shared by some of von Neumann's Hungarian-born colleagues.[53] However, since the passage cited was from a talk von Neumann gave while a member of the Atomic Energy Commission, it must be seen as not only a personal view. Whereas Wiener could speak in the role of an independent intellectual, von Neumann spoke in the role of a captive of the government establishment, who had forfeited the privilege of open discussion of political issues. Still, his wholehearted endorsement of the nuclear arms race, including the development of what he described as "nuclear weapons in their expected most vicious form," is entirely consistent with his other actions. On the whole, von Neumann's analysis seems shallow compared to Wiener's because he does not consider alternatives to an arms race. He does not ask how a nuclear holocaust might best be avoided. However, he does anticipate the possibility of the military strategy of limited war as a means of maintaining in reserve the power to counter the enemy's surprise: "You may have to hold this trump card [power stepped up to the limits of your capabilities] in reserve."[54]

317
Political
Power,
Human
Nature,
and
Society

Recently Victor Weisskopf, a distinguished mathematical physicist and president of the American Academy of Arts and Sciences, born in Austria a few years later than von Neumann, asked why the arms race between the two superpowers continues despite the knowledge that the use of the weapons would "mean certain destruction of a large part of the world, making it unfit for human habitation, with little chance of recovery," and answers, "Because neither side knows where to stop and goes on producing nuclear weapons intended for all sorts of imagined missions. Because both parties are under the grip of an unrealistic measure-countermeasure syndrome. It is the apotheosis of irrationality and anti-logic; the triumph of crazyness."[55] The Weisskopf comment provides perspective: von Neumann's thinking, which the dominant political group in the 1950s saw as the epitome of logic, rationality, and sanity, is in retrospect identified as "the apotheosis of irrationality and anti-logic; the triumph of crazyness." Both concepts, that of rationality and that of craziness, enter frequently into the debates concerning global nuclear strategy. A particular kind of self-delusion arises when a nuclear "strategist" (who once would have been called a policymaker)[56] attempts to ensure rationality by formulating the strategic decision problem in a strictly logical form, as in game theory, but builds into that formulation paranoid prem-

318

John von
Neumann
and

Norbert
Wiener

ises about the political world in the nuclear age. Adding to the linguistic confusion, one analyst has concerned himself with "the rationality of irrationality."[57]

The US government's advisers on weapons policy and global strategies had an awesome task. Historical precedents seemed to have at best limited applicability, and quite naturally the government looked more to the technologist and the mathematician for advice, rather than to the humanistically trained scholar.[58] The analytical approach incorporates besides game theory the use of the new computers and also "systems theory" and quantitative "operations research," all of which were within von Neumann's repertoire, but it is beyond the scope of the present essay to discuss these techniques.[59] While the situation called not so much for innovative techniques as for a new fundamental mode of perceiving the world situation, the military-political thinking associated with the new techniques was akin to Bismarck's Realpolitik, traditional European power politics, and the war gaming in which the German general staff in particular had engaged during the nineteenth and early twentieth centuries.[60] Von Neumann's own thinking about international relations followed nineteenth-century lines, but he brought new tools to bear. While von Neumann indeed had considerable knowledge of military history, there was little appreciation of the importance of social and political history. The technologists, scientists, and mathematicians enjoyed a new status as government advisers. The decision makers, like decision makers in an earlier era who relied on astrology, hoped for a prescience that the tools of science were unable to deliver. They didn't reckon with the selective blindness that is built into the application of game-theoretic abstractions to major policy decisions. There is no science here, only pseudoscience. However, for the executive branch of a government it is useful to have a set of "experts" whose recommendations support its policies. Solutions to policy problems may be deliberately treated as merely technical, in order to "minimize differences at the policy level and to blunt criticism from without."[61] Policy experts are useful to the government bureaucratic structure, since "the cult of the expert is so accepted in the United

States that the resort to expert advice is useful as a political maneuver to allay public fears, to lift the issue above partisan debate, and as an attempt to reduce complex problems to manageable proportions. . . . The Eisenhower Administration, particularly, recognized the advantage of this tactic."[62]

319
Political
Power,
Human
Nature,
and
Society

Game theory's usefulness to government was enhanced because it promoted a thought pattern characteristic of the postwar military-industrial bureaucracy. It favored thinking in terms of "them and us"; was as mechanical and impersonal as possible; had a simplistic model of purposes and a simple, one-dimensional, quantitative view of human nature; emphasized efficaciousness; and was conservative and uncritical of existing institutions.

Although game-theoretic analysis of policies can be complex and imaginative,[63] it is shot through with preconceptions and biases.[64] There is a tendency to revert to the "ideal" case with a unique solution—the zero-sum two-person game—which presumes total opposition of interests; anything more realistic and complex (such as the famous "prisoner's dilemma") leads to no clear conclusion. There is a tendency in the spirit of the minimax solution to base policy not on what the other parties are likely to do but rather on the assumption that they will attempt to do you all the harm they are capable of. This leads to the biased conclusion that "national security" requires more and more armament and innovation in armament. Game-theoretic thinking played a major role in evolving and justifying the policy of deterring attack by the threat of massive retaliation with nuclear weapons which was espoused by the Eisenhower administration.[65]

American foreign policy in the cold war era can be interpreted in several ways. Some regard it as merely the effort of one large power to compete with another. Others see it as aiming primarily to suppress social revolution throughout the world.[66] The focus on social revolution calls attention to the human dimension. It also reminds us of the central role played by the cooperation or noncooperation of the people with official government decision makers. The abstractions of the strategic analysts, including in particular the abstractions of game theory, protect them from facing the human dimension of the policies they espouse. In government circles the methods of the analysts seemed vindicated by the Cuban missile confrontation between President Kennedy and Premier Khrushchev. Subsequently, however, the analysts lost prestige because of their complete misjudgment of the war in Vietnam. Raymond Aron asked in 1969 if there isn't a tendency to "overestimate the technical aspect of the diplomatic or military problems, and underestimate the importance of the psychological, moral and political data—*which is different in each situation*—and allow . . . decisions to be influenced by people acquainted with strategy but not Vietnam?"[67] Game-theoretic thinking has also been used to inhibit efforts at nuclear disarmament by suggesting that disarmament might destabilize the "balance of terror" that presumably is the basis for security. P. M. S. Blackett, the leading innovator in early operations research, gave a cogent critique of these kinds of arguments as far back as 1961.[68]

The strategic analyst Thomas Schelling, who views his subject as falling strictly within the field of (non-zero-sum) game theory,[69] has also reflected on the limitations of game theory.[70] He finds, for example, that someone whose "only interest is in making the 'right' choice—a good or moral or ethical choice," regardless of its consequences, does not fit the utilitarian game-theoretic model. If a player cares about the welfare of his opponent, a paradoxical situation arises in the game-theoretic model; "that is to say, the functions relating the welfare of each to that of the other may be incompatible, or may result only from some dynamic psychological process."

Wiener emphasized the element of "control," rather than destructive power, in a social or political system. Control in any society rests not only with the elite decision makers but also with the people, who can if they choose, withdraw consent from the government and refuse to cooperate with it. As Gene Sharp has shown in his compendium of nonviolent struggles,[71] the history of political conflict is usually grossly distorted by overemphasizing the role of weapons and ignoring the role of noncooperation and other nonviolent tactics. For example, the American Revolution was won primarily by nonviolent tactics. In the Vietnam War planners erroneously thought of the decision-making process as monolithic, without properly understanding the likelihood of popular revulsion and noncooperation with the war by both Americans and South Vietnamese. No nuclear weapons policy is realistic that ignores the likelihood of a popular revulsion against weapons escalation and a consequent large-scale noncooperation on the part of crews of airplanes or submarines or missile silos. Moreover, the history of such movements shows that a mechanism of political jiujitsu causes repressive measures to stimulate even more resistance. Thus application of the theory of nonviolent action, as Sharp's pioneering study shows, would seem to amend and even reverse many of the conclusions of traditional analytical studies of war and armament. From Wiener's frame of reference the tactics of nonviolent action in a political conflict, as codified by Sharp, would be preferred in principle, because they are highly homeostatic: honest and open communication is promoted on many levels, rigidities are broken down, and control is decentralized.

321
Political
Power,
Human
Nature,
and
Society

What happens when someone deeply imbued with game-theoretic analysis of strategies opens himself up to the psychological, moral, and political data? To do so would require an entirely different mode of cognition and a renunciation of the analyst's convenient distance from the painful human implications of policy recommendations. Here, however, might be a mechanism for correcting that selective ignorance in which the modern strategist is professionally trained, and in which he is reinforced by his colleagues and employers.

One such instance has come to the attention of the public. It involved some personal courage and made political history. Daniel Ellsberg was a serious student of game theory who became a leading government analyst and deterrence theorist.[72] He liked to formulate issues like nuclear-weapons policy in game-theoretic language.[73] He had worked in the Department of Defense and had been one of the leading analysts under Rand Corporation auspices. However, he spent two years in Vietnam, during the last six months of which (December 1966–June 1967) he was special assistant to the deputy US

322
John von
Neumann
and
Norbert
Wiener

ambassador to Vietnam, William Porter, and was assigned to making field evaluations of progress and programs in the war for Porter. In this capacity he encountered and responded to Vietnamese villagers. One might say that he obtained psychological, moral, and political data firsthand. Already then his views were coming to vary from the official optimism about the war.[74] Ellsberg was explicitly concerned with the process by which major policy decisions were made, and found that in relation to Vietnam the "U.S. Government, starting ignorant, did not, would not, *learn*."[75] He studied the history of the Vietnam War and became conscious of the moral blindness of the American strategists, which he himself had once shared. In 1972 he wrote,

As I reread now my analyses written before mid-1969—and the writings of other strategic analysts, as well as official statements—I am struck by their tacit, unquestioned belief that we had had a *right* to "win," in ways defined by us (i.e. by the President); or, at least, a right to prolong a war, to "avoid defeat"; or at very worst, to lose only gracefully, covertly, slowly: all these, even the last, at the cost of an uncounted number of Asian lives, a toll to which our policy set no real limits. That belief ended with me in August and September 1969.[76]

Ellsberg had abandoned the premises of the game theorist and analyst, and the political consequences of his defection included the US government's attempt to prosecute him (which, however, backfired). It upset President Nixon's "game plan," brought out new government improprieties, and may indirectly have helped to force his resignation from the presidency.

Not only does the Ellsberg conversion show that it is psychologically difficult to overcome selective blindness (what Veblen calls "trained incapacity"); to break down the barriers that prevent unpleasant realities from undermining the biases on which abstract theories are constructed; to "come to one's senses"; it also shows that government officials are generally hindered from doing so by strong bureaucratic and political pressures and act on their knowledge only at considerable personal risk.[77]

The ready acceptance of game theory by government policymakers within a few years of its development reflects its political usefulness in helping make government policies appear "scientific." It is unusual for a theory of society to be adopted and utilized by a government within a few years after its development and for a government to promote actively further development—as the US government did game theory. Undoubtedly the ready willingness of von Neumann and Morgenstern to act in an advisory capacity facilitated the interplay between theory and government policy. Moreover, game theory and other mathematical models provided something solid to cling to, helping to protect strategists from inundation by the sea of horror, fear, and depression to which they may be subject when reflecting for prolonged periods about nuclear war. Nuclear strategy became a "game" involving the ingenious use of computers and mathematical reasoning, which lead to computing numbers describing the value of the relative "payoff" of a first strike versus a second strike, of massive reprisal against urban centers, of flexibility in the counterforce, and so on. The military-political theory based on this structure was the basis of US government thinking about global strategy not only throughout the 1950s, but also through the 1960s and the 1970s.[78]

323
Political
Power,
Human
Nature,
and
Society

Even those who would refute the conclusions of the analysts can hardly extricate themselves from the now-conventional theoretical framework. As Blackett wrote in 1961,

When I come to study in detail some of the arguments of these new military writers about nuclear war, I will necessarily have to adopt many aspects of their own methods and terminology, that is, I will have to meet them on the methodological ground of their own choosing. I want therefore to apologize in advance for the nauseating inhumanity of much of what I will have to say.[79]

But what social theory can deal with the possibilities of nuclear war, remaining detached enough from its horrors to allow clear reflection and yet fully recognizing the human miseries that the use of nuclear weapons would create? Daniel Ellsberg was apparently able to encompass the social realities of the war in Vietnam after spending two years there

324

John von
Neumann
and
Norbert
Wiener

and subsequently delving deeply into the historical background of the war. We have no information on the human effects of thermonuclear weapons. Our only data on the psychological, biological, and social realities of nuclear weapons derive from the "Hibakusha," as the survivors of the Hiroshima and Nagasaki bombs call themselves. It is notable that in May 1978 petitions urging total nuclear disarmament signed by over twenty million Japanese were presented to the United Nations; no other country has shown comparable interest in nuclear disarmament. More explicit data on these two small "tactical" explosions was reported at the International Symposium on Damage and After-Effects of the Atomic Bombing of Hiroshima and Nagasaki, held in Japan in July 1977,[80] and further studies are in progress.

Any adequate language or conceptual framework for discussing nuclear weapons policies needs to be formulated so as to take realistic and detailed account of the fact that it is not only deaths that matter, but also environmental damage and the condition of human, animal, and plant survivors. In addition, such an analysis must be informed by historical understanding of how the present situation has come into being, and what its moral components are. Systems-theoretic analyses, computer simulation, operations research, and game-theoretic models need not be excluded, for each of these techniques has a domain of usefulness. However, none of these has provided either language or conceptual tools appropriate for defining policies of nuclear strategy, whether of armament or disarmament. The object lesson of the Vietnam War deserves to be remembered: heads of governments, far from being the players who hold all the cards, are helpless in the face of popular noncooperation and especially noncooperation by members of the military or the government bureaucracy.[81] Threats, even with the most devastating weapons, may be futile. Mere destructive power is vastly different from control of a situation. Basing international relations on increases in the means of destruction is surely the result of a highly distorted model of reality.

Von Neumann's and Wiener's social theories, like any such

theories, contained ideas about the nature of human nature, of what a human being is. On one level both von Neumann and Wiener made models of organisms that called attention to patterns of organization, to internal patterns of the flow of messages and points of control or decision. These models are associated with a heuristic according to which the human being is seen as functioning through an extraordinarily complex pattern of organization. A characteristic feature of such models is that affective states and private experiences are belittled: they are of no interest in themselves. It is only indirectly, through information they provide to the organism leading to changes in its purposes and actions—learning—that they play a role in the models at all. Both men were interested in the design of relatively simple simulacra of organic systems in the form of mathematical models or electronic circuitry. These designs served to highlight the huge gap between even the most elaborate of the man-made models and the enormous subtlety and complexity of a living person. Although von Neumann in his optimism anticipated that some day empirical science and mathematical logics would advance to the point where the formal pattern of organization of a complex organism could be portrayed, he fully acknowledged this discrepancy. His brother observed that John von Neumann's preoccupation with the models seemed to lead him to experience feelings akin to awe,[82] if not religious feeling—emotions all the more noteworthy because they were entirely out of keeping with von Neumann's normal sense of total mastery and superiority. Here he was dramatically confronted with his own limitations and those of his scientific tools.

For Wiener these models became a stimulus to a more irreverent, playful, philosophical imagination.[83] He joked about the sex habits of automata and played with the idea that in principle the instruction for the design of a particular human being could be spelled out and transmitted by telephone. To Wiener, the human was distinguished from other animals by being the "talking animal," with marvelously rich possibility and variety of communication. Philosophically, Wiener re-

325
Political
Power,
Human
Nature,
and
Society

326
John von
Neumann
and
Norbert
Wiener

garded an organism as a metastable pattern of organization in opposition "to chaos, to death, as message is to noise. . . . Certain organisms, such as man, tend for a time to maintain and often even to increase the level of organization, as a local enclave in the general stream of increasing entropy, of increasing chaos and de-differentiation. Life is an island here and now in a dying world."[84] But to Wiener this made organic life, and especially human life, very precious, to be valued and loved. Wiener saw the essential meaning of his own existence in his struggle to, in his own way, increase the level of organization in opposition to that "general stream of increasing entropy." Homeostasis achieved its exalted place in Wiener's philosophy because it is the preserver of organizational patterns. Learning has a central place, for it too serves to strengthen and elaborate patterns of organization, and to raise their level of sophistication. In short, Wiener's affection for life generally, especially in its more complicated forms, underlies his social theory.

Wiener was also a strong advocate of such virtues as honesty, dignity, courage, flexibility, intellectual integrity, and refusal to exploit. This advocacy gives his social theories an ethical dimension that is an integral part of his philosophy. Thus Wiener's implicit model of human nature involves both an "is" and an "ought." His autobiography, *Ex-Prodigy*, published in 1953 when he was nearing sixty, is a remarkably candid, conscientiously written document of the pains and conflicts suffered by a gifted, precocious child. *Ex-Prodigy* displays Wiener's capacity to appreciate people as they are—although his novel exposes his limitations in this regard.[85]

Von Neumann, on the other hand, would never have put his own experience on public display, as Wiener did. He always protected his private thoughts and feelings, especially his failings and weaknesses, which he would not expose even to his close colleagues, much less a large lay public. When Wiener sent him a copy of *Ex-Prodigy*, von Neumann sent him a note describing it as "a very unusual documentation of a process that occurs rarely, and is adequately described even more rarely." Typically, he appraised the book in scientific terms—"the documentation of a process." Von Neumann mentioned no echo from his own childhood evoked by reading *Ex-Prodigy*; it would have been out of character for him to do so.

327
Political
Power,
Human
Nature,
and
Society

Von Neumann did take an amused interest in human foibles, including gossip.[86] He was not malicious, but his outlook on people did not have a moral dimension: "It is just as foolish to complain that people are selfish and treacherous as it is to complain that the magnetic field does not increase unless the electric field has a curl. Both are laws of nature."[87] Von Neumann tended to see people's psychological characteristic as essentially fixed, not subject to "improvement"; though a technological optimist, he was a pessimist about human evolution. Human society might resemble the proverbial "law of the jungle," but it could be dealt with if one understood the situation and knew how to play the game. The "ought" underlying von Neumann's game theory places the highest value on instrumental intelligence, unsqueamish ruthlessness, and forethought in pursuing self-interest. Thus, he focused on certain human traits and talents, dismissing the rich multitude of human possibilities. In spite of his personal gregariousness, his social and political-military views were consistent with this rather restrictive valuation of human qualities, and a certain contempt for the majority of human beings who have neither extraordinary intellects nor positions of power.

The sharp contrast between Wiener's and von Neumann's personal attitudes toward political power and their relation to it could not help but be reflected in their social and political views: Wiener disapproved generally of the style of action of the power elite but thought it futile for himself to attempt to influence them directly toward more humane conduct. For von Neumann, nearly the reverse was true. To Wiener, loyal to an intellectual tradition with Jewish roots, the power elite was not to be trusted. He avoided sitting with the powerful, the more so as he felt himself vulnerable in their world, as well as politically clumsy. He lived in protest against the world in which "power and money had won the day." The invitation to partake constituted a temptation to be resisted. A perceptive

328
John von
Neumann
and
Norbert
Wiener friend characterized Wiener's attitude toward the powerful very well on an occasion when Wiener had refused to have an early essay on cybernetics published in a symposium volume specifically addressed to educating the political and industrial leaders of the country:

> I suspect that if you pushed him to it, Wiener would think of mathematicians and scientists in general, including the philosophers of science, as a feeble folk immemorially pushed about by politicians and the military, to say nothing of big business. All of these would doubtless make use of any technical advantage to be had from applying science to engineering, propaganda, weapons and money-getting, but certainly they would not condescend to glean from science the idea of the good.[88]

In other words, Wiener viewed politicians, generals, and presidents of large corporations as unwilling to examine their value premises, as not educable on the deeper issues. Wiener remained a stranger to political or economic power over people.

Von Neumann, by contrast, found himself emotionally in tune with men of power. Unlike many modern intellectuals, he was at home with their thinking. When the political theorist Hans J. Morgenthau examined the personal origins of political power, he concluded that out of frustration and failure to achieve "the totality of the commitment that characterizes the pure phenomenon of love," there arises "in the great political masters a demonic and frantic striving for ever more power . . . which will be satisfied only when the last living man has been subjected to the master's will"[89]—although even then the underlying desire for "the pure phenomenon of love" remains unsatisfied. Morgenthau views this lust for power as intrinsic to human nature, especially among individuals with strong political aspirations. To Wiener this would be grounds for pessimism about the future. Morgenthau, however, reasons that consequently power politics is the only realistic kind of politics. His characterization of the great political masters dovetails neatly with von Neumann's comment to Lewis Strauss about "achieving togetherness by cosmic suicide": nuclear weapons open up a whole new range of possibility for "the great political masters" to achieve their ends! While Lifton when he speaks of nuclearism[90] and Morgenthau when he speaks of the lust for power are very close in that they identify such behavior with an underlying desire to transcend human limitations, Lifton sees nuclearism as a pathology to be cured, while Morgenthau sees the lust for power as a normal one that is particularly strongly manifested in great political leaders.

329
Political
Power,
Human
Nature,
and
Society

One of von Neumann's favorite classics, which he knew by heart and often quoted to his friends,[91] is the dialogue between the Athenians and the Melians in Thucydides' *History of the Peloponnesian War*. During the period when he was advocating a policy of preventive war against the USSR, von Neumann came back often to the Athenians' arguments. For example, he liked the rationality of their advice to the weaker Melians:

We recommend that you should try to get what it is possible for you to get, taking into consideration what we both really do think; since you know as well as we do that, when these matters are discussed by practical people, the standard of justice depends on the equality of power to compel and that in fact the strong do what they have the power to do and the weak accept what they have to accept.[92]

But as de Santillana notes, speaking of the complete dialogues, one can spot in the argument "a cold, brilliant, scientific cruelty, not of the passions but of the rational intellect, the devilish element which runs through history isolated in its pure form."[93] Elements of the Athenians' cruel logic often surface in the practical world of power politics.

13 WIENER, THE INDEPENDENT INTELLECTUAL: TECHNOLOGY AS APPLIED MORAL AND SOCIAL PHILOSOPHY

Whether or not it draws on new scientific research, technology is a branch of moral philosophy, not of science.
—Paul Goodman

Intellectuals are men who never seem satisfied with things as they are, with appeals to custom and usage. They question the truth of the moment in terms of higher and wider truth; they counter appeals to factuality by invoking the "impractical ought." They consider themselves special custodians of abstract ideas like reason and justice and truth, jealous guardians of moral standards that are too often ignored in the market place and the house of power.
—Lewis Coser

The "intellectuals" described by Lewis Coser are a special breed.[1] Most academics and professionals are not intellectuals. The professionalization and the bureaucratization of science helped to domesticate scientists into "highly specialized personnel"[2] avoiding the risks and inconveniences of the intellectual's role. Wiener, however, did function in the traditional role of an intellectual. He did not avoid the kind of activity envisaged by Thomas Jefferson in 1779 when he was planning the University of Virginia, namely, to provide an independent criticism of the powers that be and to "unmask their usurpation, and monopolies of honors, wealth and power."[3]

Wiener felt vulnerable to cooptation, exploitation, and corruption, and to retain his originality and autonomy it would be necessary for him to eschew too much integration into intellectual, bureaucratic, or political establishments.[4] To this end his rage, disappointment, and grief at the timid mentality dominating the great universities was undoubtedly helpful.[5] His conviction that the leaders in the field of mathematics were a self-chosen elite additionally served to protect his autonomy. In spite of honors and recognition from the American mathematics community, he always viewed himself as an outsider.[6] When he resigned from the National Academy of Sciences in 1941, he described himself as "profoundly suspicious of honors in science, and of select, exclusive bodies of scientists,"[7] and he retained the same attitude for the rest of his life.[8] Wiener withdrew completely from service to the government and the military toward the end of World War II, criticizing their intentions. He did not even participate in political organizations whose principles he shared. His first loyalty appeared to be to specific values rather than to institutions. As a close friend described him, "Wiener was and remained the individualist who could not and did not want to identify himself permanently with any other cause than that of rigorous honesty with himself and with others. He was even skeptical of those who were making cybernetics a cause."[9] Wiener saw it as the intellectual's duty—as his own—to be the "custodian of a tradition of honesty and sincerity."[10]

After World War II Wiener took on the role of the independent, technologically knowledgeable, humane intellectual. He tried out his political thinking on a few of his colleagues whom he particularly trusted and liked, such as Giorgio de Santillana and Dirk Struik. Because of his Marxist views Struik was an outsider as well, and Wiener remained closely in touch with administrative and political threats to Struik during the harassment of communists in the United States. In 1949 he was pleased with MIT's support of Struik, and even urged his British Marxist friend J. B. S. Haldane to join them: "MIT is an oasis in the political madness."[11] On September 13, 1951, however, Struik was indicted under the laws of Massachusetts on charges of advocating, advising, counseling, and inciting "the overthrow by force and violence of the government of the Commonwealth of Massachusetts." On the same day Wiener, in Mexico, wrote of Struik to President Killian, "If his relations with MIT suffer, unless there is far

332
John von
Neumann
and
Norbert
Wiener

more damning testimony . . . I shall regretfully . . . submit my resignation from MIT."[12] MIT suspended Struik from teaching while the case was pending, but with full pay. Five years later the charges were dropped.

Giorgio de Santillana was another whose personality and mind were congenial to Wiener. The two were also part of a small community of gifted men, including Jerome Lettvin, Warren McCulloch, and Walter Pitts, working at MIT at the interface of science and philosophy. Together with Pitts, de Santillana wrote a garbled biography, full of fantastic plays on words (which Pitts was inventing), a spoof of Norbert Wiener. It was intended to put the History of Science class on guard "against an excessive reliance on classical sources and also on the learned labors of the scholars who give us the critical editions" of ancient Greek philosophers, to show how distorted these works are likely to be. All the students knew Wiener, or at least knew who he was. Pitts and de Santillana explained that this biography was "based on the figure of a great scientist whom we both knew and loved, and who pleasantly assented to this prediction darkened and deformed in the glass of time to come." It shows something of the good-natured humor that prevailed between Wiener and some of his coworkers in 1950:

Norbertos Vindobonensis, also called Wiener of Columbus . . . paid special honor to a goddess called Jeeby, who, he said, comes after Hebe in rank, and some explain that she was the deity of Awakening ("Morning-After"). But elsewhere he says explicitly that there is no one god, but that the world is ruled by Chance and Information and that "it befits them to move hither and thither." For such are the appellations he ascribed to the American Hermes, whom he held in peculiar reverence. . . .[13]

The long history of chronicles of the lives of thinkers shows that absentmindedness is perennial and ubiquitous. The nearly otherworldly quality of being lost in his own thoughts, an apparent unawareness of surroundings intensified by his poor eyesight, was also part of the Wiener legend. Yet at MIT he was a figure of affection to janitors, machinists, students, and many of his colleagues, especially those who appreciated his gentleness: "A sweet, caring guy with a strong sense of human fellow-feeling," one MIT professor said of him.[14] Perhaps no biographical work about Wiener is complete without some sampling of the stories about his extraordinary absentmindedness. In one, a student encounters Wiener around midday on the MIT campus and engages in a conversation with him. At the end of the conversation Wiener seems confused and asks, "Do you remember the direction I was walking when we met? Was I going or coming from lunch?" The student kindly provides the needed information. On another occasion, in class, while presumably deriving a theorem on the blackboard, Wiener in his intuitive way, thinking of the problem but forgetting the students, skips over so many steps that by the time he arrives at the result and writes it down on the board, it is impossible for the students to follow the proof. One frustrated student speaks up and tactfully asks Wiener if he might show the class still another proof of the theorem. Wiener cheerfully indicates, "Yes, of course," and proceeds to work out another proof, but again in his head. After a few minutes of silence he merely places a check after the answer on the blackboard, leaving the class no wiser.

Of course Wiener heard these stories. Some students made it a habit to regale him with the latest story about him, which he would usually enjoy goodhumoredly.[15]

One of Wiener's preoccupations since the early 1940s had been to take a first step, with a clear vision, toward constructing a practical philosophy of technology for the modern age, a field nearly unworked hitherto, but one that made high demands on the thinker in order to break through the confused veil of conventional lies and apologetics for vested interests that surrounded it.[16] In this field of endeavor Wiener's basic ideas, scattered throughout his later writings, are seminal—especially for all those who share his value orientation.

His first clear-cut action—after the initial shock he experienced in the wake of the Hiroshima and Nagasaki bombings and his pondering over how that decision had been made—was his refusal to provide information for military purposes.[17] The reasoning behind his refusal was that in work useful for weapons design, "the scientist ends by putting unlimited

334
John von
Neumann
and
Norbert
Wiener

powers *in the hands* of the people whom he is least inclined to trust with their use. It is perfectly clear also that to disseminate information about a weapon *in the present state of our civilization* is to make it practically certain that that weapon will be used."[18] The italicized phrases imply that the relation between the scientist and the potential user of his investigations is that of one putting tools or powers "into the hands" of the other, and trust it is the scientist's responsibility as creator to evaluate to the best of his ability the social and political circumstances so as to judge how or if the product of his work, whether information or technology, is likely to be used, and accordingly to judge whether or not to release it for such use. These two general concepts are cornerstones of Wiener's philosophy of technology. They may serve as a framework for discussion of scientists' and engineers' responsibility generally, and they transcend Wiener's more specific and more subjective evaluation of the trustworthiness of political-military leaders and the state of international affairs, which fall into a different category. In other words, the same conceptual framework would permit the opposite conclusion for someone who judges the state of our civilization and those who rule over it differently from the way Wiener did. His refusal to be helpful served to identify noncooperation as a possible course of political action for scientists and engineers. Through noncooperation scientists, engineers, and technologists, especially if they act in concert, can overrule an irresponsible government.

The open announcement of Wiener's decision not to cooperate and his first efforts to act on that decision required him, to his own embarrassment, to renege on his agreement to participate in a military-sponsored Harvard symposium on high-speed computers.[19] He found that he had placed a new barrier between himself and the majority of the scientific and technical community. He had asserted that his function as a self-respecting person and intellectual would be very different from that of an academically "domesticated" mathematician with primarily professional concerns. Another independent intellectual and man of science, Albert Einstein, when asked to state his opinion of Wiener's noncooperation, said, "I greatly admire and approve the attitude of Professor Wiener; I believe that a similar attitude on the part of all the prominent scientists in this country would contribute much toward solving the urgent problem of national security."[20]

Norbert Wiener's new role brought him a new and large audience: the educated public. His first book for this audience, *Cybernetics* (1948), turned out to be a best seller in spite of its partly technical and mathematical content. *The Human Use of Human Beings* (1950), with technical and mathematical language eliminated, was even more popular. Wiener says that with these books he was also trying to write himself out of a financial hole, clearly as an alternative to accepting military contracts.[21] He would write four more books for the general reader. In the two books mentioned, one of Wiener's objectives was to inform the public about the potentialities of the communications and computation technology invented during World War II and to share his thinking concerning their likely implications for society. Note the contrast with von Neumann, who increasingly sought private behind-the-scenes communications with powerful men and was little interested in conveying his thinking to a broad audience.

Wiener's credibility with the public rested on his unquestioned credentials as one of America's leading mathematicians and, curiously, on his technical contributions to military engineering during the war. When he spoke about science and technology he spoke with authority. Of course he had many interesting things to say, but what distinguished his books about technology was the combination of intelligence, compassion, and independence that informed his thinking. Weekly news magazines that reviewed his books and reported on his lectures liked to describe him—with his little white beard and large paunch—as a Santa Claus figure.[22] He gave many lectures, usually to overflow audiences. His style as a speaker was to mount the podium and launch into his subject without the usual preliminaries. He spoke in a fluent, discursive manner (without notes), throwing out a multitude of ideas for an hour or two until somebody signaled him that it was time to stop. The talk, amplified over a loudspeaker system, would be punctuated by the sound he made as he drew

336
John von
Neumann
and
Norbert
Wiener

on his ever-present cigar stub. Afterwards, he would but-tonhole some unsuspecting members of the audience to ask if they thought his talk was all right.

Wiener's last two books were written after he had tired of popular speaking and too much media attention. One was a novel based on the historic role of the unappreciated, highly intuitive, and independent British scientist Oliver Heaviside and the expropriation of his discoveries by weak and unprincipled individuals seeking fame and fortune.[23] His final book returned to the philosophy of technology, with special emphasis on ethics and creation.[24]

Wiener viewed technology from two complementary standpoints: that of the scientist, engineer, or technologist concerned with moral philosophy and that of a member of the general public who wants to assess technology in terms of its beneficial or harmful effects. The second standpoint is in a sense preliminary to the first, even though in the actual development of a technology the scientist is of course involved before the public. Wiener did not regard any particular innovation as intrinsically good or bad; whether it would be beneficial or harmful could be considered only in the context of the purposes of the institutions or the society into whose hands it would be given. Though the originator loses control over his creation, Wiener does not absolve him easily from responsibility; if an inventor regards his government as irresponsible in its military policies, then it is his moral responsibility to withhold his ideas on weaponry from that government. Although in these terms the situation is somewhat analogous to an adult's moral responsibility in giving a loaded pistol to a young child or the design for a new type of weapon to an organization planning to use it for criminal purposes, Wiener does not raise the issue of legal culpability. Implicit in his formulation is the necessity that engineers and scientists come to understand the society they inhabit by extending their education beyond narrow technical training into the social sciences, philosophy, literature, and history. "It is this wholeness, this integrity, that a considerable group of us at MIT are trying to evoke and to render conscious in our students," Wiener wrote.[25] He told other scientists that even when the individual believes that science contributes to the human ends which he has at heart, his belief needs a continual scanning and re-evaluation which is only partly possible. For the individual scientist, even the partial appraisal of this liaison between the man and the [historical] process requires an imaginative forward glance at history which is difficult, exacting, and only limitedly achievable. . . . We must always exert the full strength of our imagination.[26]

Thus, Wiener had redefined the function of a scientist or engineer from mere expertise to competence and sophistication in the difficult, exacting task of anticipating the social effects of his work. He was generally skeptical of the benefits of technical innovations, whether they occurred in the United States or the USSR, because he saw that neither American industry nor the Soviet government was primarily concerned with the people, and that both were likely to use technology for exploitation or repression.

Wiener himself came to practice the "imaginative forward glance" in respect to technology. His most original and fully elaborated prophecy was of course that the technologies of communication, organization, and control would occupy center stage during the second half of the twentieth century. But his view was still broader. Writing in 1949, he drew from history to criticize sharply the then-conventional belief in the beneficence of rapid technological progress and pointed to its other side, "an exploitation of natural resources; an exploitation of conquered so-called primitive peoples; and finally, a systematic exploitation of the average man."[27] He speaks of the "process of exhaustion of our resources" and the likelihood of shortages in tin, lead, and copper. He predicts a shortage of oil in "a few decades" and suggests that the value of nuclear fission as a source of energy may have been "greatly and permanently endangered" because of its connection to weapons.[28] And he warns that it would be a foolish gamble to count on new inventions to bail us out of our difficulties. He calls attention to the impressive progress in medicine since the mid-nineteenth century, but he also warns that even "at the very scene of our greatest triumph, the conquest of infectious diseases, we must never forget that we are

338
John von
Neumann
and
Norbert
Wiener

fighting an enemy who is multiform and resourceful," referring to the new varieties of bacteria and viruses that come to the fore as other varieties are destroyed. He anticipates the increased popularity of synthetic foods or food additives but observes that they may have "slow poisonous effects" or "minute quantities of carcinogens" and warns "that the processing of foods is subjecting us to many risks . . . which may not show themselves until it is too late to do anything much about them," and he ponders how changes in the age distribution might affect the viability of the human population.

What is of interest here is that Wiener more or less correctly identified many major issues that were largely ignored in his day but have finally become prominent and public about a quarter of a century later.[29] Those who gave thought to these matters in 1947 and availed themselves of the needed information then would probably have emerged with conclusions similar to Wiener's. In retrospect his anticipations give some confirmation to his belief in the fruitfulness of the imaginative forward glance: like discovering the symptoms of a disease in its early stages, it permits simpler and milder remedies than are possible later.

Understandably, Wiener's own technical innovations concerned him particularly. He knew all too well that his mathematical work in communication theory, especially his theories of prediction, filtering, and extrapolation, would be useful to the military for the guidance of missiles, and that he had given it "into their hands" during World War II. In his position, there was nothing he could do, although someone working with the military establishment on a high level, as von Neumann did, might exert some influence. His cybernetic theories would contribute to increasingly sophisticated automation in industry, and he foresaw automation as possibly leading to a further degradation of the worker and also producing large-scale unemployment.[30] He first expressed his concern about unemployment in print in 1947, decades before the anticipated effects showed up prominently in labor statistics. He came to think of automatic control devices as "mechanical slaves" but worried that "any labor which is in competition with slave labor, whether the slaves are human or mechanical, must accept the conditions of work of slave labor."[31] Still, he wondered how the relation between humans and machines could be made noncompetitive, and a few years later he became cautiously hopeful about the possibility of skirting some of the dangers.[32] Wiener played endlessly with ideas concerning the proper relationship between humans and machines. For example, what might be the respective roles of a translation machine and a human translator in a cooperative endeavor to translate from one language to another? What should be the respective roles of computerized medical diagnosis and the physician in diagnosing a patient? These topics, as Wiener describes them, invoke the image of harmonious collaboration between people and machines. Shortly before his death he thought of his own reflections as the beginnings of a new kind of study: "What we now need is an independent study of systems involving both human and mechanical elements. This system should not be prejudiced either by a mechanical or antimechanical bias. I think that such a study is already under way."[33]

Wiener left no doubt that he regarded scientists, engineers, and businessmen engaged in the production of automatic devices and technology generally as bearing a moral responsibility. He viewed them as holding a power analogous to powerful magic or, in particular, the power of performing the Mass in Christianity: "So long as we retain one trace of ethical discrimination, the use of great powers for base purposes will constitute the full moral equivalent of Sorcery and Simony."[34] Wiener's "putting into the hands" metaphor, his urging of the "imaginative forward glance" and the kind of education that would make it possible, and his comparison of the misuse of knowledge to sorcery and simony evoke the vision of a new type of scientist and technologist, not merely an expert or hireling, but a person who insists on making his work consonant with a humane social and moral philosophy.

Scientists and engineers, although they constitute a knowledge elite, represent only one element of the complex network in society that controls technology. Everyone has a stake in it and makes some choices. Wiener, like many of the physicists who had worked on the Manhattan Project, was eager to share his knowledge of the new technology with a wider public.[35] But Wiener went further, in that he sought to convey his understanding of the significance of technology to people's lives. His central and harshest theme was the cautionary story *The Monkey's Paw* by W. W. Jacobs, which Wiener repeatedly paraphrased in his books and lectures:

In this tale, an English working family is sitting down to dinner in its kitchen. The son leaves to work at a factory, and the old parents listen to the tales of their guest, a sergeant-major back from service in the Indian army. He tells them of Indian magic and shows them a dried monkey's paw, which, he tells them, is a talisman which has been endowed by an Indian holy man with the virtue of giving three wishes to each of three successive owners. This, he says, was to prove the folly of defying fate.

He says that he does not know what were the first two wishes of the first owner, but that the last one was for death. He himself was the second owner, but his experiences were too terrible to relate. He is about to cast the paw on the coal fire, when his host retrieves it, and despite all the sergeant-major can do, wishes for £200.

Shortly thereafter there is a knock at the door. A very solemn gentleman is there from the company which has employed his son. As gently as he can, he breaks the news that the son has been killed in an accident at the factory. Without recognizing any responsibility in the matter, the company offers its sympathy and £200 as a solatium.

The parents are distracted, and at the mother's suggestion, they wish the son back again. By now it is dark without, a dark windy night. Again there is a knocking at the door. Somehow the parents know that it is their son, but not in the flesh. The story ends with the third wish, that the ghost should go away.[36]

With a technology we can magically achieve some desired objective. But it does more than we ask: it changes our lives and the world about us in many other ways. Aside from the forward glance that might allow us to anticipate the overall long- and short-term changes, Wiener urges a constant feedback that would allow an individual to intervene and call a halt to a process initiated, thus permitting him second thoughts in response to unexpected effects and the opportunity to recast wishes. In effect, Wiener is asking the user of powerful automated tools to reflect upon what his true objectives are—to appreciate that multiple objectives usually conflict with each other and that to be able to articulate what one truly wishes implies a profound and sophisticated understanding of things and people, including oneself. This constitutes an important shift from the traditional view of technology: instead of thinking of a new technology merely as something that enables you to do such-and-such (the attitude of the "gadgeteer"), you come to realize that by making it part of your ecological system you grant it the power to alter your future, for better or worse. Just what part you wish it to play in your life and what relation to it you wish to have are the choices at issue.

The usual relation of man to machine is patterned on that of master to slave. The possibility that machines would come to dominate people troubled Wiener. He was interested in machine-slaves such as the checkers-playing machine that could learn from past successes and failures and eventually outplay its designers. Wiener asserts that slavery is not only cruel but self-contradictory:

We wish a slave to be intelligent, to be able to assist us in the carrying out of our tasks. However, we also wish him to be subservient. Complete subservience and complete intelligence do not go together. How often in ancient times the clever Greek philosopher slave of a less intelligent Roman slaveholder must have dominated the actions of his master rather than obeyed the wishes![37]

But intelligence is not the only way in which the machines may in effect become dominant. It is often simply the machine's speed of operation that causes human beings to lose control: "By the very slowness of our human actions, our effective control of our machines may be nullified. By the time we are able to react to information conveyed by our senses and stop the car we are driving, it may already have run head on into a wall."[38] Wiener's recommendation, then, is that we individually and collectively make technological choices in the light of the cautions contained in the metaphor of the master-slave relation and the story of the monkey's paw, in-

342
John von
Neumann
and
Norbert
Wiener

stead of letting the market mechanism operate without con-
straint.

It was Wiener's style to begin his thinking with metaphors and images and then go on to rational analyses, as in the analysis of the decision problem in nuclear war described in chapter 12. All of these reflections were accompanied by attempts to translate his evolving philosophy into concrete actions he himself could take. The *Atlantic Monthly* letter and the personal decision it reflected had been his first such action, and the most dramatic one. When asked by the editor of the *Bulletin of the Atomic Scientists* to comment on his position just two years later, he reaffirmed the earlier letter and talked scathingly of "the degradation of the position of the scientist as an independent worker and thinker to that of a morally irresponsible stooge in a science-factory,"[39] which he claims had intensified in the previous two years.

As to his concern for the injurious social effects of industrial applications of automata, after unsuccessful attempts to arouse the active interest of labor union leaders, he vented his alarm publicly in the kind of dramatic language he so liked to use:

This new development has unbounded possibilities for good and for evil. . . . The first industrial revolution, the revolution of the "dark satanic mills," was the devaluation of the human arm by the competition of machinery. . . . The modern industrial revolution is similarly bound to devalue the human brain at least in its simpler and more routine decisions.[40]

This statement by Wiener was widely quoted by the large-circulation magazines of that era, such as *Time, Business Week,* and the *Saturday Review.*[41] When one of the leading industrial corporations solicited Wiener's advice concerning the possible production of artificial control mechanisms, he was even more disturbed and made another attempt to contact labor. This time he wrote a long letter to progressive labor leader Walter Reuther, president of the United Auto Workers (UAW-CIO), and received a quick reply by telegram: "Deeply interested in your letter. Would like to discuss it with you at

earliest opportunity following conclusion of our current negotiations with Ford Motor Company."[42] Finally, on March 14,
1950, the two men had a breakfast meeting in Boston. Reuther
proposed to Wiener a project for organizing a labor-science
council; Wiener was soon talking with likely participants,
especially Jerome Wiesner. Wiener and the labor leader continued to exchange political views and express mutual support,[43] and Reuther began to cite Wiener and make practical
political use of his ideas.[44] (In the 1940s the labor unions
seemed to Wiener to be more humane in their objectives than
management, generally speaking, and for this reason he
wanted them to be at least as knowledgeable as management
about the new technologies. By 1954, however, Wiener had
been pleasantly surprised to find that the ranks of management also included men with a sense of social responsibility,[45] although he did not alter his prediction of large-scale
unemployment in the service sector of the economy.)

All in all, Wiener's approach to a philosophy of technology
was both radical and original. Technology was subject to a
range of narrow views, from the prevailing cult of technological progress to a blanket condemnation of technological development. On the whole, the reaction among Wiener's colleagues was what one expects when an independent thinker
challenges the customary ways of a professional establishment: rejection and ridicule.[46] In the years 1968–1972 I asked
a considerable number of mathematicians and scientists
about their opinions of Wiener's social concerns and his
preoccupation with the uses of technology. The typical answer went something like this: "Wiener was a great mathematician, but he was also eccentric. When he began talking
about society and responsibility of scientists, a topic outside
of his area of expertise, well, I just couldn't take him seriously." This dismissal on the grounds of "expertise" was not
surprising, for the issues Wiener raised, if taken seriously,
would demand reexamination of a whole set of received
ideas.

Wiener's approach seems particularly isolated because it
was in the context of his personal, unique formulation of a

344
John von
Neumann
and
Norbert
Wiener

philosophy of technology. Einstein had appreciated it, as has already been noted, and so had Bertrand Russell and Lewis Mumford, among others.[47] Quite independently of Wiener an organization was founded in 1949 through the efforts of a Quaker engineer, Victor Paschkis, and the well-known pacifist A. J. Muste, which required a commitment of its members to the concept that "every scientist is personally responsible for the foreseeable consequences of his work, and science and technology should be used only for constructive purposes."[48] This was the Society for Social Responsibility in Science, which in 1949 could boast thirty-five founding members. Wiener the individualist did not join.[49]

Some of the seventeenth-century founders of modern science had recognized the social or moral responsibility of the scientist: "Descartes stated a 'scientist's oath' of classic simplicity: I would not engage in projects which can be useful to some only by being harmful to others."[50] On the other hand, science has always been the handmaiden of warfare.[51] But after the sciences became professionalized in the nineteenth century, any consideration of social responsibility in connection with scientific research became a direct violation of the standards and values of the profession, a betrayal of the cause.[52] Thus, falling into a false innocence was the price scientists paid for the "benefits" of scientific progress during the period of autonomy and professionalization. Even most of the scientists who had indicated social concern in 1945 to the extent of joining the Federation of Atomic Scientists or the later Federation of American Scientists had dropped out of those organizations by 1950.[53]

The main themes of the philosophy of technology formulated by Wiener have, however, found articulate new spokesmen. In a remarkable essay in 1969 Paul Goodman—an independent intellectual, though not a scientist—examined the philosophy of technology from the point of view of young scientists, engineers, and technologists. Goodman envisaged a "reformation" of the technological and scientific professions, a housecleaning:

Whether or not it draws on new scientific research, technology is a branch of moral philosophy, not of science. . . . As a moral philosopher, a technician should be able to criticize the programs given him to implement. As a professional in a community of learned professionals, a technologist must have a different kind of training and develop a different character from what we see at present among technicians and engineers. He should know something of the social sciences, law, the fine arts, and medicine, as well as relevant natural sciences. . . . Often it is pretty clear that a technology has been oversold. . . . Then even though the public, seduced by advertising, wants more, technologists must balk, as any professional does when his client wants what isn't good for him.[54]

Goodman observes that we do not expect a physician to give a patient any medicine he may ask for but to act on his own judgment. The crisis Goodman saw in 1969 was that young people were alienated, and his suggestions were meant to aid the new crop of scientists-technologists to "recover lost integrity." Goodman also spells out his moral criteria for a "philosophic technology" that, while different from Wiener's, is not incompatible with it and is based on an outlook that appears fundamentally similar.

Wiener's specific reflections on computers and their relation to their designers or users were examined more fully and brought up to date in 1976 by Joseph Weizenbaum.[55] The general question of the possibility of an "autonomous" technology—that is, mechanical "slaves" with a measure of control over themselves and their surroundings—has been considered on a new level of sophistication by a political scientist, Langdon Winner.[56] Wiener's own books have continued to have a wide sale. They became part of the literature of the "movement" in the late 1960s and early 1970s, which resulted on March 4, 1969, in "a 'work-stoppage' and teach-in initiated by dissenting professors at the Massachusetts Institute of Technology, followed at thirty other major universities and technical schools across the country, against misdirected scientific research and the abuse of scientific technology"

346
John von
Neumann
and
Norbert
Wiener

(especially in relation to the war in Vietnam).[57] This same movement sprouted organizations like Computer People for Peace and Science for the People. In the mid-1970s the idea of noncooperation (resignation) or "whistle blowing" was put into practice by a number of scientists and engineers engaged in government or in industrial work related to nuclear reactors as a protest against the lack of adequate safety precautions.[58] The social and moral philosophy of technology has only begun to develop.[59] Meanwhile, narrowly conceived "technology assessments" and "cost-benefit analyses" continue to prevail in practice.

As everyone recognizes, a thousand salutary instruments
came out of the new technics; but from the seventeenth century on the machine served as a substitute religion, and a vital
religion does not need the justification of mere utility.
—Lewis Mumford

Eugene Wigner, reflecting a widely held view of exceptional
mental powers, called the brain of his childhood classmate
and lifelong friend John von Neumann "a true miracle."[1] In the
same spirit, the word "genius" is associated with an extraordinary scientific mind, rather than its older meaning of the
prevailing character of a place or a person or a time.[2] In this
way the extraordinarily talented individual is set off from ordinary humanity. Not only did von Neumann know himself to be
such an individual, but beginning early in life he had experienced success after success in his career, in his many mathematical achievements, in his relations with others and the
ease with which he escaped being trapped by repressive
political regimes, through the recognition accorded his "genius." All this seemed to confirm the efficacy of his talisman,
which additionally provided him with the solitary pleasure
of mathematical thought. The technological optimism that characterized von Neumann's youth was preserved by the heady
experience of the collective success of building a plutonium
bomb—an astounding achievement that culminated in his
viewing of the first test, which was only to be surpassed by
the unequaled spectacle of the hydrogen-bomb explosions,
which he traveled to the South Pacific to witness.

Although Von Neumann was a foreigner, he was recognized

348
*John von
Neumann
and
Norbert
Wiener*

as one of the most knowledgeable weapons experts in the world, and he was to sit with the powerful in the American government, helping to ensure "national security"—that is, invulnerability—through innovations in weapons technology and their implementation. Invulnerability required unslackening technical innovation and a kind of ruthless thinking that could shrug off the cancer deaths and other consequences of nuclear weapons tests. To von Neumann it was an acceptable price to pay to secure the future against any conceivable threat, and perhaps achieve a feeling of personal immortality as well.[3] But the pattern of living and thinking engendered in von Neumann by this cumulative experience of success and the accompanying sense of invulnerability, while persistent, was not completely stable, because it could be brought up short if von Neumann's actual human vulnerability were to obtrude itself in some form that could not be ignored or denied.

This chapter deals with von Neumann as a man, rather than with his theories or his political activities. Information on von Neumann's personal qualities is drawn from the people who knew him. There is a considerable difference between the impression von Neumann made on acquaintances and the character he revealed to those who knew him well. Thus, for example, acquaintances and others who encountered von Neumann described him as "amiable," "bland," "low-key," "easy," and "agreeable," while several who were close to him personally describe him as "intense," "high-strung," "ambitious," "proud," "has to be at the center," "has to be the best," "highly competitive." Such opposite descriptions suggest a man whose personality, more than most, was colored by social circumstances. However, impressions are in each case filtered through the informant's own personality and are most meaningful when something of that person's relationship to von Neumann can also be described.

Eugene Wigner and John von Neumann had been high-school friends in Budapest, but it was when they were both in Berlin after leaving Hungary that, "being foreigners, and not feeling part of the social structure, they became especially close."[4] They had shuttled between Princeton and Berlin together over a period of years, had taught courses together, had collaborated on research, and had emigrated to the United States at the same time, where at Princeton they were again "foreigners" together. During World War II they worked with each other at Los Alamos and as late as 1954 they published a joint paper in mathematical physics.[5] Wigner also, on the whole, shared von Neumann's hard-line attitude toward the Soviet Union and has himself put in considerable energy and time working with government committees since the war. It was almost as if the two were brothers, although with different personalities. At Princeton, however, von Neumann spent more and more time with other associates, such as mathematical colleagues Oswald Veblen and James Alexander at the Institute, and later Morgenstern and the computer group. Wigner had enormous admiration for von Neumann and could not help measuring himself against him. Consider the following revealing exchange between Wigner and historian Thomas Kuhn:

349
Von Neumann:
Only
Human
in Spite
of Himself

Wigner: I loved inorganic chemistry and still know it better than most present-day chemists. . . . I liked it because I like facts.

Kuhn: And you have a good memory.

Wigner: Not like von Neumann's.

Kuhn: That, I must say, must have been a shattering experience—to have grown up with Neumann, however bright one is.

Wigner: I am not a great man and I never have any illusion about that; on the other hand, I am a modest man and I don't want to excel. I'd much rather be a soldier than an officer.[6]

In fact it seems that everyone who was close to von Neumann, including his wives and his brothers, had somehow to come to terms with their own sense of self in relation to the impressive Johnny. In turn John had not only to demonstrate his gift but also to avoid making others uncomfortable.

When I interviewed Wigner on August 10, 1971, at his summer place at Lake Elmore, Vermont, he began to muse, reaching for some balance and perspective in his view of von Neumann. He wondered aloud about what Johnny's weak-

350
John von
Neumann
and
Norbert
Wiener

nesses had been. Johnny was entirely unathletic and could not do anything physical, he noted. I knew that von Neumann dealt with this limitation by cultivating a fine contempt for athletics.[7] Furthermore, Johnny ate too much, physicians' advice notwithstanding. I had read of Klara von Neumann's comment on this failing, to the effect that "he likes sweets and rich dishes, preferably with a good nourishing sauce, based on cream. He loves Mexican food"—chicken mole was a favorite—to the extent that "when he was stationed at Los Alamos . . . he would drive 120 miles to dine at a favorite Mexican restaurant."[8] Mrs. Wigner joined us just as her husband was commenting on von Neumann's inability to drive a car well, and recalled riding with him on one occasion, barreling down a one-way road the wrong way at great speed, cars whizzing past on either side, but with von Neumann undeterred and seeming not to notice. His reckless driving—he destroyed a car almost annually—became part of Princeton legend.

Wigner went on to speak of the seeming limits to von Neumann's capacity for feeling in human relationships.[9] Wigner thought that von Neumann's mother was his real love.[10] It was she who held the von Neumann clan together once Johnny had brought them all to America. Indeed John and his mother were close until the very end: she remained his confidante in personal matters.[11] But in relation to others, Wigner reflected, von Neumann was devoid of sentiment to an extraordinary degree. ("But," Wigner added, "I still miss him.") Johnny's relations with his brothers were cold, Wigner said, noting that while they greatly admired him, admiration is not the same as reciprocal love and affection. Listening to Wigner, I conjectured that in his own relations with von Neumann he had been pained by von Neumann's coldness. Wigner went onto say that Johnny believed in having sex, in pleasure, but not in emotional attachment. He was interested in immediate pleasures but had little comprehension of emotions in relationships and mostly saw women in terms of their bodies. But then, Wigner conceded, he may have had some feeling and sympathy for Marina, his daughter.

351
Von Neumann:
Only
Human
in Spite
of Himself

The adjective "cold" has been applied to von Neumann by others, among them the Hungarian-born mathematician Paul Halmos, who, as von Neumann's young assistant and collaborator at Princeton for two years (1940–42), saw him several times every day. Although Halmos "liked and admired" von Neumann, he "found him a cold man, formal and precise."[12] Many regarded von Neumann as "deficient in emotional development."[13]

Wigner and his wife, sitting at the table in their simple, secluded house overlooking a valley in Vermont, both commented on von Neumann's desire for luxury. Mrs. Wigner stressed von Neumann's fancy silver and china and Eugene Wigner emphasized his great desire for money itself, perhaps as a symbol of status or influence: "Johnny was so very materialistic!" He remembered that as a schoolboy Johnny, admiring his father, was attracted to a life of acquiring great riches before his father persuaded him to follow a career in mathematics instead.

To put this picture of von Neumann in a somewhat different light, he was a man who liked eating out, liked sex, liked parties, liked being rich and becoming richer, enjoyed having political influence, and especially enjoyed doing mathematics. All agree von Neumann was a cheerful man. He is remembered variously by other colleagues as "very friendly and accessible,"[14] "warm-hearted and friendly with people,"[15] "easy to get along with . . . didn't give himself airs."[16]

In fact many of his associates remember von Neumann chiefly as a sociable, witty, party-loving mathematician, and they remember his parties. Some recall the grand celebration at the completion of the Princeton computer, for which Klara von Neumann had arranged a replica of the computer in ice.[17] Wigner said that von Neumann was a marvelous raconteur,[18] and according to L. J. Savage, a younger mathematician, "He loved filthy limericks. He knew all the filthy limericks in the world and told them with a sort of detached gusto. If specifically filthy words were used he might mumble the word, rather than say it. There was a certain refinement . . . it was just a little game."[19] Yet some of his colleagues found it

352
John von
Neumann
and
Norbert
Wiener

disconcerting that upon entering an office where a pretty secretary was working, von Neumann habitually would bend way over, more or less trying to look up her dress. Nevertheless, this seems in line with the view of women von Neumann expressed to his friends.[20]

Parties, intellectual conversations, and urbane social life were von Neumann's favorite recreations. Both of his wives also relished parties. When von Neumann left a party to go upstairs and work, he nevertheless seemed to like to hear the sound of it in the background.[21] When alone, he would often turn on the radio while working. It has been conjectured that von Neumann's intense gregariousness served a deeper need, providing him with a constant confirmation that despite his distinctiveness he was part of the human community. He went to the movies, even if sometimes he fell asleep. But he was constantly so active that his ever lounging on a beach in the sun seemed improbable. He was not interested in listening to music except a marching band or musical comedy show. He had little artistic sensibility—except of course in abstract mathematics—and his interest in art was that of a historian rather than a connoisseur.[22] Ever since childhood, he had had a strong interest in knowing and memorizing historical facts, having systematically read the twenty-one volumes of the *Cambridge Ancient History* and the *Cambridge Medieval History,* and he retained an excellent knowledge of ancient history especially. In addition, he was "a major expert on all the royal family trees in Europe. He can tell you who fell in love with whom, and why, and what obscure cousin this or that czar married, how many illegitimate children he had and so on."[23] He was strong in Byzantine history as well. Wiener also had an interest in history, but with a very different focus; he was more concerned to understand the social circumstances of a particular period than the life led by the upper classes.

Although to a much lesser extent than Wiener, von Neumann also was forgetful, and there are some family stories about von Neumann as the absentminded professor.[24] Stories are also told of his lecturing while writing on the blackboard and erasing what he had written so quickly that students never got a chance to see it.[25] Still, with his fastidiousness and his conservative, bankerlike attire, he appeared unlike the usual professor. When he appeared in his class one morning still in formal evening clothes and somewhat bleary-eyed, no one attributed it to absentmindedness.[26]

353
Von Neumann:
Only
Human
in Spite
of Himself

John and Klara von Neumann lived in a large house in Princeton. Some of the time his daughter Marina and his mother, whom he called by the diminutive "Gittush," lived with them. The household also had hired servants, at times including a maid and a butler. With the servants von Neumann was often very imperious. Frequently there were houseguests. An important part of the wife's function was to be hostess and to manage the household. The gatherings of people at the von Neumann house provided something like the intellectual and social excitement that the hostess of a lively rococo salon in eighteenth-century France may have experienced. Still, the rule of both of von Neumann's wives was essentially to be his handmaiden, and they both found it difficult to be individuals in their own right with Johnny.

Klara—small, vivacious, warm, and witty—worshipped her husband but also suffered under the circumstances of living in the shadow of the great man, aggravated by his need to be the center of attention. Their normal language for conversation was Hungarian, and when Johnny talked with Wigner it was usually also in Hungarian. Von Neumann often had too many important preoccupations to be aware of his wife's or daughter's ordinary needs, and his lack of interest in personal feelings made him all the more oblivious. But he had helped Klara become a computer programmer, when they were at Los Alamos, where the wives of scientists had little to occupy themselves. He was also concerned lest his daughter's intellectual development be hindered by a too-early marriage. (It is noteworthy that Marina von Neumann Whitman, who has become an economist of considerable standing, and is now a Vice President of General Motors Corporation, was not intimidated or overwhelmed by her father's intellect.) Both von Neumann and his wife were intense, ambitious, and high-strung, but their marriage has been described to me as on the whole a good match, although occasionally he would make fun of her in a cruel way in front of other people. Stanislaw

354
John von
Neumann
and
Norbert
Wiener

Ulam explains that von Neumann "may not have been an easy person to live with—in the sense that he did not devote enough time to ordinary family affairs. . . . Von Neumann was so busy . . . he probably could not be a very attentive 'normal' husband. This might account in part for his not-too-smooth home life."[27]

Von Neumann's intellectual energy was prodigious. After spending a day in his office, he would often work in his study until two in the morning, perhaps while a party went on downstairs, and be up again at six, continuing to work with powerful concentration. In the 1950s a stream of visitors, especially military technologists, came to consult with him during the evening hours. His wide range of interests, however, did not include household details: these were for the women and servants. According to a family story, one morning Klara felt sick and asked John to get her a glass of water; he returned after five or ten minutes to ask where the glasses were kept in a house in which they had been living for seventeen years. He was determinedly ignorant in these matters. His wife said that when von Neumann "is not interested in what is going on around him, he relaxes and falls asleep."[28]

Like many other creative people, including Wiener, von Neumann had a special relation to the world of childhood, which includes both what is usually referred to as "childish" and what is seen as "childlike" (both terms may be misnomers). His omnivorous curiosity and his intellectually playful view of the world have already been noted in connection with his style of work. According to an article based on information from his wife Klara, "Dr. von Neumann loves ingenious children's toys. His friends give him a vast assortment of these on his birthdays, partly as a gag, partly because they know he really adores them. He has been observed, unaffectedly scrapping with a five-year-old over who was to have priority in using a new set of interlocking building blocks. He never patronizes children."[29] Mrs. von Neumann recalled playing games with him, such as betting on which car would reach the corner first in a traffic jam. For von Neumann work and play were of a piece. Herman Goldstine, speaking of von Neumann's computer group at Princeton, recalled a game played at meetings, with von Neumann the initiator or at least an active participant: "We used to listen at meetings to serious presentations, and we would count the number of times in a sentence or a paragraph that the speaker said something which we would deem to be 'nebech.' Then we would flash the number at each other by holding up fingers. We would be flashing numbers back and forth at each other throughout the lecture."[30] In addition to toys, von Neumann was intrigued with every new gadget, the more complicated the better. Mrs. Natasha Artim Brunswick of Princeton recalled that in the early 1940s he was one of the first to have a windshield sprayer on his car and that, with boyish enthusiasm, he drove around to his friends' house to show it off.[31] According to Oskar Morgenstern, he loved to be looked at and show off in front of women, and it greatly amused him to put on toy paper hats at a party.[32]

355
Von Neumann:
Only
Human
in Spite
of Himself

Von Neumann was a child in other ways: he wanted attention, he wanted to be at the center of things. He insisted that his own stature be recognized and properly appreciated by everybody. He was justly proud of the unmatched speed with which he could work out mathematical problems in his head. This left him open to scientific practical jokes.

A friend once spent a week working out various steps in an obscure mathematical process. Accosting von Neumann at a party he asked for help in solving the problem. After listening to it, von Neumann leaned his plump frame against the door and stared blankly, his mind going through the necessary calculations. At each step in the process the friend would quickly put in, "Well, it comes out to this, doesn't it?" After several such interruptions von Neumann became perturbed and when his friend "beat" him to the final answer he exploded in fury. "Johnny sulked for weeks," recalls the friend, "before he found out it was all a joke."[33]

His friends also puzzled about his childish awe of government officials and medals. "He was much attracted by the honor of being a Commissioner," and he "liked the whole smash of the helicopter coming down on the Institute grounds to pick him up."[34] "Odd how impressed he was by uniforms and generals, the government honors."[35] And his deference

356
John von
Neumann
and
Norbert
Wiener

toward government officials: "Von Neumann was Chinese in his courtesy towards the Weather Bureau. He asked their permission where it wasn't needed."[36] Some wondered at his conception of adulthood, for when he testified before the AEC Personnel Security Board in the Oppenheimer case he had all but conceded to them that the world of military security and political maneuvering represented maturity, which "foolish" scientists ("little children") needed to face up to.[37]

He had a polite, cosmopolitan manner. Moments of anger were usually restricted to private settings, although in the 1950s he became more impatient, typically looking at his watch every few minutes in conversations with his younger associates,[38] who remember some emotional outbursts on occasions when he was pressing them to get on with the job.[39] The eighty-three-year-old Richard Courant thought that one of Johnny's salient characteristics was his "enormous intensity" and that, aside from his considerable mathematical skill and scope, he in particular "was a really great salesman and had a persuasive personality" when it came to getting funds for computers.[40]

Von Neumann was so smoothly adapted to the social world around him that one quite forgets how in some ways he was very different from other people and was indeed an outsider, although he himself could hardly have escaped being conscious of this circumstance. Whereas Wiener protected his sense of himself as an outsider, von Neumann seemed effectively to minimize it, doing what he was able to become part of the inner circle. These are very different attitudes toward marginality in relation to elite groups, and toward one's own atypicality. Colleagues at Princeton caught something of von Neumann's style of adapting in the story that made the rounds about him that he was not human but a demigod who "had made a detailed study of humans and could imitate them perfectly."[41]

Recalling von Neumann's youth, it is noteworthy that he went through his schooling together with his age-mates, and while his mathematical ability distinguished him, it in no way made him an outsider or excluded him from ordinary schoolboy activities. Moreover, upper-middle-class

357
Von Neumann:
Only
Human
in Spite
of Himself

Budapest families such as his had generally put great emphasis on the development of social skills; a courteous and charming manner was nurtured as an essential tool for advancing in the world and for surviving the reverses an uncertain future might inflict. The ambition to become acceptable and break into the social and political elite, the ruling stratum of Hungary, was normal in the community in which von Neumann was raised, so that his later style of relating to elites had an early precedent.

I have already mentioned Wigner's reference to his own and von Neumann's sense of isolation as young foreigners in Germany. Von Neumann had learned German early in Budapest, as it was to some extent the language of the upper classes, and he practiced it further as a student in Switzerland. Adopting the title "von" helped him to appear as an insider, a member of the non-Jewish German aristocracy, in title-conscious, nationalistic Germany. His mathematical work followed Hilbert's orientation and was very much in the mainstream of contemporary work in the field. All this may have served to help him disregard any disturbing sense of marginality he felt.

Von Neumann's first wife was a Hungarian-born Catholic; his second wife was also Hungarian, the daughter of a wealthy Jewish business family. This suggests that however much he liked America and regarded himself as an American, he was still most at home in a Hungarian atmosphere, and where he could speak Hungarian. Moreover, as Ulam notes, von Neumann was particularly comfortable with "third- or fourth-generation wealthy Jews," people with a background similar to his own, and his repertoire of witticisms included Yiddish humor.[42] Although apparently totally uninterested in religion (nominally he had become a Catholic during the time of his first marriage), he did not hide his Jewish origins. In addition, the prevalence of immigrants among the American scientific community after 1939 meant that von Neumann's slight accent was anything but out of place at Los Alamos and at scientific gatherings. With his superpatriotism, his promotion of the armaments race, and his advocacy of a first strike against the Soviet Union, von Neumann in the America of the

358
John von
Neumann
and
Norbert
Wiener

1950s seemed to have escaped completely from every vestige of the central European Jew's heritage of victimization, and to have crossed over to the other side. Others of his background came to reject with particular disgust the use of violence by nations who saw themselves as the lords of the earth; von Neumann made the military strength of the most powerful nation on earth his metier.

Bochner, a fellow mathematician, put his finger on the characteristic that provides the clue to von Neumann's special relation to the scientific community: "By intellectual faculty he was a mathematician, but by species *he was a scientific man within all the major connotations of today,* from esoteric ones to those of popular appeal."[43] In other words, von Neumann was a paragon of science, a view of him shared by various members of the younger generation of mathematicians.[44] Some of the most unreserved admiration for von Neumann came from physicists with whom he had worked during World War II and from scientists he had met at the cybernetics conferences, even though their comments do not refer primarily to his mathematical skill. Nobel laureate Hans Bethe's wondering whether "a brain like von Neumann's does not indicate a species superior to that of man" was inspired by Bethe's familiarity with von Neumann's work related to physics, not pure mathematics, and the Manhattan Project.[45] Similarly, historian Laura Fermi's comment that "he was . . . one of the very few men about whom I have not heard a single critical remark" describes the physicists' view of von Neumann— Fermi's husband, Nobel laureate physicist Enrico Fermi, was also one of the leaders of the Manhattan Project.[46] The highly independent younger scientist John Stroud, who came to know von Neumann only after World War II, in particular at the Macy conferences, has developed a theory according to which the next stage in civilization will be one in which scientific objectivity, scientific attention to detail, and rationality will be the bases for almost all human choices. He makes a point similar to Bochner's: "I was much impressed by von Neumann's mathematical prowess. . . . I was even more impressed by the fact that Johnny was such a damn good generalist. . . . As a matter of fact just about anything that I noticed coming to Johnny's attention got about the same kind of treatment."[47] What Stroud means is that von Neumann responded like a scientist to all kinds of information and issues, not merely to topics defined by a particular scientific discipline. Ralph Gerard, an outstanding neurophysiologist, a contemporary of von Neumann, and a regular participant at the Macy conferences, who described his own commitment to science as based on "devotion to Reason as a way of life," contrasted von Neumann with Warren McCulloch:

359
Von Neumann:
Only
Human
in Spite
of Himself

McCulloch was not a typical scientist; he was never really accepted in science, he was more a poet and philosopher. . . . For Johnny von Neumann I have the highest admiration in all regards. I would suppose that he is one of the two or three men whom I knew contemporarily who were my ideals. He was always gentle, always kind, always penetrating and always magnificently lucid. I first met him at the Macy Meetings. . . . There he managed to make extremely complicated things crystal clear. . . . He was very eager to work things through intellectually, extremely able to do so, and would defuse a confused situation.[48]

These views of von Neumann are far more interesting than the primitive worship of technical virtuosity to which he was often subject. One profound scientific value is the capacity to respond with cool, creative rationality to every circumstance, to respect facts, to be curious and enthusiastic, yet at the same time dispassionate and objective, devoted to understanding the world in logical terms. These qualities suggest something of an ideal "scientific attitude," which for some scientists represents not merely a philosophical position, but a habit and a value governing a portion of their behavior. Von Neumann seemed to personify this ideal to a high degree even in areas outside of science proper. Where Wiener may have been passionate in his enthusiasms and evocative (even dramatic) in his assertions, von Neumann was generally detached and precise in his evaluations, appearing neutral and dispassionate. Even in his computer-brain analogies he made no exaggerated claims. Where others tended to get carried away by their own enthusiasm, von Neumann seemed always to appraise a situation as it actually was, and it was the com-

360
*John von
Neumann
and
Norbert
Wiener*

bination of his habitual scientific attitude and his extraordinary mental-mathematical abililties that earned him such widespread and unreserved admiration among scientists. Also, unlike Wiener, von Neumann always remained decorously within the rules of behavior governing the scientific-mathematical community.

Any historical appraisal of von Neumann, then, demands consideration of scientific activity itself and of the values and ideals upheld by its practitioners. The ethos of science rests on two pillars, the politically useful myth of "value neutrality" and the article of faith most conducive to the growth of scientific bureaucracy, namely, that scientific innovations ("progress") and science-based technological innovations are a priori beneficial.[49] While these pillars clearly knock against each other, they continue to hold up the practice of science.

The first pillar, the myth of value neutrality, has had an important social function for the growth of modern science since it was first institutionalized by the establishment of the Royal Society in England in the seventeenth century: scientists needed and wanted the support of the theological and political powers, and this support could most easily be obtained if science in no way challenged religious or political authority.[50] The Society conveniently outlawed the subjects of theology and politics from its meetings, emphasizing useful practical knowledge, and was granted a royal charter by Charles II. Throughout the ensuing centuries the increasingly professionalized and specialized scientific community depended for employment and well-being, and for a measure of autonomy, on being regarded with favor by the political powers. Being value-neutral, the scientists posed no threat; as a source for useful inventions, they were clearly an asset. In the seventeenth century the rhetoric of value neutrality tended to obscure the fact that the new science was the beginning of a radical subversion of the status quo through a scientific-technological-industrial revolution. In the twentieth century, however, the claim of value neutrality has on the whole tended if anything to hide the fact that scientific work has directly or indirectly strengthened the hand of the already dominant centers of economic or political power.[51] This is hardly sur-

361
Von Neumann:
Only
Human
in Spite
of Himself

prising when one considers, for example, that scientific research in the United States and the Soviet Union since World War II has been largely financed by government, although in the United States a few large private corporations have hired scientists for nongovernmental research and some foundations have also provided research funds.

Aside from the social function of the value neutrality of science, its philosophical basis is a topic that has evoked controversy. However, the conviction that knowledge based on the accepted methodologies of science has some kind of absolute character as "truth," or the more extreme belief that it is the only kind of acceptable truth, has no rational foundation. In the twentieth century these convictions engender what has variously been described by philosophers as scientific "obscurantism" or scientific chauvinism,[52] a dogmatic attitude leading us astray toward nonhuman chimerical absolutes.

The second pillar, the progress of science, is the central value on which the rules of professional scientific behavior are based, and it defines special virtues accorded to scientists.[53] It was indispensable historically to the growth, professionalization, and institutionalization of science. The professional merit of a mathematician is gauged in terms of how far he contributes to progress in mathematics, and similarly in the other sciences. The idea of progress as defining the character of human history can be traced from the early eighteenth century,[54] and the apparently cumulative nature of scientific knowledge made the growth of science into a symbol of this optimistic faith.[55] Progress was believed to be unidirectional, leading toward "improvement," although the nature of the improvement was seen very differently by different thinkers, like Kant, Marx, Comte, and Spencer.

I have already noted von Neumann's special kind of scientific optimism in connection with his early attempts to prove mathematics self-consistent, his hope for a fully mathematical economics based on game theory, his mathematization of quantum theory to include even the observer, and his envisioning of the future possibility of logical models for the central nervous system. Thus it is significant that in the 1950s von

362

John von
Neumann
and
Norbert
Wiener

Neumann said that he had learned through his own experience that the standards for identifying mathematically valid statements "fluctuate" from decade to decade. Moreover, he identified himself with the large group of mathematicians who will nevertheless work in those areas of mathematics—especially "when very beautiful theories could be obtained"—in which the foundations are lacking.[56] This attitude recognizes limits to absolute rigor and yet preserves a practical optimism that facilitates the growth of mathematics. Aside from the problem of the theoretic validity of science, von Neumann feared a future in which the basis for optimism would be undercut by a change in society's values: he once "expressed an apprehension that the values put on abstract scientific achievement in our present civilization might diminish: 'The interests of humanity may change, the present curiosities in science may cease, and entirely different things may occupy the human mind in the future.'"[57] In the 1950s we find von Neumann taking various opportunities to defend "science for science's sake," against the intrusion of philosophical considerations[58] and against the imposition of mission-oriented objectives.[59] He was doing this even as his own work had become nearly entirely mission-oriented.

The possible future diminution of interest in science, which made von Neumann uneasy, was of course seen by others as promising liberation from scientific obscurantism and chauvinism, possibly through the intrusion of humanistic, social, and ethical dimensions into science. The two pillars of value neutrality and faith in the beneficence of all technological innovation could hardly hold up after World War II, and in fact Wiener had been one of those who emphasized that. If they crumbled, the true scientific virtues would also lose priority. These virtues had been promoted at the cost of others among the wide range of human possibilities. They had risen to prominence in the wake of the Newtonian and industrial revolutions. Both Wiener and von Neumann were specialized beings trained for mathematical or scientific work, and their limitations in other respects are also apparent. One may wonder whether the post-Hiroshima world does not require different virtues and an emphasis on development of other human possibilities. The rise of the "third world" indicates strong pressure in the direction of social justice; the growing likelihood of nuclear war exerts a pressure in the direction of nuclear disarmament; the limits of global resources create a pressure for conservation, awareness of ecological patterns, and restraints on the production of material luxuries; and political instabilities arising from rivalry among nation-states exert simultaneous pressures for global planning and decentralization. The assertion of the political and social dimension brings a focus on people instead of things, and subordinates science and technology to their human context. Precisely what the nature of the emerging values and virtues are or need to be is a topic deserving of a major inquiry of its own.

363
Von Neumann:
Only
Human
in Spite
of Himself

Some of von Neumann's scientific admirers have seen in him the "new man," the ideal type of future person, implied by his name. The burden of the vision as well as the historical evidence informing this book is that the direction of human development most valuable to the species in the late twentieth century is likely to be very different from that of the ideal scientific man. Von Neumann can with considerable justification be regarded as a superb and highly developed representative of a nineteenth-century vision of modern man, but if indeed a shift in the direction of civilization is occurring, it is inappropriate to view him as representative of what man should become in the future. The virtues and virtuosities of the ideal scientific man are particularly conducive to productive scientific work, but as a generalized ideal of conduct—and they have become that in a Western civilization—they are in conflict with a wider, more comprehensive humanness.

We are still in the midst of the scientific-technological ideals, values, and myths that have come to dominate Western civilization since the seventeenth century, and it is difficult to gain historical perspective. In the early feudal period in medieval France, a standard of masculine conduct emphasizing bravery, loyalty, fearlessness in battle, and gallantry toward conquered enemies was part of the code of conduct for warriors. In subsequent centuries this code of conduct evolved into the romantic ideal of chivalry, embodying these and still other virtues, and became a cultural myth.

364

John von
Neumann
and
Norbert
Wiener

Whether or not many people were true to that ideal, it surely influenced their thinking and behavior. We hardly need Cervantes to show us, or the modern feminists to point out to us, that the chivalric ideal was in fact narrow, ludicrous at some points and cruel at others, and capable of being converted to the barbaric purposes of the Crusades. The "virtues" of knighthood do not constitute a comprehensive human ideal, but a very special, historically determined one. Similarly, the scientific virtues are limited, and their dominance may well be overcome by a more wholesome, comprehensive pattern of cultural values. Those who admired von Neumann as a "new man," more highly evolved than other men because of his logical-mental ability, were under the persistent spell of a scientific-technological system of values. Similarly, those who see the new postindustrial man in those experts such as von Neumann who wield influence in government through their professional advice are also thinking within the context of current technocratic, technological ideals.[60] The idealization of these two kinds of "new man," derived by direct extrapolation from the presently dominant value structure, reflects a deep conservatism and a failure of imagination. Perhaps a modern Cervantes is needed to help our own age gain perspective.

The possibility that the scientific ideal will increase its dominance cannot be discounted. The cultural pressures in that direction, however, appear to contain self-destructive and dehumanizing elements, awareness of which tends to generate broader visions of what it means to be human. Thus the scientific ideal is likely eventually to become incorporated into a broader vision of man and subordinated to it. It might go the way of the ideal of knighthood, surviving as a minor and increasingly decadent value within the culture even though a few individuals continue to be inspired by it.

Although the line separating science from technology is a blurred one, especially for someone like von Neumann who crossed back and forth frequently and easily, some people take distinctly different views of progress in "pure" science and progress in technology. I have suggested earlier that the coincidence of von Neumann's youth with Hungary's period of economic-industrial expansion helped to determine his optimism about and emotional commitment to technological progress, which his personal talents served to intensify. Moreover, we have seen that the events surrounding World War II—the Nazis' efficient extermination techniques and the use of the atomic bomb—did not suffice to undermine this optimistic outlook. What, then, was von Neumann's implicit or explicit philosophy of technology in the 1940s and 1950s? How about computer technology, which could be expected to alter society and with which von Neumann was so intimately involved?

365
*Von Neumann:
Only
Human
in Spite
of Himself*

It is clear from von Neumann's activities from late 1944 on, when he was first brought in as a consultant for the development of the ENIAC computer prototype, that he favored and helped bring about the most rapid possible rate of innovation and development in this field. In 1945, while the ENIAC was being built, von Neumann and his group were already designing a more advanced machine, the EDVAC, and von Neumann was proposing building a computer at Princeton that would leapfrog over the advances of the EDVAC.[61] Moreover, he was eager to see his own machine surpassed. He complained to IBM executives and Navy brass that the rate of innovation is usually determined only by considerations as to "what the demand is, what the price is, whether it will be more profitable to do it in a bold way or in a cautious way, and so on" and recommends instead that it is sometimes desirable to write "specifications simply calling for the most advanced machine which is possible in the present state of the art."[62] Von Neumann meant what he said, according to Herman Goldstine:

He and I were discussing ways to encourage industry to make bold new strides forward when suddenly he asked me whether it would not be a good idea to give contracts to both IBM and Sperry Rand to build the most advanced type computers that they would be willing to undertake. Out of this conversation grew a quite formal set of arrangements between the Atomic Energy Commissioner and each of these companies.[63]

366
*John von
Neumann
and
Norbert
Wiener*
Thus, some of von Neumann's desire for the maximization of progress in computer technology was propagated in the industrial world.

In any history of technology, certain major specific innovations in computer design must be attributed to von Neumann. But it would constitute a misunderstanding of the nature of technological change to regard his role as essential to computer development. For here, as is typical in technical change,[64] numerous patent suits testify to the near-simultaneity of similar innovations by different research groups.[65] In technological development, innovation must be supplemented by hardware implementation and by government or industrial interest in applying it; otherwise it lies fallow. Through his energetic advocacy, von Neumann did far more than most mathematicians would to bring about the hardware implementation and he also affected the field indirectly, in that "he was the man who, with his great prestige in the mathematical world, taking such an interest in these things, had the effect of stimulating many, many others to regard it as a very serious discipline."[66] Von Neumann's real impact on the computer was to speed the rate of advancement of the technology—more reliable and faster computers were available sooner because of him. This was something he wanted to and unquestionably did achieve. The same could be said concerning his role in weapons development and the nuclear armaments race. What is different in the case of weapons development is that the speed with which it took place precluded the possibility that gradually changing political conditions could obviate it.

In a 1955 article in *Fortune* magazine Atomic Energy Commission member von Neumann articulated a more general philosophy of technology as a basis for dealing with the crises created by technological progress.[67] It is possible that in such an article for the general reader his primary aim was to make official AEC policy (the nuclear arms race) palatable rather than to expose his personal views. Yet even with this qualification, the article is of some interest. He defines the crisis in terms of the finite size of the earth, which ultimately limits expansion into new geographical regions. For someone else this awareness might have led to a Wienerian emphasis on homeostasis or what today would be called an ecological consciousness, but for von Neumann it did not. He did concede that impact of some large-scale phenomena (he gives the melting of the polar ice caps as an example and alludes to nuclear weapons) might be felt worldwide and not limited to smaller political units. Nevertheless, he viewed the march of technological progress as inevitable and beyond human control, and dismissed as irrelevant efforts to impose any prohibitions on technical developments. Human beings simply had to make the necessary adaptations. Of course he knew firsthand how the US government was deciding which technologies to develop. Even when they contain dangers, von Neumann insisted, new technologies are always ultimately constructive and beneficial. The scientist or technologist is consequently exonerated from responsibility. Although conceding in a vague, general way that major changes in political institutions would eventually be needed to accommodate the then-new weapons, he was adamantly opposed to any institutional changes that might slow down new technical developments. Von Neumann seemed to be religiously holding to the traditional view of the beneficence of technological progress, incorporating the new facts of nuclear weaponry without in any way significantly altering the classical views. As we have seen, in more private communications he had argued frankly that the sacrifice of human lives is an acceptable price to pay for technological progress.[68]

Von Neumann's apologetics for government policy seem to have coincided with his own views. Most of his erstwhile scientific friends and colleagues who were critical of his military involvement saw it only in personal terms or as a loss to mathematics or pure science. One of them, Warren McCulloch, wrote to von Neumann, "I regret, personally as well as scientifically, that honors and duties have ascended to your entertaining head. It would be good to see and hear you again—just for the fun of it."[69] From having read *Cybernetics*, if not from more direct conversation, von Neumann was aware of

367
Von Neumann:
Only
Human
in Spite
of Himself

368
John von
Neumann
and
Norbert
Wiener

Wiener's views on technological progress, but he was obviously unconvinced. His *Fortune* article can be regarded as a public answer to Wiener.[70]

Von Neumann's all-embracing enthusiasm for scientific and technological progress seems relatively primitive for such a complex mind; the open vista may well have had a profound personal meaning to him. The idea of unending progress implies an optimistic view of time: it neglects limits, decay, demise. Von Neumann's whole commitment was to effective development of what the sociologist Jacques Ellul has labeled *la technique,* that which in English is suggested by the word "techno-logic." It specifically includes hardware technology as well as software techniques (such as decision theory and computer programs). It refers to a focus on instrumental rationality to achieve well-defined objectives and the creation or use of all kinds of mechanical and formal tools for this purpose. It has been persuasively argued by Ellul that when patterns of thinking and acting along the lines of techno-logic dominate an individual, he is deprived of essential elements of consciousness concerning what it means to be human.[71] Does von Neumann the individual confirm Ellul's analysis?

The view of von Neumann as a paragon of science and as a technologist par excellence raises fundamental issues concerning the scientific community, technology, and our advancing but simultaneously deteriorating civilization. This is what makes von Neumann such a fascinating historical figure. On a personal psychological level he was deeply committed to unlimited technological progress, to be achieved with the greatest possible speed, a commitment related to the maintenance of an attitude of cheerful optimism, as if innovation could rejuvenate us, could save us from old age and death, that cyclical, unprogressive aspect of time. This was the irrational foundation that underlay von Neumann's sophisticated application of reason. On a sociological level his commitment was to providing high technology for that powerful group sometimes called the military-industrial complex, in effect strengthening the already great power of that group. Uninhibited by ethical considerations in his drive toward innovation, for better or for worse he chose to be "hard-boiled" concerning the exploitation of nuclear weaponry. By putting himself at the service of the conservative powers that be, he elaborated and helped to advance existing trends in weapons policy and could not help but become a symbol for these very trends.

369
Von Neumann:
Only
Human
in Spite
of Himself

Mircea Eliade has characterized the aspiration that lies at the heart of the West's technique- and technology-oriented civilization:

We must not believe that the triumph of experimental science reduced to nought the dreams and ideals of the alchemist. On the contrary, the ideology of the new epoch, crystallized around the myth of infinite progress and boosted by the experimental sciences and the progress of industrialization which dominated and inspired the whole of the nineteenth century, takes up and carries forward—despite its radical secularization—the millenary dream of the alchemist. . . . On the plane of cultural history, it is therefore permissible to say that the alchemists, *in their desire to supersede Time*, antici-pated what is in fact the essence of the ideology of the mod-ern world.[72]

But the potential for fulfilling the dream of superseding time through science or technology is limited. Some mathematical theorems may be nearly timeless, but personal immortality is not the same as the long life of a theorem. Nevertheless, the ideal is dear in personal terms, and it is only human to main-tain it as long as possible, however illusory it may be. If some circumstance, perhaps a painful confrontation with reality, destroys this illusion in someone whose attachment to it is especially strong, the victim of the experience of the harsh contradiction between illusion and reality may literally go crazy. If fortunate, the victim recovers from the experience.

In the summer of 1955 von Neumann slipped on a corridor in an office building and hurt his left shoulder. The injury led to the diagnosis of a bone cancer in August. It had already metastasized. His first response to the situation was optimis-tic, even stoic, and he worked harder than ever. To a remark-able extent he continued to meet the many demands made on him in his capacity as AEC commissioner and to pursue his other interests. There was a bitter irony in the situation in that von Neumann, who had waved away the cancer-producing effects of nuclear weapons tests, would himself contract this

370

John von
Neumann
and
Norbert
Wiener

awful disease. He had increased his own chances of getting cancer by personally attending nuclear weapons tests and by staying at Los Alamos for long periods.

Before long his accustomed four or five hours of sleep did not suffice, and his enormous activity was forced into a slower pace. Yet death was unthinkable, inconceivable—so many plans and projects were still unfulfilled. The psychological stress was enormous, the more so as his special sense of invulnerability, which many of us share but to a lesser degree, was being challenged. To ease his spiritual troubles von Neumann sought not a Jewish Rabbi but a priest to instruct him in the Catholic faith. Morgenstern, among others, was shocked. "He was of course completely agnostic all his life, and then he suddenly turned Catholic—it doesn't agree with anything whatsoever in his attitude, outlook and thinking when he was healthy."[73] Whether his choice was prompted by his fondness for ritual, or his longing for immortality, or by some nostalgia for a world from which he had been excluded in childhood, or by a spiritual impulse to renounce the dynamo for the Virgin, or by entirely different reasons, in the spring of 1956 Father Strittmatter began coming to see him regularly.

But religion did not prevent him from suffering; even his mind, the amulet on which he always had been able to rely, was becoming less dependable. Then came complete psychological breakdown; panic; screams of uncontrollable terror every night. His friend Edward Teller said, "I think that von Neumann suffered more when his mind would no longer function, than I have ever seen any human being suffer."[74] And Eugene Wigner wrote,

When von Neumann realized that he was incurably ill, his logic forced him to realize also that he would cease to exist, and hence cease to have thoughts. Yet this is a conclusion the full content of which is incomprehensible to the human intellect and which, therefore, horrified him. It was heart-breaking to watch the frustration of his mind, when all hope was gone, in its struggle with the fate which appeared to him unavoidable but unacceptable.[75]

Von Neumann's sense of invulnerability, or simply the desire to live, was struggling with unalterable facts. He seemed to have a great fear of death until the last, at least for as long as he could still communicate with Father Strittmatter.[76] No achievements and no amount of influence could save him from extinction now, as they always had in the past. Johnny von Neumann, who knew how to live so fully, did not know how to die.

371
Von Neumann:
Only
Human
in Spite
of Himself

Physiologically his brain was fully intact, not touched by the cancer, but he had been suffering steadily from physical pain.[77] And still the United States government depended on his thinking. As Admiral Lewis Strauss describes it, speaking of sometime after April 1956,

Until the last, he continued to be a member of the [Atomic Energy] Commission and chairman of an important advisory committee to the Defense Department. On one dramatic occasion near the end, there was a meeting at Walter Reed Hospital where, gathered around his bedside and attentive to his last words of advice and wisdom, were the Secretary of Defense and his Deputies, the Secretaries of the Army, Navy and Air Force, and all the military Chiefs of Staff. The central figure was the young mathematician who but a few years before had come to the United States as an immigrant from Hungary. I have never witnessed a more dramatic scene or a more moving tribute to a great intelligence.[78]

Nevertheless, von Neumann was assigned an aide, Air Force Lt. Col. Vincent Ford, and air force hospital orderlies with top-secret security clearance, lest in his distraction he should babble "classified information." He died February 8, 1957, at the age of fifty-three.

But searching deeper into the Jewish phenomenon, we come upon a paradox that constitutes the actual uniqueness of this ethnic community: no other particular people has been so constantly and immediately involved and concerned in the destiny of humanity at large: no other people's individuality was so intrinsically interwoven with genuine *universality*.
—Erich Kahler

"The story is shot through with pain," one reviewer commented on the first volume of Wiener's autobiography,[1] a manuscript first entitled *The Bent Twig* but eventually published under the title *Ex-Prodigy*. Indeed, his father's efforts to mold Norbert, even though they remained a source of stimulus and direction throughout Norbert's adult life, also left him with a large residue of conflict and pain. Not only had his father seen his own educational function as that of a sculptor, but Norbert himself had described his relation to his father as resembling that of the statue come to life, Galatea, to her creator Pygmalion. At times he preferred the metaphor provided by the legend of the rabbi and the golem: the creator creates in his own image, as God created Adam in the story of Genesis.

   This is an unusual experience that few people could weather successfully. It contains paradoxical, mutually contradictory elements that required Norbert to have the wit and strength to avoid an intolerable double bind—for example, to interpret appropriately for himself an insistence on obedience to the injunction to be intellectually independent and original. His friendship dating from 1909 with his coeval, mathematical prodigy William James Sidis, who had been the subject of his father Boris Sidis's effort to demonstrate a theory of education somewhat similar to Leo Wiener's, again and again reminded Wiener of the doubtful merit of such educational experiments: Boris Sidis, like Leo Wiener an immigrant Russian Jew, was a leading New England psychiatrist. The son eventually broke under the psychological pressure and turned against his whole upbringing. Norbert arranged for a routine job for Sidis at MIT. Eventually, William Sidis killed himself. In 1950 Wiener expressed his private feeling when he wrote: "Sidis failed, and I have had worldly success, but we are cut out of the same cloth. There, but for the grace of God, go I."[2] Other elements of pain in Wiener's life derived from his belated discovery of himself as an "unloved Jew," his physical and social clumsiness, his sense of naiveté and vulnerability to exploitation, his persistent sense of himself as a misfit and outsider, and his prolonged dependency on his mother for so many practical and social necessities.

Wiener's later years bring us full circle. The childhood origins of the unresolved inner conflicts of the mature Norbert Wiener are indeed recognizable, but during the course of his career he had himself become the actor, the creator, the sculptor in the medium of mathematics. His adult sense of identity is expressed in the title he gave to the second volume of his autobiography, *I Am a Mathematician*, although in his later years he drifted away from mathematics into other disciplines, and one of his younger friends suggested that he ought also to have written a third volume, one entitled *After-Math*.[3] Wiener's autobiography was more than a literary achievement; the reexamination of his life was one means to alleviate personal discord and unresolved pain. Not surprisingly, other family members resented the book. Yet in *Ex-Prodigy* Wiener "achieves an almost Olympian objectivity and gentleness"—as Margaret Mead put it in her review. Von Neumann referred to the autobiography as the "documentation of a process," suggesting that despite, if not because of, the strong emotions with which it was infused, it transcended the merely personal.

Occasionally Norbert Wiener consulted a psychiatrist. De-

374
John von
Neumann
and
Norbert
Wiener

pressions, fantasies and impulses, or conflict in personal relations could severely disturb his peace of mind. There is no sense of urgency in the letter to a New York woman psychiatrist dictated to his daughter Peggy in 1950:

Since I have seen you last, I have the testimony of my family that I have adhered to a reasonable standard of equanimity and rational conduct. However, in view of various annoyances, these are wearing a bit thin, and I think the time is not far off when I had better consult with you again. I can, of course, come down to New York, but if you are to have this summer's vacation in the same idyllic spot in which we found you last summer, we might as well look you up there.[4]

A faint echo is heard here of the young man reporting from England to his parents about his managing acceptable behavior, including baths, haircuts, and personal relations. However, Wiener never permitted himself to be drawn deeply into any psychotherapy or psychoanalysis. It was his habit to utilize his personal conflicts and tensions, to turn them away from the "merely personal" toward creative intellectual work.

For the historian the consideration of Wiener's relation to the personal or impersonal energies represented for him by the symbol of the rabbi of Loew and the golem, the creator and his creature, are of considerable interest, because it sheds some light on the question of why Norbert Wiener, in particular, would be the scientist who would occupy himself with the social and moral philosophy of technology, and radically challenge outmoded assumptions. As a mathematician Wiener was a creator, but it was when he began to work with mechanical analogies between organisms, or the nervous system of organisms, with formal or mechanical automata and simulacra, that he came to resemble the builder of a golem. The building of artifacts to simulate organisms is a familiar theme that can be traced from ancient Egyptian figurines through Leibniz and Pascal to the modern designers of cybernetic machines like von Neuman and Wiener himself.[5] Historian Derek Price even places the legends of Pygmalion and the golem in the same broad tradition with the development of mechanical contrivances. He describes this tradition as motivated by man's eternal preoccupation with the problems of creation. . . . The making of tangible artifacts showing the nature of the material universe and the nature of a creature was . . . the two movements of a "do-it-yourself creation kit." By playing God, man could know God. . . . For almost five thousand years this urge dwelt in the minds of ingenious men, fostering their highest ingenuities and calling forth a wealth of mechanical skills and scientific understanding.[6]

Wiener, a modern scientific atheist,[7] was indeed preoccupied with the problem of creation. But to "play God," to design an electronic "creature" and help it to learn, might lead some to an alarming sense of omnipotence, which Wiener plainly rejected. Wiener's unresolved tensions could not in any case be resolved by the crude psychological process of unthinkingly shifting from the passive role of creature to the powerful role of creator. He was himself forever learning and creating: he was both rabbi and golem. What intrigued him both in itself and as a means to mitigate the conflict was the achievement of a "proper relationship" between the two. Thus Wiener's unresolved pain stemming from his boyhood relation to his father became transmuted and was given explicit form in the mature Wiener's development of a comprehensive moral and social philosophy of technology. According to Wiener, "The machine . . . is the modern counterpart of the Golem of the Rabbi of Prague."[8] The modern inventor and his team of engineers play the role of the rabbi, or that of the creator of Adam. They are building machines in their own image.[9] What continued to trouble Wiener and remained a preoccupation in his writing and activities was the painful awareness that the Garden of Eden at second remove, much of it created by inventive and industrious man, is far from harmonious.

Since the time when Wiener had belatedly discovered himself to be a Jew, how had he lived out his conscious Jewishness? Had he come to terms with the phenomenon of anti-Semitism? Wiener's hypersensitivity to anti-Semitism never left him. When he heard of some anti-Semitic episode, it usually upset him considerably, and during the years when his eyesight was particularly bad and he preferred to have his

376
John von
Neumann
and
Norbert
Wiener

mail read to him, his wife, enlisting his secretary's cooperation, went so far as to "protect" him from hearing or reading
such mail. His daughter Barbara thought this unfair, believing
her father needed no such protection—and doubtless father
and daughter would have agreed. As to his Jewishness, it is
worth noting that Wiener chose a non-Jewish German wife,
fond of things German, which may have complicated matters
for him during the Nazi years. Wiener was proud of his purported medieval Jewish ancestor Maimonides—an earlier
man of science also concerned with moral law. Wiener's
placing science and technology within a moral universe fits
that tradition. However, Wiener did not profess Judaism or
learn to speak and read Hebrew. In his novel, *The Tempter*,
the immigrant hero is an Armenian; had Wiener made him a
Jew he would have more directly confronted the specific
American Jewish experience in the novel, and perhaps in
himself.

Wiener perceived himself as coming out of and sharing in
the historical experience of diaspora Jews. Wiener's aloof-
ness from the Boston area's Jewish community can be viewed
as consistent with a major stream of European Jewish tradition: relative disregard for local loyalties in favor of more universal values. Wiener's style and outlook, while rooted in the
experience of his youth and the historic experience of Jews,
had no particular connection to the condition of American
Jews in the 1950s. Wiener's self-definition shows a preference
for marginality—not as a response to anti-Semitism, which by
the 1950s, especially in the scientific and intellectual communities, was no longer a major factor,[10] but because it was
congenial to his originality and creative power and to the
sense of his own dignity and manhood, helping to protect him
from corrupting compromise. Besides, the conditioning of his
youth, to which he had after all assented (unlike Sidis), may
have put bounds on his flexibility in this regard.

Wiener had pleasant childhood recollections of his mother
reading Rudyard Kipling's stories to him.[11] When in his sixties
he reexamined Kipling, he recognized the British writer's
alienation but also his attempt to escape from marginality
through an absolute loyalty to the British colonial elite in

India, a narrow nationalism, and compulsive in-groupishness.
In an essay on the subject, Wiener and his coauthor Karl Deutsch derided that choice, pointing out that "just as marginality itself can be a curse or an opportunity . . . so Kipling's own insights may merely speed men in that panicky flight from loneliness into the nearest ingroup. . . . The relief found in belonging to a group may become the rigor of fear-rooted conformity."[12] Wiener with his strong and incorrigible sense of marginality recognized in it an opportunity that must not be lost, one that from a sociological point of view has a particular asset, since

it is in the mind of the marginal man that the moral turmoil which new cultural contacts occasion manifests itself in the most obvious forms. It is in the mind of the marginal man—where the changes and fusions of culture are going on—that we can best study the process of civilization and progress.[13]

And again, "the marginal man is always relatively the more civilized human being."

Wiener spoke as a respected mathematician yet also as an intellectual outsider when, beginning in 1946, he sounded the call to what would be a high historical undertaking—placing science and technology within a social and ethical universe. A quarter century later it could be stated that "we now see emerging a 'critical science,' in which science, technology, politics, and ultimately the philosophy of nature are involved, and which may be the most significant development in the science of our age."[14]

Wiener did not live to see this development, the germ of which was in his own writings. It remains to be seen to what extent the philosophy of technology and critical science will become coopted or turned into apologetics for the powers that be, or to what extent they will serve as an effective challenge to reform in the direction Wiener envisioned.

Wiener did see, however, that his work in pure mathematics had proved seminal, as had his application of mathematics to electrical engineering. It was too soon to be sure about the future of cybernetics, but much in it looked promising. Indeed, Wiener was fortunate in being able to realize so many of the

378
John von
Neumann
and
Norbert
Wiener

possibilities of his various vocations. If he saw the circumstances of his childhood as a mixture of unusual advantages
and unusual handicaps as he looked back on them in his sixties, he must have felt triumphant and grateful that it was the
advantages and not the handicaps that had proved decisive.
In fact such sentiments do emerge in the second volume of his
autobiography.

Norbert Wiener's family consisted of his wife Margaret, who
with a practical intelligence devoted herself to being the wife
of a brilliant, moody, absent-minded man with more social
needs than social skills, and their two daughters, Peggy and
Barbara. They lived modestly in suburban Belmont and sum-
mered in their cottage in North Sandwich, New Hampshire.
The cottage, according to Wiener's secretary in the 1960s, was

a charming farm house, high up on a hill and surrounded by
first of all meadows, hills, and then huge woods. It's absolutely pastoral; it's just beautiful. He was a very simple man.
He shunned a great many luxuries, he was a vegetarian, he
didn't like cars, he didn't like big social functions. So this
suited him perfectly. He loved to walk through the greenery,
along paths in the woods . . . and to visit a few friends.[15]

Still, Wiener usually wanted company. He, who defined the
human being as the "talking animal," always needed people
to talk with. In New Hampshire he had a few old friends and
hiking companions. On his walks in Belmont he would stop to
talk to everyone—the grocer, the shoemaker, and other shop-
keepers. He would speak classical Greek to the Greek grocer,
who was hardly in a position to respond. Because of his extreme nearsightedness, even after a cataract operation, extended periods of reading were strenuous for him, although it
helped that he could read exceptionally fast. He dictated his
books to a secretary and did mathematics on a blackboard
with a younger coworker looking on, taking notes. He liked a
responsive listener and onlooker while working. The nontechnical books he dictated in nearly finished form; they required
very little editing.[16] His mathematics done on the blackboard
tended to require more repetition, elaboration, and clarification, but even here he did a good portion of the work in his
head.

At MIT Wiener launched himself from his office frequently, if not daily, on a passegiata known as the "Wienerweg." The route—pursued when Wiener was working out a problem, was stuck and depressed, was euphoric with a new insight, or was concerned about a political situation—included a cross-section of the MIT population, from janitors to graduate students and professors. According to Julius Stratton, a former student of Wiener's, "He'd get an idea and start wandering through the halls, went into anyone's office, or those of his friends—and I was very proud that I was on the 'Wiener-weg.' "[17] Stratton later became president of MIT, but his attitude toward Wiener's interruptions didn't change:

After I was in the President's office, I was very pleased that Norbert would come in . . . and of course it is customary when you come to the lobby of the President to ask if he was busy—not Norbert. He would just walk right in—whoever was there—he would interrupt and talk. This was not rudeness —he was just carried away.

Stratton recalled fondly the zest of Wiener's enthusiasms, his warm personality, his strong feelings. He recalled their early joint effort to transform MIT from a training school for engineers to a university engaged in fundamental mathematics and physics research,[18] and how Wiener, the resident genius, became the institute's pride. Stratton recalled some long conversations with Wiener around 1949 concerning MIT's Research Laboratory for Electronics (RLE), which had evolved out of the wartime Radiation Laboratory, and in which many of the interdisciplinary ideas originating in Wiener's cybernetics were being explored.[19] These conversations concerned the advisability of permitting the major funding for RLE to come from the military services, as it did then and has continued to do. Wiener was deeply concerned. As Stratton recalls, "We asked the question: Is there a will, a direction imposed upon us by the funding, or are we free?" Wiener withdrew from participation at the RLE for a time, "but there was never any anger between us and we always remained friends."[20] Wiener himself rejected direct financial support from military sources, but over the years he provided much of the inspiration for the

380
John von
Neumann
and
Norbert
Wiener

fundamental scientific work at RLE (work unrelated to military purposes), and spent time with his colleagues at the laboratory. Nor did he stand in the way of his students' receiving support through RLE. In defense of his own views, Stratton added casually, "Some of my Harvard friends get mad when I remind them that in the early part of this century a major source of funds for Harvard University came from investments in the red-light district of Boston. Still, it's a great scholarly institution."

Another stopover on the Wienerweg was the American historian Neal Hartley, who spoke about Wiener's friendliness and sympathy toward fellow human beings, even though his overbearing manner sometimes made him insufferable— egalitarian, and yet so vain![21] Wiener had consulted Hartley, among others, on his dilemma whether or not to include his mother's anti-Semitism in his autobiography, especially as his mother was still alive and would probably be hurt if she read it.

When Wiener reached retirement age, MIT appointed him an Institute Professor. Henceforth he would no longer be specifically identified with the mathematics department but would be completely interdepartmental.[22] In fact this had been his style for some years already, although Hartley recalled that Wiener happily wandered through the institute that day, announcing that he was now "legitimate."

Hartley called to my attention that Wiener even in his later years sometimes visited his old school friends in Ayer, Massachusetts; Alfred Richardson of Ayer confirmed this.[23] Local people seemed pleased that their most distinguished alumnus had not forgotten them, and they were fond of him. They also recall that when Wiener was in his fifties on one occasion he wanted to join in when some other men were pitching horseshoes but seemed unable even to aim the horseshoe in the right direction. Some things hadn't changed much.

Some, including linguist Morris Halle, felt something missing in Wiener when he came by to talk with them. His involvement in his subject was so intense that to Halle and some others Wiener seemed hardly aware of the person he was talking with.[24] There are stories of Wiener walking into the wrong classroom with the wrong group of students but apparently not realizing it and giving his lecture anyway, with the students enjoying it without being able to follow and the instructor unwilling to interrupt.[25]

Halle also recalled the occasion of Wiener's trip to Russia in 1960. In Russia the reception of cybernetics had undergone a cycle of rejection as capitalist science followed a few years after Stalin's death by enthusiastic acceptance as a set of ideas entirely consistent with dialectical materialism.[26] The Soviets expected cybernetics to be a useful tool for clarifying patterns of political organization based on reconciling decentralization of control with overall central purposes. Halle recalls Wiener, in anticipation of his trip, inquiring anxiously about conditions in Russia and whether he should expect any unpleasant political experiences. When Wiener arrived, cybernetics was on the upswing, and he in fact had a triumphant tour. He even gave an outspoken paper discussing his own ideas on science and society, rather than restricting himself to technical matters.[27]

One can imagine Wiener at MIT, walking the long, industrial halls, leaning backward in his ducklike fashion, looking up as he tossed peanuts into the air and caught them in his mouth —a skill he had perfected. If sometimes his unannounced entrances were an annoyance, the chances of hearing something interesting from him outweighed this for most people, particularly for Jerome Wiesner, a close associate who was a member of the Research Laboratory for Electronics after World War II and became president of MIT:

Many can remember Norbert's daily visits around the Institute from office to office and his conversations that always began with "How's it going?" He never waited for the answer before sailing into his latest idea. "By the way," he would say, "have you heard about what Arturo and I have been thinking?" Or, "have you heard about what Walter Rosenblith has said?" Sometimes he was disturbed: someone had challenged one of his ideas; or he had become convinced that some foolish action of the President or Secretary of State was going to

382
John von
Neumann
and
Norbert
Wiener

catapult the world into oblivion. "Do you think there is going to be a war?" was his standard question on those blue days. Whatever was on his mind, Norbert Wiener's visit was one of the high points of the day at MIT for me and many others. For those of us working in the lonely isolation of Building 20, Norbert's visits were especially welcome, for he was one of our best links with the main building.[28]

Electrical engineer Robert Fano recalls Wiener, when *Cybernetics* had just been published, entering his office, proudly showing him the book and letting him look at it for a few moments, and then immediately asking, "What do you think of it?"[29] A younger friend of Wiener, interested in understanding him, explained that when Wiener said, "What do you think of X?", he normally meant, "Do you get the point of X?"[30] Henry Zimmerman, an electrical engineer who was also director of the Research Laboratory for Electronics for a number of years, recollected Wiener as a "warm, caring person of great charm." He tells of an incident when in his official capacity he was entertaining a visiting Greek dignitary at the MIT Faculty Club and happened to notice Wiener standing by himself. Zimmerman introduced the guest to Wiener, hoping that the famous mathematician might somehow help MIT's public relations. After a brief conversation, he heard Wiener sing the Greek national anthem for the visitor, who was thrilled.[31] Wiener generally let MIT use him for public relations purposes, although he was not always predictable and could snub an important visitor from the military-industrial establishment or insult him in scathing language. One summer MIT set up a program to "retread" working engineers and industrial administrators. "MIT trotted out Wiener for a lecture—he of course just talked about what interested him at the moment, with all the doubts, blind alleys, difficulties and so on," which would have hardly been comprehensible to the majority of that audience. "He was picking his nose energetically throughout the lecture, presumably peripherally conscious of that but feeling that it didn't matter a bit."[32]

Not everyone looked forward to encounters with Wiener, especially in the period immediately preceding World War II, when, according to his mathematical colleague and former student Norman Levinson, Wiener was troubled by self-doubt:

His usual words of greeting became, "Tell me, am I slipping?" Whether one knew what he had been doing or not the only response anyone ever made was a strong denial. However this was usually not enough and it was necessary to affirm in the strongest terms the great excellence of whatever piece of his research he himself would proceed to describe sometimes in the most glowing terms. Altogether such an encounter was an exhausting experience.[33]

At times Wiener, intrigued by a favorite notion of his in biology or physics or some other discipline, would periodically visit MIT scientists in the field, such as the theoretical physicists Herman Feshbach and Victor Weisskopf, to discuss his ideas. While Feshbach and Weisskopf appreciated Wiener's ability, they found his ideas on quantum theory anachronistic and his visits distracting. As a solution, they assigned a young theoretical physicist, Armand Siegel, to work with Wiener.[34] As it turned out, an interesting reformulation of quantum theory as a stochastic process resulted from the Siegel-Wiener collaboration, although it was out of the mainstream, and the friendship that developed between the two men long outlasted their collaboration, which began in 1951.

Another group of MIT researchers working with brain waves also viewed Wiener's ideas as on the wrong track, and Wiener was inclined to waste their time. So they contrived to place a man where he could see Wiener coming. He would alert the others, who would then scatter in all directions, even hiding in the men's room. In the MIT idiom of the 1950s this arrangement was known as the Wiener Early Warning System.[35]

By contrast, students seeing the familiar rotund figure of Wiener waddling through the halls would often accost him and inveigle him into conversation. When students asked him to give a special lecture, he usually complied, giving an enlightening and stimulating talk in his picturesque and lucid English.[36] Asim Yildiz took a graduate seminar with Wiener dealing with nonlinear stochastic processes, one of the last Wiener taught. Although he recalled that in many ways Wiener

384
John von
Neumann
and
Norbert
Wiener

was "like a child"—when he had been invited to go some-where interesting, to Mexico or Denmark, he would excitedly share the happy news with his class—he was very generous in helping Yildiz, an immigrant from Turkey, to find an aca-demic position in America. Yildiz also noted his legendary absentmindedness: "Sometimes when the class was sup-posed to begin, we would have to look for Wiener, drag him out of somebody's office where he was engaged in conversa-tion, and bring him to class."[37] Yildiz added that the lectures contained some new material that has never been published.

Through the winding, tree-lined streets of suburban Bel-mont, Norbert created another Wienerweg, since it was not just during his weekday hours at MIT that he felt the need to seek out companionship. Dirk Struik, Lawrence Frank, and Armand Siegel were among the accessible neighbors. Frank, with a lively interest in recognizing emerging cultural values, made a good audience for Wiener's ideas and also re-sponded compassionately to Wiener's periods of "depression about himself and discouragement about the human race."[38]

I interviewed Struik, Wiener's close friend for over forty years, and Mrs. Struik later joined the conversation. The Struiks' house is comfortable and unpretentious, with a Breughel festival production and plates of Dutch design on the livingroom wall. Struik mentioned that the Wieners' house had been similar to their own. The tall and slim Struik, origi-nally Dutch, was friendly, laughed easily while reminiscing, yet was personally considerate and restrained in his manner. He must have formed quite a contrast to the short and corpu-lent Wiener, volatile and socially awkward. Struik said that Wiener one year had taught a course on seventeenth-century philosophy, especially Spinoza, Descartes, and Leibniz, and had brought to it a great depth of understanding of these philosophers and their relation to history. Wiener thought Leibniz the most relevant for the modern age. Struik also re-called Wiener's encyclopedic memory; he listened to others more than he seemed to, as shown by the complete under-standing he would afterward show of their ideas. Struik mentioned that Wiener especially loved to talk with Johnny von Neumann.

Other notable Wienerian traits were his friendliness with the farmers and his other neighbors in New Hampshire; his loud voice and the way he had of looking around to see people's responses to him; his generosity with students whom he liked; his constant dependence on the approval of others; and his deep appreciation of his wife, who was hospitable, a good cook, good-natured, and at ease socially, and who also felt great admiration for him. Wiener's family disliked the autobiography he wrote and refused to talk about it.

In the 1930s, the Struiks recalled, Norbert would sometimes come and talk for hours with Dirk, unburdening himself, until sometimes it became too much and they sent him home. In the 1950s the two men talked a good deal about questions of conscience and politics. Wiener was getting offers from various companies and the Defense Department, and he would seek out Dirk to discuss them. In the end, of course, he always refused such offers.[39]

Discussing Wiener's novel, in which the characters are wooden and the interpersonal relations lack subtlety, Mrs. Struik ventured that it reflected his own lack of awareness of others, an insensitivity to the feelings that govern human relationships. In fact the origin of The Tempter was Wiener's anguish at the historical circumstance of the exploitation of Oliver Heaviside's ideas by hucksters, together with Wiener's romantic view of the unappreciated but undaunted scientist following a deep and correct mathematical intuition. As early as 1941, after seeing the film Citizen Kane, directed by Orson Welles, Wiener had sketched the historical plot of The Tempter and sent it to Welles with the invitation to use it as the basis for a film.[40] So the 1959 novel was for Wiener simply another means to express his frustration with the way things were. His descriptions of people in his autobiography are much more convincing.

For a scientist, Wiener took an exceptional interest and pleasure in poetry and literature. In situations where neither his

386

John von
Neumann
and
Norbert
Wiener

frustrations nor his anxieties were aroused, he would respond to people with spontaneous emotion, intuitively understanding a situation. The cause and effect of his insensitivity was his fear of his own gullibility and vulnerability, and much of his volatile anger, which sometimes led to the rupture of relationships, seemed to emanate from such a fear.

Wiener had a particular weakness for the genre of the detective story. He was a member of the "Speckled Band of Boston,"[41] a group of Sherlock Holmes enthusiasts, and was prone to picking up detective stories in drugstores.[42] Using the pseudonym W. Norbert, he wrote one fiendish little story in which a brain surgeon through circumstance has the opportunity to use his talent to take delicious revenge against a mobster.[43] In another little story, published in the same MIT magazine, he teases the scientific reader with the possibility of the supernatural influencing experimental results.[44]

Wiener encouraged younger people close to him to follow up strong scientific intuitions or interests of their own and not to be intimidated by what the recognized leaders in any particular discipline happened to consider important. Thus he could be deeply encouraging to a young man with independent ideas but who felt self-doubt in the face of what was generally accepted; Wiener bolstered his confidence in his own powers, letting him know that he and Wiener were alike in viewing themselves as outsiders, and in pursuing what might be unfashionable.[45] In following such a direction, a young scientist is taking risks with his career, but probably those who modeled themselves after Wiener in this respect were men whose personal sense of identity was confirmed by taking these kinds of risks. Among them was Armand Siegel, a generation younger than Wiener and also a Belmont resident. Siegel, after having worked with Wiener, became a professor of physics at Boston University. His special field was statistical mechanics, but later he followed in Wiener's footsteps in a shift of interest to neurobiology, becoming attracted to extending A. N. Whitehead's philosophy of science and to Freudian psychoanalytic theory, which he believed was undervalued by most scientists. The Wiener and Siegel families became close friends in the 1950s.

It is evident from the picture of Wiener that has emerged that he had a penchant for self-dramatization, which extended to his very choice of words. Even his social obliviousness left his acquaintances wondering to what extent it was not just an expedient response he had made into a habit. Siegel recounted the following interchange with Wiener:

It is a story kind of against him. He was saying something about his inability of dealing with certain human situations, and I said "Oh, go on, Norbert. You're a ham from the word *go*." You know, I could always kid him, I've kidded him any number of times . . . but this time he was terribly hurt (laughter)—as if I had no right to say that he was a ham. Usually even a joke against himself he could accept, but maybe this time it was a little too close to home.[46]

The extent and quality of Wiener's social sensitivity are hard to evaluate, since a person's state of awareness is not open to direct inspection. But Wiener's genuine devotion to honesty and sincerity need not blind us to his poses and his showmanship on a different level.

Struik has said that "Wiener was a foreigner wherever he was,"[47] but MIT was a second home to him. He cared about the institution and was very grateful to its people for having created the conditions that allowed him to thrive. In turn he contributed enormously to MIT, leaving a powerful imprint on the institution. One immediate impact was on his mathematics students, such as Ted Martin and Norman Levinson, each of whom later became a distinguished mathematician and chairman of the MIT mathematics department. As Levinson noted, Wiener encouraged not only young scientists striking out on their own, off the beaten path, but equally young mathematicians "doing work in established areas close to his own."[48]

In the decades before World War II, according to the present provost and the president of MIT, "the period during which the Institute transformed itself from a technical school into a university . . . 'polarized around science,' Wiener's in-

388

John von
Neumann
and
Norbert
Wiener

tellectual virtuosity, curiosity and integrity contributed importantly to that transition."[49] But even after the new emphasis on mathematics proper and basic science had been established at MIT under Karl Compton's presidency, not only did Wiener continue to maintain "the high and hard mathematical standards that he demonstrated in every piece of work he did,"[50] but in the 1930s set a pattern of continued collaboration with the electrical engineering department, including the design of computers. Through Wiener the level of sophistication in mathematics applied to electrical engineering generally was raised, but particularly and most directly at MIT.

In the postwar years, Wiener's incorrigibly cross-disciplinary style helped to generate new kinds of administrative entities at MIT other than the traditional departments. The Research Laboratory for Electronics became a prototype organization. Although Wiener could not bring von Neumann to MIT, he helped to add neurobiologists and then linguists to the physicists and electrical engineers at the laboratory. Others contributed to widening the influence of Wiener's own thinking; Yuk Wing Lee became the outstanding expositor of Wiener's mathematical ideas in communication engineering, effectively translating them into the form of a course offered to budding electrical engineers; Jerome Wiesner, who was an enthusiast for many of Wiener's ideas about cybernetics, was effective in bringing about the administrative action needed to favor the new field of inquiry. As Wiesner recalled the early days of RLE,

Fired up by Norbert Wiener's cybernetics, we explored the far-ranging implications of the concepts of information and communication theory; our interests ranged from man-made communication and computing systems to the sciences of man, to inquiries into the structure and development of his unique nervous system, the phenomena of his inner life, and finally his behavior and relation to other men. . . . Norbert Wiener drew the communication sciences together at MIT . . . [he was] the catalyst, though catalyst is a lukewarm description of his role. Ordinarily one thinks of a catalyst as a passive participant in a reaction. The platinum surface to which atoms cling to form compounds gives nothing of itself to the new

creation. Wiener, the catalyst, sought out his atoms and drew
them together. He did this job almost without recognizing his
role, for his interest was in ideas. But his action did bring together people with related interests. He did this in many
ways—by personal contact, by stimulating meetings, and
particularly by writing and speaking . . . he organized dinner
meetings for a diverse (the label "interdisciplinary" stuck)
group of scientists and engineers.[51]

Wiener's active outgoingness, his sharing of his own thinking,
apparently served a useful function at MIT. A perusal of a
current (1979) MIT catalogue reveals that the work carried out
or initiated by Wiener still pervades numerous courses in
many departments as well as special interdisciplinary laboratories throughout MIT. His objection to carrying out work
for military purposes, however, seems to have had relatively
little, if any, impact on the institute, and the current group in
Technology Studies dealing with critical science and the
philosophy of technology has not yet been brought fully into
the mainstream of MIT education. But the fact that over a period of forty-five years Wiener the idiosyncratic individualist
and MIT the educational institution had such beneficial impact on each other does credit to both parties.

Reflecting on this unlikely character, Norbert Wiener con-
jures up a double image: on one side an amusing figure—
childish, bombastic, absentminded, emotionally vulnerable,
and extremely near-sighted. Joined to it we see a bringer of
truth, and not only in mathematics; a purveyor of profound insights on the state of contemporary civilization, providing the
larger society with some of the basic wholesome intellectual
nourishment it needs to protect it from itself.

With the reader's indulgence I preserve here two more
small but typical personal reminiscences.

He'd come around Sunday morning when nobody was
dressed and we hadn't finished breakfast. He was patient.
He'd sit in the living room and muse to himself or wander off.
. . . We lived in Belmont just a half mile from his house—ours
was one of the places he would stop at on his rounds. Then he
would talk about everything, world events, things that were
bothering him about the government, and things that bothered

390

John von
Neumann
and
Norbert
Wiener

him about the trends in science. His insights were always fascinating. He loved Scrabble, and we'd get out the Scrabble board. Our kids loved Scrabble too. Andy, at the age of eight or nine, would be a member of the Scrabble game. . . . Andy was very taken with Norbert, and it seemed to be mutual. Andy would take his time to find his word and Norbert would be as patient as could be. To him a child was just as good as anybody else.[52]

One day I wanted Wiener to meet Yilmaz (another physicist). He had actually already met him, but had forgotten. And we decided to go to a Chinese restaurant because Wiener was a vegetarian. We went to his favorite restaurant, namely Joyce Chen. I'll never forget, all the way to the restaurant he was telling Yilmaz the history of Turkey, because Yilmaz is Turkish . . . so I heard him describe one of the Turkish Sultans, and things about Turkey all the way to Joyce Chen. We sat down, and then Wiener ordered in Chinese and held an animated conversation with the waiter in Chinese. After that interlude, and after the meal was brought, Wiener stopped and turned the conversation to Yilmaz. Now people say that Wiener wasn't good at listening to people. Wiener was perfectly quiet as Yilmaz expounded his color theory . . . then we went on to invariance, and science and epistemology; the conversation then took on a very general tone. I remember it was a beautiful, sunny day; it was a very heady experience; we were really getting along perfectly. Then at the end Wiener said to me: "Was I a good listener?"[53]

Wiener's philosophy had always stressed the finite. Referring to both Gibbs and Kierkegaard, he had described life as "swimming upstream against a great torrent of disorganization," tragic because of inevitable death and defeat by the second law of thermodynamics. Yet its great glory consisted of the establishment of enclaves of order in the face of these conditions, and of the "declaration of our own nature" through our efforts. The sense of how finite are the most exalted human efforts, and how finite the duration of any created order, had long been part of Wiener's consciousness and existential outlook. He would have liked to extend his own life, but not forever.[54]

He had heart attacks. His attitude toward his own possible death was one of matter-of-fact objectivity, like the detached interest of a physician in the mechanical malfunctioning of the human body. "He would take digitalis and various drugs, and would talk about it, as if he were a laboratory animal, and he regarded his own body as a laboratory."[55] In his last years he envisaged the technical feasibility of a compact mechanical device that could be attached to the human heart, sense the first symptoms of a coronary attack, and immediately release an anticoagulant into the bloodstream, thus preventing a fatal consequence.[56]

In the spring of 1964 he was again traveling in Europe with his wife Margaret. He had visited Scandinavian friends, one of whom was the versatile Piet Hein. He and Margaret were planning a little celebration with the Struiks in an Amsterdam restaurant. But Wiener didn't get to the celebration; he died of a heart attack in Stockholm on March 18. He was sixty-nine years old.

Wiener (far right) with other mathematicians in ballistics unit, Aberdeen Proving Ground, 1918. Courtesy of MIT Historical Collection.

Von Neumann (front, center) at International Conference of Mathematicians, Istanbul, August 1952.
Photograph from Library of Congress.

Quantum theorists, Zurich, 1929.
From left: J. Robert Oppenheimer, I.
I. Rabi, L. M. Mott-Smith, and
Wolfgang Pauli. Courtesy of AIP
Niels Bohr Library.

Wiener (front, center) with Electrical
Engineering Department, Tsing Hua
University, Peking, 1936. Courtesy of
MIT Historical Collection.

Wiener, Jerome Wiesner, and Yuk Wing Lee with MIT Autocorrelator. Courtesy of MIT Historical Collection.

Wiener and Max Born, 1925. Photo by George H. Davis, Jr., courtesy of MIT Historical Collection.

Von Neumann and J. Robert
Oppenheimer with the Institute for Advanced Study computer. Courtesy of
Princeton University Archives.

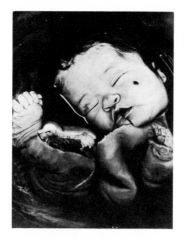

The mothers of these babies survived the bombing at Nagasaki but their babies were born deformed and dead. After autopsy their bodies were preserved to show others what may happen when genes are damaged by nuclear radiation. No mere photographs can convey the nature of atom bombs. The reader may appreciate social scientist Robert Lifton's writing that "despite considerable previous research experience with people subjected to extreme situations, and a certain amount of beginning knowledge of the atomic bomb exposure. I was not prepared for the things that I heard. I found that the completion of each of these early interviews (with Hiroshima survivors) left me profoundly shocked and emotionally spent." Photographs by Teruaki Tomatsu, from Felix Greene. *Let There Be a World* (Palo Alto, Calif.: Fultan Publishing Co., 1963).

EPILOGUE

Wiener's and von Neumann's shifts of primary activity from pure science to work in the service of explicit social and political values signals and symbolizes what now appears as the cardinal cultural transformation of the second half of the twentieth century. This transformation makes scientific and technological progress obsolete as a primary goal; considering which aspects of science and technology to adopt in our lives becomes paramount. This shift involves a fundamental psychological reorientation. Instead of faith in the progress of science and technology, we rely on people as pivotal in all constructive changes and on the political and social processes that determine the place of science and machines in our lives.

The transformation, which in von Neumann's and Wiener's time had barely begun, is now in full swing. It was and is forced upon us by external circumstances made possible by ideas emanating from technically and scientifically inventive minds:

Modern people have filled the world with the most remarkable array of contrivances and innovations. If it now happens that these works cannot be fundamentally reconsidered and reconstructed, humankind faces a woefully permanent bondage to the power of its own inventions. But if it is still thinkable to dismantle, to learn and start again, there is a prospect of liberation. Perhaps means can be found to rid the human world of our self-made afflictions.[1]

The present task is to examine, reconsider, and reconstruct. One approach is to identify inappropriate hardware: tech-

nologies that make our environment less congenial or habit-
able, including weapons and especially nuclear weapons,
and technologies that violate the human scale, interfere
with the fulfillment of basic human needs, or generate
alienation. But honest men differ as to which technologies are
appropriate and beneficial; the issues become political and
social rather than technical. Another approach, which appears altogether different, is to focus on the personal qualities
presumably needed by the coming generation to deal with
future exigencies. Here one can only hope that the pressure of
circumstances will stimulate the combined wisdom of old and
young to recognize and develop the traits needed to guide the
transformation in progress in a humane direction. But these
circumstances are contingent on social structures. Thus,
whether one is concerned with appropriate hardware or with
appropriate human talents, it is useful to begin by considering
patterns of social organization, institutions, and collective attitudes and beliefs.

The Practice of Science

Professionalism and careerism are characteristic of modern
science, although in no intrinsic way connected to scientific
exploration (in fact, Wiener was highly conscious of the threat
to the originality of young scientists posed by the profession
itself). Some of the most creative work in science has been
carried out in other social contexts: Einstein did his revolutionary work in theoretical physics as an avocation while
employed in a patent office. In retrospect he viewed this circumstance as fortunate, because he believed that the
professional-careerist-bureaucratic conditions prevailing
among university physicists would have hampered his intellectual independence.[2] He went so far as to advocate that
young scientists take this option for the sake of their own independence and profundity. The early membership of the
Royal Society formed in seventeenth-century England was
made up largely of scientific amateurs whose sources of income were unrelated to their avocation, but that period was a
highly creative one in science in England.

A major reason for maintaining science within a manage-

400
John von
Neumann
and
Norbert
Wiener

able bureaucratic context in a country like the United States is the consensus among scientists, business executives, and political leaders that a strong science establishment can be utilized directly or indirectly to bring about all kinds of technical developments. In the weapons field especially, scientific manpower and direct access to the latest in scientific knowledge are reckoned as part of the indispensable arsenal for furthering the "national interest."

The modern scientist is typically employed by a research institute, a large corporation, a university, or a government laboratory, and consequently his research is governed— aside from his professional associations—by the bureaucratic structure of his place of work and the source of his research grants. Of course there are exceptions: von Neumann, after leaving the academic world, continued to work on his automata theory in what spare time he could find. The bold interdisciplinary group of scientists who in recent years formed the New Alchemy Institute on Cape Cod work in a cooperative nonbureaucratic arrangement.[3] A question that deserves debate is whether the intrinsic spirit and pleasure of science may not be heavily compromised by the usual context imposed by the profession, the employing institution, and the granting agency. But then one must also inquire which alternative patterns of social organization might effectively serve to keep alive whatever is judged to be of greatest value in scientific research.

As with poetry or mountain climbing or baseball, some people for one reason or another find scientific research satisfying. A mathematician may enjoy the aesthetic and creative experience, as both Wiener and von Neumann did; an empirical scientist may enjoy the process of exploration and discovery; an experimenter may delight in his ingenuity and craftsmanship. For some, science, like other scholarly pursuits, offers a happy escape from a harsh world or painful personal relations. For others, scientific research, by participation in the enterprise of making the structure of the universe more comprehensible, provides a means to ease existential anxieties and feel at home in the world. Nor is scientific research an isolated activity. It is necessarily related to a scien-

tific community capable of sharing one's appreciation of a
new theorem or discovery. One might say that scientists continually give gifts to each other of their own insights and knowledge.[4] It is a shared enterprise, and the community extends worldwide. Much of the knowledge, however, is not readily codified and put into books but has the character of the tacit knowledge of craftsmen, passed on personally from investigator to investigator, from teacher to student.[5] An important element in the enterprise of science is the continuity of the knowledge and craft skills that have evolved over the past centuries.

We know of other crafts and craft communities that have been lost through specialization, mechanization, and industrial organization but have more recently been revived as serious avocations. Whatever the future development of scientific professions operating within bureaucratic social structures may be, it would be desirable to promote alternative social structures especially suited to the craft character of scientific practice. Since it is apparent in American society that the available labor pool far exceeds the actual full-time work force needed for producing the necessary goods and services—especially if full use is made of automation—the possibility of pursuing serious avocations is in principle available to an increasing number of people. The freedom of choice and pleasure inherent in a serious avocation is entirely unlike any alienating type of work or the negative condition of mere unemployment. Scientific activity as a serious avocation, and the creation of social and economic patterns conducive to it, is one particular example of a constructive and practical alternative to prevailing patterns. Perhaps not many will follow their scientific muse as artists and poets are expected to. Some may seek association and collaboration with like-minded individuals in their scientific pursuits, but they need not make themselves into members of a bureaucracy on that account. It is a fairly simple matter for a small group of people to arrange to meet regularly to report and hear of each other's research and to discuss scientific questions.

The freedom afforded by a diversity in social patterns would surely liberate new energies for scientific inquiry. With the

402
John von
Neumann
and
Norbert
Wiener

pressures for superficiality, conformity, and narrow specialization lifted, more fundamental scientific issues might come
to the fore, such as the question of the proper context in which
to view scientific knowledge.

As the reader may recall, von Neumann sought to place
scientific knowledge within a formal logical structure which
he viewed as primary and universal. Wiener, by contrast,
placed scientific activity and natural events all within the fundamental and comprehensive notion of "process," in which
science was a process for describing process. These are
certainly possible contexts for science, but it is worth emphasizing that mathematical and scientific knowledge do not
include a knowledge of the context in which they are to be regarded.[6] The contexts within which a scientist perceives his
research express his relationship to certain abstract ideals, to
nature, and to society. The value of the human activity of scientific research is contingent upon the character and validity
of these relationships, an issue outside the realm of science
itself. Scientists who have seriously reflected upon the matter
may differ greatly. For example, G. E. Hutchinson, the distinguished ecologist, a contemporary of Wiener and von
Neumann and a personal acquaintance of both men, sym-
pathized with the view of Simone Weil:

She rejected completely the social significance of science as
it is commonly understood. Her whole life was an intense and
desperate exploration of a psychological wilderness of afflic-
tion, and from this wilderness she cried stridently that we
should take in the entire universe in an act of intellectual love
extending infinitely far into the future and past and excluding
nothing but our momentary sins. It was as a part of this act that
she conceived science.[7]

A naive but conventional conviction is that scientific knowledge, confined to the accepted methodology of science, has
some kind of absolute character as "truth," even as the only
kind of truth acceptable. Although this view is extremely vulnerable to refutation by philosophic criticism,[8] it is a possible
choice of value context, as is the similarly vulnerable opposite view that scientific knowledge is pure illusion. Besides
the "search for truth," the other conventional value context is

the "power over nature" that scientific knowledge provides.
The power achieved over nature is often viewed as convincing evidence that the reality selected by scientific investigators is the only reality that matters, the one we should live by. One might compare the mastery over nature in science to a Hindu yogi's impressive mastery over his own bodily needs and functions. In the latter's metaphysics science is traditionally relegated to the veil of maya, illusion. Both instances reflect the all-too-human vulnerability to being impressed by magic, and in many a culture this vulnerability has led to disastrous consequences. In particular, the power-over-nature ideology assumes a particular kind of relationship between the technologist and the natural world, one that obscures the fact that people themselves belong to nature and are part of ecological systems.[9]

Power over nature may have seemed appealing in the kind of world in which Francis Bacon found himself, but in the current era the objective of enhancing viable ecological patterns within which people are part of mutually regulating complex systems is more appropriate. Such a shift of objectives would mean a reshuffling of research priorities. Similarly, in place of the abstract ideal of truth, one might regard scientific knowledge as a part of one's understanding and wisdom concerning the way things are, an adjunct to practical wisdom. All these are reasonable possibilities worthy of consideration. When context is emphasized, content will be seen in a new light. Even a highly skilled scientist, if he is unreflective or unconcerned about the context of his work, is poorly qualified to evaluate research programs. These considerations lead me to the specific suggestion that the open and skeptical exploration of alternative contexts for science should be viewed as an essential part of general education, and in particular of the education of scientists.

Nuclear Weapons
Part of the tension in the scientific community after the successful completion of the Manhattan Project—a tension that pervaded the lives of Wiener and von Neumann—resulted from diverse responses to the possibility of further nuclear

weapons development. This tension persists. In 1945 many were shocked by the new powers of destruction; now an apparent complacency prevails because these powers have not been used since then except as threats. But just below the surface one finds a widespread sense of futility and powerlessness concerning the course of international events, and even the expectation of a nuclear holocaust in the not-too-distant future. The failure of all efforts at nuclear disarmament and the increased development, production, and stockpiling of nuclear weapons, especially by the United States and the Soviet Union, constitute the material basis for this expectation.

Many will concede the irrationality of the arms race, in which the United States has been leading the rest of the world. The large number of cancer deaths resulting from the mining, processing, storing, and testing of nuclear materials and the diversion of the limited supply of some natural resources merely to create armed camps are evidence against the advisability of nuclear arms development which supplement the primary reason, the likelihood of world catastrophe. Arrayed against these rational arguments are the economic interests of industrial oligopolies, the inertia of powerful governmental and industrial bureaucracies, the internal logic of international relations in a world of competing nation-states, and the emotional investment of some in the arms race as a means to national security or national self-assertion.[10]

When all the momentum lies with the armaments race and none with disarmament, what practical steps can be taken to develop the option of disarmament?

Unfortunately, the group of men who have the political and economic power to advance the disarmament option appear to support the present trend toward increased nuclear weaponry, even though publicly they may even espouse disarmament. Dwight Eisenhower's hypocrisy in this regard is typical. The kind of nonsensical doublethink that calls for more nuclear weapons while avowing interest in disarmament does not deserve credence. Instead, individuals must confront the either-or question: Do you view increased armament or disarmament as the better route to security for yourself and whoever you care about? If one comes out clearly for the disarmament option and all it implies, issues like national and ideological rivalries and economic interests have to be subordinated to the requirement of disarmament.

My first practical suggestion is that not only the reader but also government officials and candidates, representatives, intellectuals, civil servants, and government advisers be challenged to identify their own position as to the either-or question on disarmament. Even a small minority with a strong commitment to nuclear disarmament can provide the needed momentum. By a commitment to nuclear disarmament I do not necessarily mean commitment to unilateral disarmament by the United States but rather to a foreign and domestic policy in which the United States exerts its leadership and political power to the fullest to bring about the dismantling of nuclear arsenals. (The SALT treaties, incidentally, have not been concerned with disarmament; they are little more than agreements between the two giants about how each will engage in further armament.) Once the commitment to disarmament has been made, the concrete questions of how to proceed will become paramount, but at present the commitment is lacking.

A clear stand on disarmament is most meaningful when it is based on more than official information and the news media. Independent historical studies on technological and political events since World War II provide another valuable source of information. Herbert York, Daniel Ellsberg, and Andrei Sakharov each engaged in such historical research as well as soul-searching introspection, and all three shifted away from the prevailing attitudes toward disarmament. Equally important is an acquaintance with the medical effects of nuclear weapons, such as may be obtained from the victims of Hiroshima and Nagasaki. The more personal and direct one's acquaintance with victims, the more fully can one grasp the nature of the impact of the detonation of nuclear weapons. I recommend that all those who take part in making nuclear-weapons decisions engage in these means to inform their understanding and wisdom.

American nuclear policies are made and carried out

406

John von
Neumann
and
Norbert
Wiener

primarily within the executive branch of the government. The Congress is circumvented as much as possible. What happens when a government official responsible for dealing with weapons-related activities comes to believe in disarmament, a policy at variance with that of his department and his government? He may suppress his dissident views so as to retain his position, he may try to influence official policies by articulating his views, he may become difficult and uncooperative or even engage in civil disobedience, or he may leave government and work to implement his views privately. His colleagues may view him as a traitor or rebel, but the defector from government service can be a particularly effective critic, for he is familiar with government procedures and customs.

A grass-roots movement can be important.[11] Aside from providing the dissident with a political constituency and an alternative to anonymity and isolation, a variety of organizations devoted to the nuclear disarmament option, which develop momentum, loyalty, vested interest, and a dynamic of their own, are invaluable in opposing a very large network of bureaucratic structures devoted to making weapons for war and planning their use. Active dissent from policy can also arise among administrators, scientists, and engineers within industry. One can envisage the snowballing of a disarmament movement. Moreover, such a flocking to the disarmament viewpoint could happen in many countries simultaneously, and its participants could make common cause. Of course, the rebellion or noncooperation so easily advocated is not easily carried out: the nuclear arms bureaucracy would seek to make victims of its challengers. The transition from a world governed by the terror induced by the nuclear threat to a world in which nuclear weapons play no major role cannot but require several generations to complete, but the political pressure for nuclear disarmament itself could initiate the process in the early 1980s. Among the changes nuclear disarmament is likely to engender is a shift of loyalties from particular large power blocs or powerful nations to either relatively decentralized political units or the human species at large. The increased appreciation of economic pluralism is also a likely concomitant. These changes would contribute to the stability of a world no longer governed by fear of nuclear bombs.

Technology for Whom and for What?

At the time of this writing, native Americans in New Mexico and in the Black Hills of South Dakota, having seen numerous Navajo uranium miners die of lung cancer and seen their sacred lands mutilated with mines, processing plants, radioactive mill tailings, power plants, and so on, are calling for help[12] to halt the continuing desecration and decimation of their world by corporate giants having US government backing. Examples of unfilled community and individual needs abound, and in many cases technology or technical skills might be helpful. But, ordinarily, available technologies are in the service of the relatively affluent and powerful, often corporations concerned with profit rather than social benefit. Even when government agencies are created to respond to human needs, they tend to serve the internal needs of their bureaucracy more than the needs of the people they are intended to serve.[13]

There is an alternate route for applying technical skills to respond to community and individual needs, which is socially of far greater significance than its small contribution to the economic balance sheet would indicate. This route typically requires both technical and social innovation. Perceiving a particular need, the innovators obtain the necessary financial support and then proceed to create both the social mechanism and the required technology. Such enterprises tend to be personal, flexible, low in cost, and small in scale.[14] At its best the participants share in a sense of community yet retain individual autonomy. This is the decentralist approach, variants of which have been advocated by Peter Kropotkin, Lewis Mumford, Paul Goodman, Martin Buber, and E. F. Schumacher. These efforts are often difficult in a world dominated by bureaucratic, centralized organizations, but they provide a promising route for a future in which technologies and methods are made appropriate to people's lives. In some instances a suitable form of organization is that of a small business, while in others backing from philanthropic founda-

408
*John von
Neumann
and
Norbert
Wiener*

tions is an appropriate resource. Currently one factor inhibiting the development of alternative institutions is that the economic rewards for participation are usually so much lower than those offered by mainstream institutions.

Decentralized alternative associations to respond to unfilled human needs have sprung up everywhere. They range from food cooperatives to alternative medical centers, solar energy technology development groups, and radio stations. Some are study groups providing needed information, such as NARMIC, the action-research project of the American Friends Service Committee in Philadelphia concerned with the facts of military and police technology, the Science for the People group in Cambridge, Massachusetts, and the public-interest law group associated with Ralph Nader—the latter engaging in legal action over and above providing information.

Wiener, von Neumann, and the Many Dimensions of Technology

The story of Wiener and von Neumann can be brought to bear on present concerns and options, even though circumstances have changed considerably since their day. The dimensions relevant to technology and the available options can be viewed abstractly and thus propelled out of the realm of history.

The process of choosing among options usually entails a kind of thinking that incorporates meanings and values. Following a particular option may be viewed as a means of manifesting or even defining who one is. It is a mode of reflection and assertion in which the scientific and technical communities are traditionally untrained. In our society, in which scientific and technical expertise is so highly developed, the capacity for this nearly orthogonal mode of thought—not specialized but incorporating fully all that is individual and human—is generally rudimentary and underdeveloped.

As a stimulus for discussion and as a bridge between systematic scientific thought and the incorporation of values and meanings, a simple taxonomy may be useful. My procedure here will be merely to identify a number of facets or qualitative dimensions pertinent to characterizing a scientist's relation to technology and society, and to locate both Wiener and von Neumann in each dimension. (They can be used as prototypes or straw men.) Each dimension is the subject of a considerable literature. This complex topic becomes much more manageable and open to more sophisticated analysis if only one dimension at a time is considered and synthesis is deferred. The mere recognition of the deep-rootedness of the issues has sharp implications for the education of people—scientists and engineers especially.

A pivotal dimension is the philosophy of technology itself: By what criterion is one to judge a technological development as beneficial or harmful? Von Neumann and Wiener had opposing views. For von Neumann new technology was always basically beneficial, even if sometimes hazardous; people simply had to adapt to the inevitable march of progress. Wiener regarded most technological developments as harmful, not intrinsically but because their control was in the hands of institutions and individuals prone to use them in ways harmful to the human community as a whole. His metaphors were "give into . . . hands," the golem, and the "imaginative forward glance." Marcuse in *Eros and Civilization* asks that technology serve to enhance vitality and enjoyment of life; Jacques Ellul, in *The Technological Society*, prefers technology to fit a view of life emanating from Catholicism, and Langdon Winner emphasizes the desirability of a technology that permits individual freedom of choice in the face of the diversity of preferences and the autonomy of technology itself. All three agree that our present use of technology is predominantly inhumane. Winner in particular raises the political question of who makes and who ought to be making the decisions concerning technology.

Underlying any philosophy of technology is a concept, however inchoate, of human nature. This is another topic deserving reflection and discussion among technologists. We have some indication of von Neumann's views from his emphasis on the fixity of human nature and the use of the utility function for decision making. A clue to Wiener's view lies in his emphasis on learning, communicating, and the fulfillment

410

John von
Neumann
and
Norbert
Wiener

of human possibilities. Wiener's model for decision making includes an explicit ethical dimension. Closely related to one's view of human nature is one's conceptualization of political events, the subject of the established discipline of political science. Both von Neumann's and Wiener's explicit political models are described: Wiener thought in terms of the values embodied in the patterns of communication and control and assumed the power elite to be unscrupulous; von Neumann uncritically accepted the Athenians' view in the dialogue with the Melians—ruthless instrumental rationality in the service of maximizing power—as realistic.

A separate facet deserving consideration is the relation of technologies to time scales, which has both a subjective aspect and a practical aspect—for example, from the viewpoint of a worker in a partially automated factory, a military strategist, or an ambulance driver. The topic of time scales as experienced by Wiener and von Neumann is scattered throughout the preceding chapters. Here I will recapitulate the highlights of their respective attitudes: Wiener's starting point is the natural time scale of organic processes, such as an organism's natural lifetime or the duration of a heartbeat or the time needed to learn something new. Wiener's emphasis concerning the element of time is ecological—the time scale of machines should be chosen so as to be appropriate for the natural time scales of the organisms with which they interact, for otherwise they will be harmful to the organisms. Thus the sensory mechanism in an artificial arm must match the time constants of the biological nervous system. In general, time constants of coupled systems must be so matched that feedbacks are monitored and responded to sufficiently quickly to prevent runaway processes, whether one is talking of a device to prevent death from heart attacks or the coupling of historical process to the growth of scientific knowledge. Applying this reasoning to the currently controversial nuclear reactors, the two time constants, that of the half-life of plutonium in the nuclear waste and the period of stability of social institutions, fail to match, indicating that reactors are likely to be a poor choice of technology. The time scales of all processes are finite and related to other time scales. The infinite has no place

in Wiener's view of technology. Von Neumann's predominant
attitude toward time seems to fit Eliade's description of the modern technologist who, like the early Chinese metallurgists and alchemists, views technology as a means of speeding up natural process, thus achieving a mastery over time and in effect increasing the human life span by comparison. The technologies of transportation and computation could serve to illustrate the concept. Much of von Neumann's applied work had the quality of a race against time, as did his philosophy of technological change.

A fruitful way of presenting the topic of time scales and technology for purposes of discussion is to consider the sequence of construction, preservation, and dismantling that is the life cycle of all humanly created objects, techniques, and institutions. Organisms experiencing birth, life, and death naturally have a similar life cycle. Thus change is unavoidable, and change requires some dismantling or destruction. Modern man, probably much more than medieval man, is conscious of his power to create, to preserve, and to dismantle or destroy; spontaneous nature is not the only actor. But when is the appropriate time to innovate, when to leave things as they are, and when to dismantle? What is it that should change: the machinery, the social patterns, or human psychology? The spectrum of answers to these questions that are possible and can be acted upon in a given situation reveals the range of options concerning the relation of time span to technologies, and each option is likely to carry particular values. The space within which each option is considered is much larger than the dimension described under the rubric of "philosophy of technology." The structure of the questions raised is the same whether one is speaking of large-scale national policies or ordinary everyday choices. To illustrate with a simple example: if someone contracts a mild infectious disease, one possible action is to see a physician, who predictably will prescribe antibiotics, which are introduced into the sick body to eradicate the unwanted bacteria; alternatively he may forgo that technology and rely on changing his habit patterns and diet, so that the resilient ecological system, the human body, will be restored to a new healthy equilibrium. A

412

*John von
Neumann
and
Norbert
Wiener*

third option is to disregard the disease altogether and trust to "time." The difference between the first two options is, characteristically, more pervasive than emphasis on the technology alone would indicate. The antibiotic is expected to give "quick relief" but increases vulnerability. In one case the sick person is defining himself as a patient in relation to the authority of the doctor; in the other case he is trusting in his own sense or in spontaneous process, or perhaps turning to alternative institutions or groups of people with an approach to health different from that of the conventional physician. Introducing the antibiotic becomes an alternative to introducing new habit patterns into his life. If many were to avoid the antibiotics and the consultation with a physician, some institutional patterns of medical practice would wither. What is typical in this illustration is the interplay between introducing (or forgoing) a technology, institutional changes, and individual changes. Generally the three are intertwined, and the choice to introduce or abandon a particular technology involves the other two elements as well.

A different dimension requiring both objective examination and introspection is that of the individual's loyalties, including in particular loyalties to social or economic class. One's commitment can be to a particular nation and power bloc or to humankind and future generations. Wiener and von Neumann differed in this respect. For many the nuclear family comes first, but for others primary loyalty is to a more extended group. Class identification is particularly important. Does one wish to serve those with political and economic power, or the middle class, or the victims of society? Often class interests conflict, but the classic hidden-hand argument does not fit and cannot legitimate the actions of major corporations and dominant political powers. The argument may have to be turned around. The disadvantaged groups have most to gain by changes in the prevailing patterns and are most motivated to bring changes about, but are often resigned to powerlessness in this regard, and sometimes are witting or unwitting accomplices in their own oppression. However, a dynamism, a vitality, is stimulated in some disadvantaged groups as they come to recognize that through organization, solidarity, leadership, and finding some allies within the dominant groups they can improve their lot while retaining their integrity. Generally this brings a cultural enrichment to all strata. The relatively disadvantaged groups more often than not favor a course coincident with generally beneficial changes. For example, in the 1950s many Third World nations advocated total nuclear disarmament, while the United States and the USSR engaged in the arms race. If nuclear disarmament is desirable per se, regardless of power conflicts between the superpowers and the nonnuclear nations, then the interests of nonnuclear nations here coincided with general interests. Similarly, the interests of independent American fishermen coincide with the well-being of the general population when it comes to protecting the ecological equilibrium of the oceans; and the respect native Americans have for their land supports the general public value of a viable environment, even if other forces in our weapons-minded and energy-minded civilization prevail.

A final dimension is that of social function. Von Neumann's function, the technologist influential in high places, fits Bell's concept of postindustrial man. (Actually it has its preindustrial counterpart in the advisers to kings on astrology and other matters.) Wiener's function can be likened to that traditionally performed by conscientious Quakers, even though Wiener functioned in the world of high technology and approached every issue intellectually. Other social functions are possible for scientists and technologists. One new concept of responsibility includes "whistle blowing": calling attention to irresponsible actions of one's employer. Some groups, like the Union of Concerned Scientists, have blown the whistle on whole industries and their associated government agencies.

To help implement the cultural transformation demanded by changed circumstances, college engineering and science students ought to be required to engage in the examination of these kinds of issues with the same degree of seriousness and discipline as they are expected to devote to the technical aspects of science and engineering proper. The radical cultural transformation that appears to be in process means that a quest for genuine wisdom must become a high priority in

414
*John von
Neumann
and
Norbert
Wiener*

society, supplanting to some extent the current dominance of the quest for scientific knowledge and technical skill.

The Irony of Progress

Contemplate now for a moment the great movement of Western science since the days of Galileo Galilei, its pioneer and quite properly its hero. The subsequent centuries of scientific activity may be viewed metaphorically as a journey of discovery and exploration, away from the medieval world, the personal and subjective, the moral, the theological, and the political, and into an objective, empirical, public reality in which measurements fit into abstract mathematical patterns with a claim to universality and the human observer is eliminated. The eye-opening insights of a Newton, a Gauss, an Einstein are among the great treasures discovered on the journey and shared by all who can appreciate them. It was part of the same journey of Western civilization to create machinery of many kinds: elaborate tools, weapons, methods of mass production and complex organization, magical and diverse gadgets—in short, modern technology. And this civilization, drunk with the power of this amazing technology and the benefits it seemed to bring, so forgot itself that it lost all perspective. It let its mode of existence be determined by science and technology. The Nazi gas chambers which came out of that civilization and nuclear bombs, its latest high technology, were like a shot of cold water in the face, awaking us to the discovery, once we had seen past the dazzling treasure, that our journey hadn't taken us as far as we had imagined. It was a familiar landscape because what dominated it, after all, was people—play and affections, politics and passions, pleasures and pains.

NOTES

Preface

1
Jerome R. Ravetz, *Scientific Knowledge and Its Social Problems* (New York: Oxford University Press, 1973), pp. 10, 44.

2
Hannah Arendt, in an article in the *New Yorker* (November 21, 1977) exploring the nature of thinking and its implications for action, asserted that thought has to do with the quest for meaning, and that it is a mental process essentially different from that involved in finding scientific or other truth.

3
A volume of essays has been devoted to each of their mathematical works by the *Bulletin of the American Mathematical Society*. For von Neumann see vol. 64, no. 3, pt. 2; for Wiener see vol. 72, no. 1, pt. 2. See also the collected works of John von Neumann (New York: Macmillan, 1963) and those of Norbert Wiener (Cambridge, Mass.: MIT Press, 1975, 1979). The latter also contain essays appraising Wiener's published work.

4
Benjamin Franklin, "A Proposal for Promoting Useful Knowledge among the British Plantations in America," May 1743, in *The Papers of Benjamin Franklin*, eds. L. W. Labaree et al. (New Haven: Yale University Press, 1959), 2: 380–383.

5
My own position is not holier-than-thou. During the late 1950s, before completing my doctorate, I worked at Ames Research Center (NACA-NASA) in the field of hypersonic flow. Although my work was not classified and applied generally to space

vehicles reentering the atmosphere, it was also clearly useful for missile design. During the 1960s I finally clarified my own views on scientific work with foreseeable military applications.

6

Compare H. F. York, *Race to Oblivion: A Participant's View of the Arms Race* (New York: Simon & Schuster, 1970); *The Advisors: Oppenheimer, Teller and the Superbomb* (San Francisco: Freeman, 1976); C. P. Snow, *Science and Government* (Cambridge, Mass.: Harvard University Press, 1961).

7

The most helpful discussion of this issue I have found is in the concluding chapter of Philip P. Hallie, *The Paradox of Cruelty* (Middletown, Conn.: Wesleyan University Press, 1969). Shaw wrote in the preface to his play *Saint Joan*, "There was a great wrong done to Joan and to the conscience of the world by her burning. *Tout comprendre, c'est tout pardonner,* which is the Devil's sentimentality, cannot excuse it. When we have admitted that the tribunal was not only honest and legal, but exceptionally merciful in respect of sparing Joan the torture which was customary . . . the human fact remains that the burning of Joan of Arc was a horror, and that a historian who would defend it would defend anything."

1

Boston Evening Transcript, March 19 and 26, April 2, 9, 16, 23, and 30, 1910. The series is entitled "Stray Leaves from My Life." See also the Harvard University Archives for information about Leo Wiener.

2

Elias Schulman, introduction to Leo Wiener, *The History of Yiddish Literature in the Nineteenth Century*, 2d ed. (New York: Hermon Press, 1972), p. ix, states that Leo Wiener's parents also left Europe for the United States in 1882.

3

Wiener, "Stray Leaves," April 23.

4

Ibid., April 30.

5

See Wiener, *The History*, cited; or Leo Wiener's introduction to Morris Rosenfeld, *Songs from the Ghetto,* Copeland and Day, 1898.

6

Wiener, *The History,* p. 10.

7

Rosenfeld, *Songs from the Ghetto*.

8

For Leo Wiener's relation to Judaism see Norbert Wiener's *Ex-Prodigy* (Cambridge, Mass.: MIT Press, 1964), hereafter referred to as EP. See also Schulman's biographical introduction.

9

L. N. Tolstoy, *Complete Works*, 24 vols. (Boston: D. Estes & Co., 1904–1905).

10

EP, p. 22.

11

Ibid.

12

Ibid., p. 27. I am especially indebted to Mr. and Mrs. Alfred Richardson of Ayer, Massachusetts, for an interview (Dec. 11,

1971) concerning their recollection of the Wieners. Alfred R. was Norbert's classmate; Mrs. Richardson is the sister of Norbert's teacher, Miss Leavitt.

13
Ibid., p. 146. Norbert Wiener hesitated for some time before deciding to include these recollections of his mother in his autobiography. Historian Neal Hartley told me (Nov. 15, 1971) of conversations with Wiener on this subject at the time he was working on it. In an earlier version of EP, an MS. entitled "A Bent Twig," Leo Wiener's rejection of Judaism and of the Jewish community is given greater prominence.

14
H. Addington Bruce, "New Ideas in Child Training," *American Magazine*, July 1911, pp. 291–292.

15
EP, p. 136. Norbert Wiener gives his own views on education in *The Human Use of Human Beings* (Boston: Houghton Mifflin, 1950), pp. 150 ff., hereafter referred to as HUHB.

16
EP, p. 82.

17
See Gershom Scholem, *On the Kabbalah and Its Symbolism* (New York: Schocken, 1969), chap. 5, for a discussion of the golem. The golem of Prague had, according to Jewish folklore, been created by Rabbi Loew and his disciples during the sixteenth century. Somewhat analogously to the biblical creation of the first man, they kneaded a figure of a man out of clay; after the appropriate incantations (however blasphemous) and rituals were performed, the golem opened his eyes and got to his feet. This was at a time of a particularly malicious persecution of the Jews of the Prague ghetto, and the golem was given a specific task that effectively protected the Jews in the ghetto. After a few years, when the danger had subsided, his task was complete. The process of the creation of the golem was then carried out in reverse, and the golem again became inanimate.

18
EP, pp. 67–68.

19
Ibid., p. 63.

20
Ibid., p. 62.

21
Interviews with Mr. and Mrs. Alfred Richardson and Henry Brown, Dec. 11, 1971. Brown, an Ayer druggist, is brother of Frank Brown, described in EP. See EP, pp. 15, 93, 101.

22
Turner's Public Spirit (Ayer, Mass.), June 2, 1906.

23
EP, p. 100.

24
The Guide to Nature (an illustrated magazine for adults, published by the Agassiz Association, Stamford, Connecticut), April 1909.

25
EP, p. 121.

26
See Barbara Miller Solomon, *Ancestors and Immigrants* (Cambridge, Mass.: Harvard University Press, 1956). For the Adams brothers specifically, see *The Letters of Henry Adams,* ed. W. C. Ford (Boston, 1938), *An Autobiography* by Charles F. Adams (Boston: Houghton Mifflin, 1916) and *The Education of Henry Adams* (Boston: Mass. Historical Soc., 1918); Henry Adams, *The Degradation of Democratic Dogma* (New York: Macmillan, 1920), in particular the introduction by Brooks Adams (New York, 1920); and the biographies by Ernest Samuels of Henry Adams and Arthur F. Beringause and Thornton Anderson of Brooks Adams. Another aristocrat expressing strong anti-Semitic views was John Jay Chapman; see Richard B. Hovey, *John Jay Chapman—An American Mind* (New York: Columbia University Press, 1959). See also Edmund Wilson, *Commentary* 22 (October 1956): 329–335, and John Hingham, *The Mississippi Valley Historical Review* 43 (March 1957):559–578.

27
Quoted in Samuel Eliot Morison, *Three Centuries of Harvard* (Cambridge, Mass.: Harvard University Press, 1965), p. 418.

28
Statistical information is from William Coolidge Lane, "The University During the Last 25 Years, 1881–1906," Harvard

University Archives. For criticism of Eliot's policy, see John
Jay Chapman, "President Eliot," in *The Selected Writings of
John Jay Chapman*, ed. Jacques Barzun, 1957; George Santayana, "The Academic Environment," in *Character and
Opinion in the U.S.* (New York: Norton, 1967), and "The Spirit
and Ideals of Harvard U.," in *George Santayana's America*,
ed. J. Ballowe (Urbana: University of Illinois Press, 1961).
Harvard's Scientific School had increased more than tenfold
from 1880 to 1906 and the graduate school similarly, while the
total enrollment had only tripled during that period.

29
Solomon, *Ancestors and Immigrants*, chap. 5.

30
Ibid.

31
EP, p. 148.

32
Norbert Wiener, *I Am a Mathematician* (Cambridge, Mass.:
MIT Press, 1964), p. 215, hereafter referred to as IAM.

33
Norbert Wiener to Leo Wiener, n.d. (Wiener Collection, MIT
archives).

34
EP, pp. 215–216.

35
N. Levinson, "Wiener's Life," *Bulletin of the American Mathematical Society* 72, no. 1, pt. 2 (1966): 7.

36
Since this chapter was written, additional material detailing
the intellectual relation of Wiener to Russell has appeared.
See vol. 1 of *Wiener's Collected Works* (Cambridge, Mass.:
MIT Press, 1977); see also I. Grattan-Guiness, "Wiener on the
Logics of Russell and Schroeder: An Account of His Doctor's
Thesis, and of His Discussion of It with Russell," *Annals of
Science* 32 (1975): 103–132.

37
Bertrand Russell to Lucy Donnelly, October 19, 1913, quoted
in I. Grattan-Guiness, "The Russell Archives: Some New Light
on Russell's Logicism," *Annals of Science* 31 (1974): 406.

38
EP, p. 190; IAM, p. 22.

39
Leo Wiener to Norbert Wiener, November 3, 1913 (MIT archives).

40
Norbert to his mother, January 31, 1913 (MIT archives).

41
Norbert to his father, October 25, 1913 (MIT archives).

42
EP, p. 240.

43
The story emerges from a half-humorous but nevertheless quite angry "proclamation" by N. Wiener that, from that time on, a state of war between Wiener and Sergeant Smith was in force.

44
He defends this thesis in *The Technology Review* 32, no. 3 (1929): 129.

45
G. H. Hardy, *A Mathematician's Apology* (Cambridge: Cambridge University Press, 1940).

46
Vannevar Bush, *Operational Circuit Analysis* (New York: Wiley, 1929); the appendix contains original work by Wiener and deals with the mathematical justification for the Heaviside Operational Calculus. See also IAM, p. 139.

47
Paul Forman, "The Financial Support and Political Alignment of Physicists in Weimar Germany," *Minerva* 12 (1974): 39; "Scientific Internationalism and Weimar Physicists," *Isis* 64 (1973): 151.

48
See for example Peter Gay, *Weimar Culture* (New York: Harper & Row, 1968).

49
This comment is based on Wiener's letters, especially to his sister and brother-in-law.

50

Dirk Struik, private communication.

51

J. Robert Oppenheimer, interview with T. S. Kuhn, November 20, 1963. Archives for the History of Quantum Physics, American Philosophical Society Library, Philadelphia, Pa. (hereafter called Quantum archives).

52

The Courant-Wiener relation is described in IAM, pp. 113–118, and in Constance Reid, *Courant* (New York: Springer, 1976), pp. 104–106, 121; see also Wiener's letters home in 1925–26 (MIT Archives). See pp. 173–174 and 180–187 of *Courant* on the Courant-Douglas relation. See also Norbert Wiener to J. R. Kline, Feb. 17, 1941, and Wiener to the Secretary, American Mathematical Society, November 10, 1942 (MIT archives). These events did not alter Hilbert's high opinion of Wiener's mathematical ability: "I can say that I assess Mr. Wiener's mathematical ability as being of a high order. He combines to a high degree great knowledge and creative power in analysis with understanding for applications, especially to modern physics" (Hilbert, August 4, 1928, in a letter recommending Wiener for a position).

53

Their joint articles are "Quantum Theory and Gravitational Relativity," *Nature* 119 (1927): 853; "A Relativistic Theory of Quanta," *Journal of Mathematics and Physics* 7 (1927): 1–23.

54

Wiener to J. R. Killian, September 13, 1951.

55

IAM, p. 105.

56

Max Born and Norbert Wiener, "A New Formulation of the Laws of Quantization for Periodic and Aperiodic Phenomena," *Journal of Mathematics and Physics* 5 (1926): 84. Born and Wiener introduced the formulation of quantum mechanics in terms of a general linear operator calculus. See Max Jammer, *The Conceptual Development of Quantum Mechanics* (New York: McGraw-Hill, 1966), pp. 220–227, for an appraisal of the Born-Wiener article.

57
IAM, p. 109.

58
Constance Reid, *Hilbert* (New York: Springer, 1970), p. 180.

2 Von Neumann's Youth: Mathematical Reasoning Power as Amulet

1
Eugene Wigner, *Symmetries and Reflections* (Cambridge, Mass.: MIT Press, 1970), p. 260.

2
Numerous quotations could be cited. The comment of Hans Bethe, Nobel Prize winner in physics, may suffice to illustrate the genre: "I have sometimes wondered whether a brain like John von Neumann's does not indicate a species superior to that of man" (*Life*, February 25, 1957, p. 90). See also Wigner, *Symmetries and Reflections*, p. 198; Oskar Morgenstern, *Economic Journal*, March 1958, pp. 170–174; Herman Goldstine, *The Computer from Pascal to von Neumann* (Princeton: Princeton University Press, 1972), p. 176.

3
Compare Stanislaw Ulam, *Adventures of a Mathematician* (New York: Scribner's, 1976), pp. 78–79.

4
Interview with William Fellner, October 7, 1971.

5
See *Hungary To-day*, ed. Percy Alden (London: Brentano's, 1909). The Budapest population was 730,000 in 1900, while total Hungarian population was 19,300,000. Magyars constituted 45 percent of the total population and Jews 5 percent. (Statistics prepared by the Hungarian government.) See also R. W. Seton-Watson, *Racial Problems in Hungary* (London: A. Constable, 1908).

6
Zoltan Horvath, *Die Jahrhundertwende in Ungarn* (Berlin: Luchterhand, 1966); David Kettler, *Marxismus und Kultur* (Berlin: Luchterhand, 1967); see also the succinct comments of Lewis Coser in *Masters of Sociological Thought* (New York: Harcourt Brace Jovanovich, 1971), p. 442, and the review of Horvath's book in *New Hungarian Quarterly* 3, no. 6 (1962): 207–213 by Peter Hanak.

7
The Universal Jewish Encyclopedia, ed. Isaac Landman (New York: The Universal Jewish Encyclopedia, Inc., 1941); *Ungarische Bibliothek*, vol. 25 (Berlin: Julius von Farnkas, 1940);

Jerome and Jean Tharaud, *Die Herrschaft Israels* (Vienna: Amalthea, 1927). For a short, readable account of Hungarian history, see Emil Lengyel, *1000 Years of Hungary* (New York: John Day, 1958); also Carlile A. Macartney, *Hungary* (Chicago: Aldine, 1962).

8
Horvath, *Jahrhundertwende*; Hanak; William O. McCagg, *Jewish Nobles and Geniuses and Modern Hungary* (New York: Columbia University Press, 1973).

9
Jean-Paul Sartre, *Anti-Semite and Jew* (New York: Grove Press, 1962), p. 132. (The original French edition appeared in 1946.)

10
This discussion relies heavily on McCagg, *Jewish Nobles and Geniuses.*

11
Ibid., p. 33.

12
Ibid.

13
Arthur Koestler, *Arrow in the Blue* (New York: Macmillan, 1952).

14
Quoted by Lewis Strauss, *Men and Decisions* (Garden City, N.Y.: Doubleday, 1962).

15
T. von Karman, *The Wind and Beyond* (Boston: Little, Brown, 1967); Franz Alexander, *The Western Mind in Transition* (New York: Random House, 1960); Koestler, *Arrow in the Blue*; Edward Teller and Allen Brown, *Legacy of Hiroshima* (Garden City, N.Y.: Doubleday, 1962). Teller is also quoted by Laura Fermi, *Illustrious Immigrants* (Chicago: University of Chicago Press, 1968). See also Eugene Wigner, recorded interviews, Archives for the History of Quantum Physics, American Philosophical Society, Philadelphia, and Ulam, *Adventures of a Mathematician*, which discusses Ulam's as well as von Neumann's views.

16
McCagg, *Jewish Nobles and Geniuses*; Fermi, *Illustrious Immigrants*, pp. 53 ff.

17
Alexander (*Western Mind*) has emphasized the familial encouragement of young men to develop themselves and their potential talents in middle-class Budapest prior to 1914.

18
Alexander, *Western Mind*.

19
Education in Hungary (Budapest: Royal Hungarian Ministry of Religion and Public Instruction, 1908); Julius Kornis, *Education in Hungary* (New York: Columbia University Press, 1932) and a somewhat different German version, *Ungarns Unterrichtswesen Seit dem Weltkriege* (Leipzig: Quelle und Meyer, 1930).

20
The foremost subject was Latin—an hour a day six days a week for all eight years of gymnasium. Hungarian language and literature was next, and then mathematics. The gymnasium had eight years of mathematics; German language and literature, history, and Greek were other required subjects, in declining order of importance. Religion was also a required subject—Lutheran, Catholic, or Jewish, depending on one's faith. In addition to all of these requirements, there were two years of general science and a year of physics. For details see Kornis, *Education in Hungary* and *Ungarns Unterrichtswesen*, and Royal Hungarian Ministry of Religion and Public Instruction, *Education in Hungary*.

21
McCagg, *Jewish Nobles and Geniuses*, p. 215. Wigner has described the atmosphere of the high school in the recorded interviews deposited at the Quantum archives.

22
McCagg, *Jewish Nobles and Geniuses*, pp. 220–222.

23
Fermi, *Illustrious Immigrants*, p. 53.

S. Ulam, "John von Neumann," in *Bulletin of the American Mathematical Society* 64, no. 3, pt. 2 (1958): 1. (My emphasis.)

25
Dennis Gabor, *Innovations* (Oxford: Oxford University Press, 1970), writes that we have now reached a stage in which technological innovation is misguidedly governed by the "compulsive" belief, "Innovate or die!" Gabor is another Budapester of von Neumann's generation, an innovator and a Nobel Prize winner.

26
Perhaps none was more deeply stirred to innovate in response to the possibility and then the actuality of the atomic bomb than Leo Szilard. For Szilard, however, innovation was not solely technological but included political organization, political action, the shifting of his career from physics to biology, and the writing of *The Voice of the Dolphins*. His emphasis was on social innovation, while Teller and von Neumann concentrated on innovations in weapons technology.

27
Teller and Brown, *The Legacy of Hiroshima*, chap. 5.

28
McCagg, *Jewish Nobles and Geniuses*, p. 31.

29
During World War I the name of the street was changed to Vilmos Csaszar-ut, and its present name is Bajcsi Zsilinszky-ut.

30
I am grateful for the recollections of John von Neumann's childhood recounted to me by his brothers Nicolas Vonneuman and Michael Neumann, his cousin Catherine Pedroni, and his schoolmates Eugene Wigner and William Fellner.

31
McCagg, *Jewish Nobles and Geniuses*, p. 31.

32
Ibid., p. 17.

33
Ibid.

34
Hanak, review of Horvath, p. 210.

35
Wigner, recorded interview (Quantum archives). Wigner also discusses Ladislas Ratz.

36
Wigner, ibid.; compare McCagg, *Jewish Nobles and Geniuses,* pp. 214–215.

37
Interview of Wigner by T. S. Kuhn, tapes 92b and 93a (Quantum archives).

38
"Über die Lage der Nullstellen gewisser Minimumpolynome," *Jahrberichte Deutschen Mathematischen Verein* 31 (1922): 125–138.

39
Wigner, *Symmetries and Reflections*, p. 258.

40
Interview of Wigner by the author, August 10, 1971.

41
Interview of Wigner by Kuhn, tape 92b (Quantum archives).

42
Based on independent recollection of Jansci's close friend, Eugene Wigner, and his brother Michael Neumann. The account in T. von Karman's ghosted autobiography, *The Wind and Beyond*, gives a different picture of Max Neumann's views on Jansci's career.

43
The generational tendency from business to academia held true for many of John von Neumann's contemporaries who like von Neumann participated in the development and early application of quantum theory; examples are Bloch, Oppenheimer, Wigner, Peirls, and Schrödinger. Typically, the fathers wanted their sons to come into the business; the sons sought scholarly pursuits instead (recorded interviews, Quantum archives).

44
Ulam, *Adventures of a Mathematician*, p. 71.

45
References on which the discussion of Hungary is based, especially that of the years of 1914–1921, include Macartney, *Hungary*, *Hungary and Her Successors* (Oxford: Oxford University Press, 1937), and *October Fifteenth* (Edinburgh: Edinburgh University Press, 1956); Michael Karolyi, *Memoirs* (London: Jonathan Cape, 1956); Oskar Jászi, *The Dissolution of the Habsburg Monarchy* (Chicago: University of Chicago Press, 1929); *Revolution and Counter-Revolution in Hungary* (Westminster: King, 1924); Rudolf L. Tökes, *Béla Kun and the Hungarian Soviet Republic* (New York: Praeger, 1967); Lengyel, *1000 Years of Hungary*; A. Kaas and F. de Lazarovics, *Bolshevism in Hungary* (London: Grant Richards, 1931); Hugh Seton-Watson, *The East-European Revolution* (New York: Praeger, 1951); John Weiss, *Conservatism in Modern Europe* (New York: Harcourt Brace Jovanovich, 1977) and *The Fascist Tradition* (New York: Harper & Row, 1967); and Koestler, *Arrow in the Blue*.

46
See Alexander, *Western Mind*, for a discussion of this parental attitude in prewar Hungary.

47
Bolshevism in Hungary, Jászi, *Revolution Etc.*, p. 42.

48
Kaas et al., pp. 220–222.

49
John von Neumann's testimony at the AEC Personnel Security Board Hearings concerning J. R. Oppenheimer, April 27, 1954; verbal accounts from members of the Neumann family given to the author.

50
McCagg, *Jewish Nobles and Geniuses*, p. 16.

51
Koestler, *Arrow in the Blue*, pp. 64–65.

52
Recollections of Eugene Wigner, August 10, 1971, and William Fellner, October 7, 1971, related to the author.

53
Macartney, *October Fifteenth*, 1: 23.

54
Tökes, *Béla Kun*, p. 214.

55
Jászi, *Revolution Etc.*, pp. 60–61, 160.

56
McCagg, *Jewish Nobles and Geniuses*, pp. 16–17.

57
The number of Jews was to be in proportion to the fraction of Jews in the population of Hungary. This required a severe reduction in the number of Jewish students.

58
Kornis, *Education in Hungary*, p. 143.

59
Eugene Wigner, interviews recorded by T. S. Kuhn.

60
Constance Reid, *Hilbert* (New York: Springer, 1970), p. 172.

61
Ibid.; see especially the appendix by Herman Weyl.

62
Ulam, "John von Neumann," p. 5.

63
Peter Gay, *Weimar Culture* (New York: Harper & Row, 1968); for photographs see Thilo Koch, *Die Goldenen Zwanziger Jahre* (Berlin: Althenaion, 1970).

64
Fritz K. Ringer, *The Decline of the German Mandarins: The German Academic Community 1890–1933* (Cambridge, Mass.: Harvard University Press, 1969).

65
Joseph Ben-David, *The Scientist's Role in Society* (Englewood, N.J.: Prentice-Hall, 1971), p. 135.

66
Ibid., p. 137.

67
Paul Forman, "Scientific Internationalism and the Weimar Physicists," *Isis* 64 (June 1973): 151; "The Financial Support and Political Alignment of Physicists in Weimar Germany," *Minerva* 12 (1974): 39.

68
Ringer, *Decline of the German Mandarins*; Ben-David, *Scientist's Role in Society.*

69
Ben-David, *Scientist's Role in Society*, p. 138.

70
See Philip H. Abelson, "Relation of Group Activity to Creativity in Science," *Daedalus*, Summer 1965, on Wigner and von Neumann as "foreigners" in Germany.

71
Lothar Nordheim to the author, July 24, 1978.

72
Reid, *Hilbert*; Richard Courant, interview with T. S. Kuhn, May 9, 1962 (Quantum archives).

73
W. Heisenberg, interview with T. S. Kuhn, February 7, 1963 (Quantum archives). His Göttingen lectures were in winter 1925–1926; he had been working both in Göttingen and in Copenhagen during 1925.

74
Maria G. Mayer, interview with T. S. Kuhn, February 20, 1962.

75
Lothar Nordheim to the author, July 24, 1978: "I tried at that time to cast the unifying Dirac-Jordan transformation theory into a simpler and more easily understandable form and to convey its essence to Hilbert. When von Neumann saw this he cast it in a few days into an elegant axiomatic form much to the liking of Hilbert. (This is the origin of the paper "Über die Grundlagen der Quantenmechanik" by Hilbert, von Neumann and myself . . .). The method used was that of integral operators. . . . This work set von Neumann on his way to his definite studies on the foundations of quantum mechanics."

76
Heisenberg, interviews with T. S. Kuhn; he suggests Pauli and Einstein agreed with his view.

77
Ulam, "John von Neumann," pp. 8–15, gives an overview of von Neumann's work on the foundations of mathematics. His original papers on this subject are reprinted in vol. 1 of his *Collected Works* (New York: Macmillan, 1961).

78
Koestler, von Neumann coeval and compatriot, who unlike von Neumann had been attracted to leftist politics, wrote in *Arrow in the Blue* that "at a conservative estimate, three out of every four people I knew before I was thirty were subsequently killed in Spain or hounded to death at Dachau, or gassed at Belsen, or deported to Russia; some jumped from windows in Vienna or Budapest; others were wrecked by the misery and aimlessness of permanent exile."

79
Shields Warren, in interview with the author at the Shields Warren Radiation Laboratory, Deaconess Hospital, Boston, Mass., Dec. 21, 1972.

1

Giorgio de Santillana, *The Origins of Scientific Thought*
(Chicago: University of Chicago Press, 1961), provides a
knowledgeable overview of ancient Greek mathematics and
science. Standard works on Greek mathematics are listed in
its bibliography. Most histories of mathematics devote considerable space to the ancient Greeks.

2

Ibid., p. 68, based on translation by Kathleen Freeman in *An-
cilla to the Pre-Socratics* (Cambridge, Mass.: Harvard University Press, 1948).

3

For the details of the development of measure theory in the
nineteenth century, see Ivan N. Pesin, *Classical and Modern
Integration Theories* (New York: Academic, 1970); Thomas W.
Hawkins, *Lebesgue's Theory of Integration—Its Origin and
Development* (Madison: University of Wisconsin Press, 1970).
See also the elementary book Henri Lebesgue, *Measure and
the Integral*, ed. K. O. May (San Francisco: Holden-Day,
1966).

4

Jean Perrin, *Atoms* (London: Constable, 1916), pp. 82–83.

5

The classic articles by Robert Brown are "A Brief Account of
Microscopical Observations made in the Months of June, July
and August, 1827, on the Particles in the Pollen of Plants; and
on the general Existence of active Molecules in Organic and
Inorganic Bodies," *Philosophical Magazine*, n.s. 4 (1828):
161; "Additional Remarks on Active Molecules," *Philosophical Magazine*, n.s. 6 (1829): 161. Reviews of the early history
of Brownian motion are contained in Jean Perrin, *Brownian
Movement and Molecular Reality* (London: Taylor and Francis, 1910); Edward Nelson, *Dynamical Theories of Brownian
Motion* (Princeton: Princeton University Press, 1967).

6

In particular, M. Gouy and F. M. Exner made careful experiments on Brownian motion late in the nineteenth century. In
addition to Perrin, *Brownian Movement* and *Atoms,* their

work is reviewed in G. L. de Haas-Lorentz, *The Brownian Movement and Some Related Phenomena* (Braunschweig: Vieweg, 1913).

7

Aside from the writings of the principals in the controversy, one might read Stephen Brush's introduction to Ludwig Boltzmann, *Lectures on Gas Theory* (Berkeley: University of California Press, 1964); C. C. Gillispie, *The Edge of Objectivity* (Princeton: Princeton University Press, 1960), chap. 2; and M. Klein, *Paul Ehrenfest* (London: North-Holland, 1970), pp. 37, 63.

8

Boltzmann, *Gas Theory*, p. 216.

9

Ibid., introduction, p. 17.

10

Nelson, *Dynamical Theories*, and Perrin, *Brownian Movement*, review the experiments.

11

Einstein was apparently unaware that Gibbs had already made such derivations. See *Albert Einstein*, ed. Paul A. Schilpp, 1: 47.

12

Ibid.

13

"On the Movement of Small Particles Suspended in a Stationary Liquid Demanded by the Molecular-Kinetic Theory of Heat," *Annalen der Physik* 17 (1905): 549; published in English in Albert Einstein, *Investigation on the Theory of the Brownian Movement*, ed. R. Fürth (New York: Dover, 1956).

14

"On the Theory of the Brownian Movement," *Annales de Physik* 19, (1906): 371; published in English in Einstein, *Theory of the Brownian Movement*.

15

L. Bachelier, "Théorie de la spéculation," *Annales Scientifiques de l'Ecole Normale Supérieure* 17 (1900): 21; M. von Smoluchowski, "Zur kinetischen Theorie der Brownschen

Molekularbewegung und der Suspension," *Annalen der Physik* 21 (1906): 756.

16
Perrin, *Atoms*, 2d English ed.

17
R. E. A. C. Paley and Norbert Wiener, *Fourier Transforms in the Complex Domain* (New York: American Mathematical Society, 1934), p. 157.

18
IAM, p. 23.

19
Ibid., p. 33.

20
Such a function was first constructed by Bilzano in 1834; other examples were given by Weierstrass thirty years later. The Weierstrass nondifferentiable function is

$$f(x) = \sum_{N=0}^{\infty} b^m \cos (a''\pi x),$$

where *a* is an odd positive integer, $0 < b < 1$, and $ab > 1 + 3\pi/2$. See E. O. Titchmarsch, *Theory of Functions*, 2d ed. (Oxford: Oxford University Press, 1939), p. 351, for proof of its nondifferentiability. Another example of such a function is the snowflake curve first constructed by Helge von Koch in 1904; it is constructed as follows:

 Equilateral
triangle

 Trisect each line;
erect new equilateral
triangle on middle of each line.

 Again trisect each line of
previous figure, and erect new
equilateral triangle on middle of each line.

Repeat this process *indefinitely,* and the curve will be without a tangent. (Based on C. Boyer, *A History of Mathematics* (New York: Wiley, 1968.)

21

G. H. Hardy, *A Mathematician's Apology* (Cambridge: Cambridge University Press, 1969), p. 84.

22

In particular, functions that are discontinuous everywhere were considered, such as $f(x)$ is 1 for all rational numbers but is 0 for all irrational numbers for $0 < x < 1$. Such a function possesses no Riemann integral. See discussion in Edna Kramer, *The Nature and Growth of Modern Mathematics* (New York: Hawthorn, 1970), p. 645.

23

Stanislaw Saks, *Theory of the Integral*, preface to the 1st ed. (New York: Stechert, 1937). Hermite is quoted by Saks.

24

Hawkins, *Lebesgue's Theory,* lists a number of problems having to do with nonanalytic and everywhere-discontinuous functions which helped motivate the Lebesgue theory of integration based on measure theory.

25

Lebesgue, *Measure and the Integral*, p. 98.

26

J. P. Daniell, W. H. Young, M. Frechet, R. Gateaux, P. Levy, J. Radon, E. H. Moore.

27

"The Mean of a Functional of Arbitrary Elements," *Annals of Mathematics* 22 (1920): 66.

28

IAM, p. 36.

29

In 1961 S. G. Brush reviewed some applications in statistical physics of the Wiener integral in "Functional Integrals and Statistical Physics," *Review of Modern Physics* 33 (1961): 79.

30

The probability for a Brownian particle to travel a (net) distance between x and $x + dx$ in time t is shown by Einstein to be

$$\frac{dx}{\sqrt{4\pi Dt}} \exp(-x^2/4Dt),$$

where D is the diffusion constant. The Gaussian form for the probability for the net distance traveled, which after all is the effect of a large number of collisions, follows from the central limit theorem. Einstein originally derived it, including the dependence on D and t, by showing that the probability density must satisfy a diffusion equation. Wiener's measure for a particle going between a_1 and b_1 at time t_1, between a_2 and b_2 at time t_2, etc., given it left $x = 0$ at time $t = 0$, is

$$[(4\pi D)^n t_1(t_2 - t_1) \cdots (t_n - t_{n-1})]^{-1/2} \int_{a_1}^{b_1} \int_{a_2}^{b_2} \cdots \int_{a_n}^{b_n}$$

$$\exp\left(-\frac{x_1^2}{4Dt_1} - \frac{(x_2 - x_1)^2}{4D(t_2 - t_1)} - \frac{(x_3 - x_2)^2}{4D(t_3 - t_2)}\right.$$

$$\left. - \cdots - \frac{(x_n - x_{n-1})^2}{4D(t_n - t_{n-1})}\right) dx_1\, dx_2 \cdots dx_n.$$

This expression is an immediate consequence of the product rule for independent probabilities, applied to the probability distribution of a Brownian particle traveling a distance between x and $x + dx$ in time t.

31
Wiener had allowed all continuous and discontinuous trajectories, although only the former make physical sense. When he imposed the Wiener (Einstein) measure, however, all the discontinuous functions, except for a set of measure zero, were automatically eliminated. This is reassuring. Nondifferentiability idealizes the circumstance that a Brownian particle undergoes about 10^{20} collisions with molecules per second, so that it changes its direction so frequently that the periods during which the Brownian particle moves without abrupt change in direction are too rare and too brief to be caught even by modern high-speed photography.

32
George Uhlenbeck and L. S. Ornstein, "On the Theory of the Brownian Motion," *Physical Review* 36 (1930): 823.

33
J. L. Doob, "The Brownian Movement and Stochastic Equations," *Annals of Mathematics* 43 (1942): 351. The curves have no second derivative anywhere.

34
The original articles by Wiener are "The Mean of a Functional of Arbitrary Elements," *Annals of Mathematics* 22 (1920):

66; "The Average of an Analytical Functional," *Proceedings of the National Academy of Science* 7 (1921): 253; "The Average of an Analytical Functional and the Brownian Movement," *Proceedings of the National Academy of Science* 7 (1921): 294; "Differential Space," *Journal of Mathematics and Physics* 2 (1923): 131; "The Average Value of a Functional," *Proceedings of the London Mathematical Society* 22 (1924): 454.

35
For an appraisal of Wiener's work on Brownian motion, see the articles by Mark Kac and J. L. Doob in *Bulletin of the American Mathematical Society* 72, no. 1, pt. 2 (1966). Wiener et al., *Differential Space, Quantum Systems and Prediction* (Cambridge, Mass.: MIT Press, 1966), which was published after Wiener's death, in chapters 1–3 contains a review of some of Wiener's early work on Brownian motion—a very readable exposition for people with some knowledge of mathematics who are not professional mathematicians. In the same category is chap. 4 of Mark Kac's *Probability and Related Topics in Physical Science* (New York: Interscience, 1959). Expositions of the modern theory of Brownian motion—for mathematicians—are Kiyoshi Ito and Henry P. McKean, *Diffusion Processes and Their Sample Paths* (Berlin: Springer, 1965), and the shorter *Stochastic Integrals* by Henry P. McKean (New York: Academic, 1960).

36
The first instance is Plancherel's "Proof of the Impossibility of Ergodic Mechanical Systems," *Annalen der Physik* 42 (1913): 1061–1063.

37
Kac, *Bulletin of the American Mathematical Society* 72, no. 1, pt. 2 (1966): 68.

1

Psychoanalysts have emphasized that both parricidal and homosexual emotions are often sublimated by preoccupation with the logical game of chess. See, for example, Ernest Jones, "The Problem of Paul Morphy: A Contribution to the Psychology of Chess," in *Essays in Applied Psychoanalysis,* vol. 1, ed. E. Jones (London: Anglobooks, 1951); Norman Reider, "Chess, Oedipus and the Mater Dolorosa," *International Journal of Psychoanalysis* 40 (1955): 320–333; Reuben Fine, *The Psychology of the Chess Player* (New York: Dover, 1967).

2

Die Philosophischen Schriften von G. W. Leibniz, ed. C. I. Garhardt (Berlin: Asher, 1875–1890), 5: 447, quoted by Ian Hacking, *The Emergence of Probability* (Cambridge: Cambridge University Press, 1975), p. 57.

3

On the history of the theory of probability, see I. Todhunter, *A History of the Mathematical Theory of Probability* (New York: Chelsea, 1949); Florence N. David, *Games, Gods and Gambling* (London: Griffin, 1962); Hacking, *Emergence of Probability*. On Pascal, see, for example, F. Strowski, *Pascal et son temps*, 3 vols. (Paris: Plon-Nourrit et cie, 1907–1922); Jean Mesnard, *Pascal: His Life and Works* (New York: Philosophical Library, 1952).

4

Apparently an earlier machine had been built by Wilhelm Schickard in 1623, but it was destroyed by fire, and Schickard died of the plague. Neither Pascal nor Leibniz knew of the machine. See Herman Goldstine, *The Computer from Pascal to von Neumann* (Princeton: Princeton University Press, 1972).

5

Hacking, *Emergence of Probability*, p. 63. I refer the reader to Pascal's *Pensées* for details of his reasoning and to Hacking for an analysis of Pascal's logic, including some subtle points.

6

Voltaire, *Lettres philosophiques* (Amsterdam, 1734), Letter 25 par M. de V. Quoted by Hacking, *Emergence of Probability*, p. 72.

7

I am indebted to Silvan S. Schweber for calling this aspect of nineteenth-century intellectual life to my attention.

8

Adam Smith, *The Wealth of Nations* (New York: Random House, 1937; first published in 1776).

9

Charles Darwin, *The Origin of Species* (New York: Mentor, 1958), p. 75.

10

General Karl von Clausewitz, *On War* (London: Kegan Paul, Trench, Trubner, 1940).

11

See chapter 34 of Charles Babbage, *Passages from the Life of a Philosopher* (London: Longman, Green, Longman, Roberts & Green, 1864); for a good introduction to Babbage and his writings, see Philip and Emily Morrison, eds., *Charles Babbage and His Calculating Engines* (New York: Dover, 1961). Games are discussed on pp. 152–156.

12

Morrison, *Babbage*, p. 153.

13

Émile Borel, "La Théorie du jeu et les équations intégrales à noyau symétrique," *Comptes Rendus de l'Académie des Sciences* 173 (1921): 1304–1308; "Sur les jeux ou interviennent l'hasard et l'habileté des joueurs," *Théorie des probabilités* (Paris: Hermann, 1924), pp. 204–224; "Sur les systèmes de formes linéaires à determinant symétrique gauche et la théorie générale du jeu," *Comptes Rendus de l'Académie des Sciences* 184 (1927): 52–53. An English translation of these articles was published in *Econometrica* 21 (1953): 97–117.

14

Here is an instance of a situation common in the mathematical and scientific community: a dispute over who was first. Such disputes reflect the importance the individual scientist as well as the scientific community attaches to innovation of an idea or theory. The innovator is honored and admired. By placing a high value on "priority" the scientific community places private motivations in the service of progress in science or mathematics. In his 1928 paper on game theory, von

Neumann's reference to Borel (note 9) made Borel's contribution appear minor, so that for a long time von Neumann indeed was regarded as the "initiator" of game theory. But in 1953 the mathematician M. Frechet rediscovered the early Borel papers, had them translated into English, and wrote a little note of appreciation of Borel's "initiating" the theory of games. The translator, the young American mathematician L. J. Savage, who liked and admired von Neumann, recalled that he and Frechet had sent the manuscript to von Neumann, seeking his opinion, "and von Neumann had been very angry. He phoned me from someplace like Los Alamos, very angry. He wrote a criticism of these papers in English. The criticism was not angry. It was characteristic of him that the criticism was written with good manners. . . . I feel that if you were going to make a serious biographical analysis of von Neumann, that you would have to take into account that his pride was hurt." (Interview, Yale University, November 10, 1968.) Both the Frechet note and the public von Neumann criticism appear in *Econometrica* 21 (1953): 95–126.

15
Mathematische Annalen 100 (1928): 295–320, reprinted in von Neumann, *Collected Works*, 6: 1–28 (New York: Macmillan, 1963).

16
A good introduction to the essential ideas of game theory is found in Anatol Rapoport, *Two-Person Game Theory* (Ann Arbor: University of Michigan Press, 1966). A more complete mathematical exposition of game theory is found in R. D. Luce and H. Raiffa, *Games and Decisions* (New York: Wiley, 1957); the original book, J. von Neumann and O. Morgenstern, *Theory of Games and Economic Behavior* (New York: Wiley, 1944) is also very readable, although the mathematics is more elaborate than in Luce and Raiffa.

17
The phrase *on the average* is taken to mean in the present context the expected result if the players were each to stick to their respective policies for many rounds of the game. For a sufficiently large number of rounds the elements of good or bad luck in the cards dealt to each person are likely to even out.

18
H. W. Kuhn, *Lectures on the Theory of Games* (Princeton: Princeton University Press, 1952), an Office of Naval Re-

search report, gives a bibliographic review. Some specific references are: J. Ville, in E. Borel, ed., *Applications aux jeux de hasard,* vol. 4 (Paris: Gauthier-Villars et cie, 1938), S. Kakutani, *Duke Mathematics Journal* 8 (1941): 457–458; L. H. Loomis, *Proceedings of the National Academy of Science* 32 (1946): 213–215; H. Weyl, in Kuhn and Tucker, eds., *Contributions to the Theory of Games,* vol. 1 (Princeton: Princeton University Press, 1950); J. F. Nash, *Annals of Mathematics* 54 (1951): 286–295; G.' B. Dantzig, *Pacific Journal of Mathematics* 6 (1956): 25–33.

19
"A Model of General Economic Equilibrium," *Review of Economic Studies* 13 (1945–1946): 1–9; originally published in German in 1937. A statement of the Brouwer fixed-point theorem is: Let S be a compact convex subset of n-dimensional Euclidian space, and let f be a continuous function mapping S into itself. Then there exists at least one $x \in S$ such that $f(x) = x$.

20
This analysis was first published in von Neumann and Morgenstern, *Theory of Games and Economic Behavior* (Princeton: Princeton University Press, 1944); as to the priority question, see fn. 2 on p. 186 of the 3d edition, published by Wiley, 1953.

21
Ibid., 3d ed., pp. 186–219; D. B. Gillies, J. P. Mayberry, and J. von Neumann, "Two Variants of Poker," *Annals of Mathematical Studies*, 1953, no. 28, pp. 13–50.

22
Put all the 2,598,960 possible hands in a sequence ordered according to strength, equally spaced from each other, so that the strongest hand is given the value 2,593,960 and the weakest the value 1. A hand is "strong" if its number on this scale is bigger than $(1 - L/H) \times 2,598,960$. Otherwise it is "weak." Thus the word *strong* depends on the ratio of the high to the low bid. The quantitative fraction of the times that are meant by *sometimes* is $L/(H + L)$ of the time. So the frequency of bluffing for a sound strategy also depends on the ratio of the high to the low bid.

5 Axioms and Atoms

1
Cyril Bailey, *The Greek Atomists and Epicurus* (Oxford: Oxford University Press, 1928); Giorgio de Santillana, *The Origins of Scientific Thought* (Chicago: University of Chicago Press, 1961); see also the Roman atomist Lucretius, *On the Nature of Things.*

2
De Santillana, *Scientific Thought*; Ettore Carruccio, *Mathematics and Logic in History and in Contemporary Thought* (Chicago: Aldine, 1964).

3
Joseph Needham, *The Grand Titration* (London: Allen & Unwin, 1962), pp. 242–243, 311–312, comments on ancient Chinese science and explores possible explanations for the difference between the character of Chinese science and that of ancient Greece and Renaissance Europe.

4
Dirk Struik, *A Concise History of Mathematics*, 3d ed. (New York: Dover, 1967), p. 50.

5
It is noteworthy that Newton could have obtained many of his results most easily by use of the methods of the calculus he had himself invented. Nevertheless, in his *Principia* he relied on the traditional style of geometric demonstration.

6
H. Pemberton, paraphrasing Newton, in *A View of Sir Isaac Newton's Philosophy* (London: S. Palmer, 1728), p. 23. Pemberton, who had edited the third edition of the *Principia*, is also quoted in E. W. Strong's essay, "Newton's 'Mathematical Way,'" in P. P. Wiener and A. Noland, eds., *Roots of Scientific Thought* (New York: Basic Books, 1957).

7
Adolphe Wurtz, *The Atomic Theory* (New York: Appleton, 1881).

8
Ernst Mach, *Die Mechanik in Ihrer Entwicklung* (Leipzig: Brockhaus, 1883), p. 396 (my translation).

9

Maupertuis' principle of least action, as restated by Euler.

10

Hamilton's formulations of Newton's mechanics. Hamilton's formulations were particularly suggestive to both Schroedinger and Heisenberg in their respective formulations of quantum theory.

11

Mach, *Mechanik*, p. 439 (my translation).

12

This is the classic Balmer series. $1 \text{ nm} = 10^{-9} \text{ m}$.

13

For a historical review, see Max Jammer, *The Conceptual Development of Quantum Mechanics* (New York: McGraw-Hill, 1966). Also see the textbook *Atomic Physics* by Max Born, 1st ed. 1935; 4th ed. (New York: Hafner, 1946).

14

Comment by M. Mayer in interview of J. Franck and H. Sponer, July 14, 1962, conducted by her with Thomas Kuhn (Quantum archives).

15

For a detailed chronology of the two competing approaches, see Jammer, *Quantum Mechanics*.

16

Ibid., p. 272.

17

J. von Neumann, "Mathematische Begründung der Quantanmechanik," *Göttinger Nachrichten* (1927), pp. 1–57. The matrix method required mathematical justification whenever one dealt with a continuous spectrum. Von Neumann also objected to the use of the "improper" Dirac delta function as an eigenfunction. Besides the lack of unity and the lack of rigor in the extant formulations, he also had a minor objection to the Heisenberg and Schrödinger formalisms (pertaining to the formal nicety of the theories). In computing the state of an atom one also calculates an in principle unobservable phase factor, although to compare results with experiment one would eventually eliminate the phase factor. To von Neumann this seemed unsatisfactory and roundabout.

18
M. Pasch, *Vorlesungen über neuere Geometrie* (Leipzig: Teubner, 1882); G. Peano, *I Principii di Geometria* (Torino: Fratelli Bocca, 1889); D. Hilbert, *Grundlagen der Geometrie* (Leipzig: Teubner, 1900). The other seminal nineteenth-century book on the foundations of mathematics is G. Frege, *Die Grundlagen der Arithmetik* (Breslau: Koebner, 1884).

19
Hilbert had not been able to prove that the axioms of geometry are free from contradiction but only that if the axioms of analysis are free of contradiction, so are the axioms of geometry.

20
D. Hilbert, "Neubegründung der Mathematik" (1922), in *Gesammelte Abhandlungen* (Berlin: Springer, 1935), 3: 161.

21
D. Hilbert, *Gesammelte Abhandlungen,* 3: 217–289; *Mathematische Annalen* 72 (1912): 562–577.

22
See for example his "Naturerkennen und Logik" (1930) in *Gesammelte Abhandlungen*, 3:378–387.

23
One could equally well use the equation $\sqrt{x^2 + y^2 + z^2} = 1$. Squaring both sides of the equation gives the equation in the text.

24
For a historical overview with extensive reference to the original literature, see Michael Bernkopf, "Development of Function Spaces," *Archives for the History of Exact Sciences* 3 (1966–1967): 1–96.

25
J. von Neumann, *Mathematische Grundlagen der Quantenmechanik* (Berlin: Springer, 1932).

26
M. Born and N. Wiener, "A New Formulation of the Laws of Quantization of Periodic and Aperiodic Phenomena," *Journal of Mathematics and Physics* (MIT) 5 (1925–1926): 84–98.

27
The use of the delta function. Dirac, *The Principles of Quantum Mechanics* (Oxford: Oxford University Press, 1930).

28
"Mathematische Begründung der Quantenmechanik," pp. 1–57, "Wahrscheinlichkeitstheoretischer Aufbau der Quantenmechanik," pp. 245–272, 273–291. See also von Neumann, *Collected Works*, vol. 1.

29
John von Neumann, "Allgemeine Eigenwerttheorie Hermitischer Funktionaloperatoren," *Mathematische Annalen*; 102 (1929): 49–131, "Zur Algebra der Funktionaloperatoren und Theorie der normalen Operatoren," 370–427; *Journal für die Reine und Angewandte Mathematik* 161 (1929): 208–236. See also von Neumann *Collected Works*, vol. 2.

30
Mathematische Grundlagen; the English edition is *Mathematical Foundations of Quantum Mechanics* (Princeton: Princeton University Press, 1955).

31
I use the word *atom* for brevity and simplicity, although the term *micromechanical system* is more precise. It refers to a particle such as an electron or any collection of particles, such as an atom or molecule.

32
Cf. Max Jammer, *The Philosophy of Quantum Mechanics* (New York: Wiley, 1974), p. 5.

33
The Hilbert space is complex, so each of the numbers is a complex number.

34
The proof is based on the Fischer-Riesz theorem.

35
The five axioms defining the abstract Hilbert space R are:
A. R is a linear space.
B. A Hermitian inner product is defined in R.
C. There are arbitrarily many linearly independent vectors.
D. R is complete.
E. R is separable.
For generality, von Neumann also included the possibility of a finite-dimensional space. Axioms C, D, and E are then replaced by C_n: There are exactly *n* linearly independent vectors.

36
Richard Courant, interview, May 9, 1962 (Quantum archives).

37
G. C. Wick, A. S. Wightman, and E. P. Wigner, "The Intrinsic Parity of Elementary Particles," *Physical Review* 88 (1952): 101; E. P. Wigner, "Die Messung quantenmechanischer Operatoren," *Zeitschrift für Physik* 133 (1952): 101. The formalism of Hilbert space can be modified by the introduction of "superselection rules" to incorporate the exceptions.

38
See for example R. Haag, "Canonical Commutation Relations in Field Theory and Functional Integration," in *Lectures in Theoretical Physics* 3, ed. Brittin, Downs, and Downs (New York: Wiley Interscience, 1961); also K. L. Nagy, *State Vector Spaces with Indefinite Metric in Quantum Field Theory* (Groningen: Noordhoff, 1966).

39
G. Birkhoff and J. von Neumann, "The Logic of Quantum Mechanics," *Annals of Mathematics* 37 (1936): 823.

40
The status of abstract Hilbert space is reviewed by R. Haag, "Quantum Field Theory," *Contemporary Physics* 2 (1969): 459.

6 Scientific Style and Habits of Work

1
Jacques Hadamard, *The Psychology of Invention in the Mathematical Field* (Princeton: Princeton University Press, 1945); H. Poincaré, *Science and Method* (New York: Dover, undated).

2
Norbert Wiener's "Mathematics and Art," *Technology Review* 32 (1929): 271, belongs to this genre.

3
Stanislaw Ulam, "John von Neumann, 1903–1957," *Bulletin of the American Mathematical Society* 64, no. 3, pt. 2 (1958;: 8.

4
IAM, p. 60. Wiener wrote: "All mathematical work is done under sufficient pressure, and its increase by such a fortuitous competitive element is intolerable to me."

5
M. Kac, "Wiener and Integration in Function Space," *Bulletin of the American Mathematical Society* 72, no. 1, pt. 2 (1966): 52–68.

6
Ibid., p. 52.

7
Ulam in *Adventures of a Mathematician* states (p. 74) that he was told at that time (1935–36) "that three persons 'owned' the American Mathematical Society: Oswald Veblen, G. D. Birkhoff, and Arthur B. Coble, from Illinois. Most academic positions were secured through the recommendations of these three."

8
See N. Levinson, "Wiener's Life," *Bulletin of the American Mathematical Society* 72, no. 1, pt. 2 (1966), on Wiener's relation to the mathematical community.

9
Ulam, *Adventures of a Mathematician*, p. 96.

10
Cf. J. D. Bernal, *Science in History* (Cambridge, Mass.: MIT Press, 1971), pp. 49–51, 1219–1231. Bernal relates this issue to social class structure.

11
Little Review, May 1929 (the final issue of the magazine), quotes the response of Russell to a questionnaire whose first two questions were: (1) What should you most like to do, to know, to be (in case you are not satisfied)? (2) Why wouldn't you change places with any other human being? Russell replied, "I must warn you . . . that I am making my answers truthful rather than interesting. I have a rooted belief that the truth is always dull. . . . (1) I should like to do physics, to know physics, to be a physicist. (2) There are about a dozen human beings with whom I would gladly changes places; first among them I should put Einstein." Quoted in Margaret Anderson, *Little Review Anthology,* 349ff. (New York: Hermitage House, 1953).

12
Constance Reid, *Hilbert* (Berlin: Springer, 1970).

13
IAM, p. 23.

14
J. Needham et al., *Science and Civilization in China*, vol. 5, part 4 (in press); N. Sivin, "Chinese Alchemy and the Manipulation of Time," *Isis* 67 (1976): 513; M. Eliade, *The Forge and the Crucible* (New York: Harper & Row, 1962).

15
See, for example, IAM, pp. 85–86.

16
John von Nuemann, *Collected Works* 1: 9 (Oxford: Pergamon Press, 1961). The essay "The Mathematician" was first published in 1947.

17
Von Neumann, "The Mathematician," in *Collected Works*, vol. 1; Wiener, "Mathematics and Art," *Technology Review* 32 (January 1929): 129.

18
Concerning von Neumann, see for example the "charming tale about himself" reported in Herman Goldstine's *The Computer from Pascal to Von Neumann* (Princeton: Princeton University Press, 1972), p. 174; R. Gerard wrote about von Neumann that "the steps of a proof commonly enter his consciousness as a linked chain of steps, obviously in proper se-

quence in the unconscious but dragged into awareness (as if from memory) in haphazard fashion. When the whole has emerged and he finally writes it as an article, he occasionally develops a strong distaste for continuing this chore beyond some point. Experience showed him that such a block almost always results in his discovering a previously unrecognized error at that step where the unreasoned blocking occurred." "The Biological Basis of Imagination," *Scientific Monthly* 62 (June 1946): 479n. Concerning Wiener, see his autobiography and the quotations in this chapter.

19
Interview with the author, December 16, 1970; Dirk J. Struik, "Norbert Wiener—Colleague and Friend," *American Dialog*, March–April 1966, p. 34.

20
Levinson, "Wiener's Life," p. 25.

21
EP, p. 211.

22
Wiener, "Mathematics and Art," *Technology Review* 32 (1929): 129. (My emphasis.)

23
Armand Siegel, interview with the author, June 8, 1971.

24
Edward Teller in the film *John von Neumann*, produced by A. Novak; Eugene Wigner, Julian Bigelow, and Hermann Goldstine in separate interviews with the author, August 10, 1971, November 12, 1968, and February 26, 1971.

25
Teller, in the film *John von Neumann*.

26
Goldstine interview; see also Samuel Grafton, "Married to a Man Who Believes the Mind Can Move the World," *Good Housekeeping*, September 1956, p. 80.

27
John von Neumann, "The Mathematician," in James R. Newman, ed., *The World of Mathematics* (New York: Simon and Schuster, 1956), vol. 4, p. 2053.

28
Ibid., p. 2062.

29
Goldstine, *The Computer*, p. 182, gives an anecdote of how some colleagues contrived to upstage von Neumann on one occasion.

30
Ulam, *Adventures of a Mathematician*, p. 78.

31
Such a rhythm—a relatively long period of preparation succeeded by quick, apparently spontaneous action—is reminiscent of the brush painters of the Zen Buddhist school, who work with black ink on highly absorbent rice paper. The emphasis is on "spiritual preparation" of the mind prior to painting as well as on technique, which must become second nature because the medium requires spontaneous brush strokes not subject to correction. The required state of mind is one of absorption and "insight" into the subject. According to the Zen literature, it is just the state of absorption and insight into the subject of the painting that gives the painter delight.

32
Bigelow was in his late twenties when he started to work as Wiener's assistant; Wiener was nineteen years his senior. He assisted in Wiener's mathematical work and incidentally built engineering devices to illustrate the mathematics. He was in his mid-thirties when he started to work for von Neumann, who was only ten years his senior, and of course he had his previous experience with Wiener behind him. Under von Neumann, Bigelow had the task of chief engineer for building the first Princeton computer; von Neumann had other mathematical collaborators at that time.

33
Interview, with the author, November 12, 1968; compare Poincaré's distinction between "logical" and "intuitive" mathematical minds, quoted by Hadamard, *The Psychology of Invention in the Mathematical Field* (New York: Dover, 1945), p. 106.

34
IAM, p. 168.

35
"Differential Space," *Journal of Mathematics and Physics* 2 (1923): 131–174; Generalized Harmonic Analysis, *Acta Mathematica* 55 (1930): 117–258.

36
Ulam, "John von Neumann, 1903–1957," p. 12.

37
Reid, *Hilbert*, p. 172.

38
Goldstine, *The Computer*, p. 171. (My emphasis.)

39
Based on interviews with several mathematicians who visited at Princeton.

40
Goldstine, *The Computer*, p. 167; see also Eugene Wigner in *Symmetries and Reflections* (Cambridge, Mass.: MIT Press, 1970) and the introduction to von Neumann, *Computer and the Brain* (New Haven, Connecticut: Yale University Press, 1958).

41
Quoted by Wigner in *Symmetries and Reflections*, p. 261.

42
Ulam, "John von Neumann 1903–1957," p. 6.

43
Niels Bohr, *Atomic Physics and Human Knowledge* (New York: Wiley, 1958); *Atomic Theory and the Description of Nature* (Cambridge: Cambridge University Press, 1961).

44
Specifically, this is the noncommutativity of position and momentum described in chapter 4. See also his reaction to Bohr's Lake Como lecture in 1927, quoted by Wigner, and Max Jammer, *The Conceptual Development of Quantum Mechanics* (New York: McGraw-Hill, 1966), p. 354.

45
Mathematical Foundation of Quantum Mechanics, p. 420. Of the original group of quantum theorists, Heisenberg's view concerning the nature of observation probably comes closest to von Neumann's. Modern textbooks tend to gloss over the

difference between Bohr's and von Neumann's interpretations, a tendency reflecting physicists' uninterest in philosophy.

46
Ibid., p. 421.

47
Wigner, *Symmetries and Reflections,* p. 186.

48
In this section I have in particular made use of the article by
Abner Shimony: "Role of the Observer in Quantum Theory,"
American Journal of Physics 31 (1963): 755.

49
Ibid.

50
This view is explicit in some of von Neumann's later writings
on the logic of quantum mechanics, cited in chapter 4.

51
Feyerabend, Paul, *Against Method* (London: Humanities
Press, 1975), p. 64.

52
A similar situation occurs in connection with the application of
the concept of probability. Von Neumann made use of the
concept in connection with quantum theory, in applying game
theory to economics, in automata theory, and in other applications of mathematics. The mathematical rules of combining
probabilities are self-consistent and free of difficulty, but the
interpretation of the notion of probability has been a subject of
long-standing controversy. Some theorists are proponents of
the "frequency interpretation," some of the "logical interpretation," and still others defend some variant of a "subjective" interpretation of probabilities. Von Neumann has in different contexts made use of all three interpretations, choosing
whichever seems most convenient. In *Mathematical Foundations of Quantum Mechanics* he mentions the frequency interpretation (n. 165) and appears to be thinking in terms of it;
in his later work on quantum logics he uses a logical interpretation; and in automata theory he also favors viewing
probability as a branch of or extension of logic [*Theory of
Self-Reproducing Automata*, ed. Burks (Urbana: University of
Illinois Press, 1966), pp. 58–59]. In applying game theory to
economics he explicitly uses a frequency interpretation but

notes that a variant of a subjective interpretation now referred to as "personal probability" might be used equally well [von Neumann and Morgenstern, *Theory of Games and Economic Behavior*, 3d ed. (Princeton: Princeton University Press, 1953), p. 19].

53
Jerome Lettvin, in a lecture at MIT on February 23, 1977, explicitly related Leibniz's monads to von Neumann's automata. I also follow Lettvin in viewing the *Monadology* as a treatise in epistemology rather than ontology. Lettvin credits Walter Pitts with having introduced him to a proper understanding of Leibniz's *Monadology*.

54
Quoted in Bertrand Russell, *A History of Western Philosophy* (New York: Simon and Schuster, 1945), p. 592. Pitts and Lettvin's view of Leibniz is sharply at odds with Russell's.

55
The comments on von Neumann in this chapter need to be tempered by the observation that his attitudes toward logic changed as he himself became older. In 1947, reflecting on the nature of mathematics, he took cognizance of the "mutability" of the criteria of rigor. At that time he also conceded that mathematics is somehow incomplete, needing "something extra" from an empirical source, or from philosophy or somewhere else. See von Neumann, "The Mathematician."

56
Norbert Wiener, "Quantum Mechanics, Haldane and Leibniz," *Philosophy of Science* 1 (1934): 479; "The Role of the Observer," *Philosophy of Science* 3 (1936): 307; "Back to Leibniz," *Technology Review* 34 (1932): 201.

57
Wiener, "Back to Leibniz."

58
Bohr, *Atomic Theory,* p. 68: "In the classical theories any succeeding observation permits a prediction of future events with ever-increasing accuracy, because it improves our knowledge of the initial state of the system. According to the quantum theory, just the impossibility of neglecting the interaction with the agency of measurement means that every observation introduces a new uncontrollable element."

59
He believed that the existence of hidden variables underlies the statistical character of quantum theory. Wiener appeared to be agnostic concerning the hidden variable controversy. Von Neumann sought to settle the controversy by a mathematical proof, which succeeded in severely restricting the kinds of hidden variables compatible with his formulation of quantum theory and was widely held by physicists to rule out the possibility of hidden variables.

60
Much later he worked with younger collaborators Armand Siegel and Giacomo Della Riccia on a stochastic formulation of quantum theory: "A New Form for the Statistical Postulate of Quantum Mechanics," *Physical Review* 91 (1953): 1551; "The Differential Space Theory of Quantum Systems," *Nuovo Cimento* 2, supp., ser. 10 (1955): 982; "Wave Mechanics in Classical Phase Space, Brownian Motion and Quantum Theory," *Journal of Mathematical Physics* 7 (1966): 1372.

61
It should be mentioned that Wiener had concerned himself with clarifying and using logic to make unambiguous the significance of measurements of the magnitude of "sensations" (e.g., loudness, brightness). He had carried out this analysis in 1919, while still under the spell of the Russell-Whitehead *Principia*, years before the epistemological issues of quantum theory had surfaced. Still, before applying symbolic logic, he had formulated the process in ordinary English; only in the second half of his article does he restate matters in formal logical notation. Norbert Wiener, "A New Theory of Measurement," *Proceedings of the London Mathematical Society* 19 (1921): 181.

62
J. B. S. Haldane, "Quantum Mechanics as a Basis for Philosophy," *Philosophy of Science* 1 (1934): 78–98.

63
Cf. also "The Mathematician," where von Neumann (who incidentally enjoyed double meanings of words) speaks about the "duplicity" of mathematics—empirical and axiomatic.

64
See for example Lynn White, *Dynamo and the Virgin Reconsidered* (Cambridge, Mass.: MIT Press, 1968), chapter 3.

65

The hidden variable controversy mentioned in n. 57 is illustrative of the belittling of philosophy and overvaluing of mathematics in the physics community. Bohr and Einstein had in the 1920s and thereafter engaged in discussions about the hidden variable question, on a level mingling physics and philosophy. However, when von Neumann's mathematical proof appeared showing that under certain assumptions no hidden variables are compatible with quantum theory, it was regarded by the physics community to have settled the question once and for all. The atypical Einstein did not agree but saw the proof as irrelevant. Only thirty years afterward did the mainstream of the physics community awake to realize the restricted nature of the assumptions in the von Neumann proof and again opened up the question of the possible existence of hidden variables. See for example John S. Bell, "On the Problem of Hidden Variables in Quantum Mechanics," *Reviews of Modern Physics* 38 (1966): 447.

7 The Foundation: Chaos or Logic?

1
EP, p. 96.

2
Ibid., p. 46.

3
Norbert Wiener, "Is Mathematical Certainty Absolute?" *Journal of Philosophy, Psychology and Scientific Method* 12 (1915): 568.

4
EP, pp. 192–193; Wiener, "Is Mathematical Certainty Absolute?" *Journal of Philosophy, Psychology and Scientific Method* 12 (1915): 568.

5
For a semipopular exposition, see E. Nagel and J. R. Newman, "Gödel's Proof," in *The World of Mathematics*, vol. 3, ed. J. R. Newman (New York: Simon and Schuster, 1956).

6
Ibid.

7
Hermann Weyl, in Constance Reid, *Hilbert* (Berlin: Springer, 1970), p. 273.

8
"Tribute to Dr. Gödel," remarks made in March 1951 (von Neumann archives, Library of Congress, Washington, D.C.).

9
Eugene Wigner, interview with the author, August 10, 1971, and with Valentine Bargman, December 4, 1972. Wigner has recalled being present when von Neumann opened a letter from Gödel giving the proof and then saying something like "This changes my whole work." Bargman was attending von Neumann's course at the time.

10
Goldstine, *The Computer*, p. 174.

11
Von Neumann's correspondence, some of which is with the Veblen papers in the Library of Congress, incidentally shows

that when many years later Gödel (a somewhat erratic and unstable character) and von Neumann were both at the IAS von Neumann repeatedly used his influence to ensure that Gödel was properly treated and valued by the administration of the institute.

12
For nontechnical historical comments on the two approaches see E. T. Bell, *Men of Mathematics* (New York: Simon and Schuster, 1937), chap. 29. Also compare N. Wiener et al., *Differential Space, Quantum Systems, and Prediction* (Cambridge, Mass.: MIT Press, 1966), pp. 12–14.

13
EP, p. 212.

14
N. Levinson, "Wiener's Life," in *Bulletin of the American Mathematical Society* 72, no. 1, pt. 2 (1966). There was an accidental element in the choice of topic, for a young mathematician from Ohio, I. Barnett, happened to stop by at MIT and suggested the problem to Wiener as possibly an interesting one.

15
M. Kac, "Wiener and Integration in Function Space," *Bulletin of the American Mathematical Society* 72, no. 1, pt. 2 (1966).

16
Gibbs's method was to *imagine* a large number of identical physical systems and then to treat this ensemble of systems statistically and predict the likely properties of the one physical system at hand. By "physical system" something of ordinary size like a glassful of water or a chunk of copper is meant. See *Elementary Principles in Statistical Mechanics* (New Haven: Yale University Press, 1902) and E. Schrödinger, *Statistical Thermodynamics* (Cambridge: Cambridge University Press, 1948). Because each physical system considered contained a large number of degrees of freedom as a result of its many atomic constituents, the statistical predictions produced near certainties.

17
IAM, pp. 85–86.

18
Norbert Wiener, "The Role of the Observer," *Philosophy of Science* 3 (1936): 307. See also comments by Julian Bigelow and Armand Siegel cited in the text.

19
H. Poincaré, *Science and Method* (New York: Dover, n.d.), chapter 3; J. Hadamard, *The Psychology of Invention in the Mathematical Field* (Princeton: Princeton University Press, 1945).

20
EP, pp. 212–213.

21
Wiener, "The Role of the Observer."

22
He had written that "quantum theory reduces the whole of physics to a form of statistical mechanics," and he clearly meant the Gibbs form of statistical mechanics: "Other possible worlds are introduced . . . [and] considered from the standpoint of probability." See "Back to Leibniz," *Technology Review* 34 (1932): 201. See also chapter 6.

23
In "The Role of the Observer" Wiener suggests that because of the discontinuities in history, one does not have sufficiently long statistical runs to permit fruitful application of the theory of stochastic processes.

24
J. von Neumann, "A Model of General Economic Equilibrium," *Review of Economic Studies* 13 (1945–1946): 1. (Originally published in German in 1937.)

25
Originally he sought to implement the suggestions of Lewis F. Richardson, *Weather Prediction by Numerical Process* (Cambridge: Cambridge University Press, 1922); subsequently he became acquainted with the physically motivated approximations suggested by J. G. Charney and worked together with Charney. See their "Numerical Integration of Barotropic Vorticity Equation," *Tellus* 2 (1950): 237.

26
Proceedings of the Third Berkeley Symposium on Mathematical Statistics (Berkeley: University of California Press, 1954). Conflict and competition between the Neumannian and Wienerian approaches to weather prediction was overt and intense among their respective adherents and was much in evidence at some conferences on weather prediction.

27
IAM, p. 260.

28
Norbert Wiener, *Cybernetics*, appendix of 2d ed. (Cambridge, Mass.: MIT Press, 1961).

29
John von Neumann to George Gamow, July 25, 1955.

30
Letter, Armand Siegel to author, February 15, 1973.

31
HUHB, 2d ed. (Garden City, N.Y.: Doubleday, 1954), p. 11. (My emphasis.)

32
The analogy is surely too literal, but note that the Brownian particle is constantly subject to bombardment from billions of molecules, so metaphorically Wiener himself, or his ego, was under constant bombardment from impulses, desires, and ideas from his own turbulent unconscious, as well as from diverse interactions with other people and with the world about him.

33
J. O. Urmson, *Philosophical Analysis* (Oxford: Clarendon, 1956).

34
IAM, pp. 324–325.

35
HUHB, pp. 11, 34–35, 190–191.

36
See Robert Merton, *Social Theory and Social Structure* (Glencoe, Ill.: Free Press, 1949).

37
"Science as a Vocation," in *From Max Weber*, ed. H. H. Gerth and C. Wright Mills (Oxford: Oxford University Press, 1946).

38
Even Wiener imposed a stronger specialization on himself as an ambitious young man than he did in middle age.

39
This difference in style of thought between Wiener and von Neumann is reflected in the relation of their thought to that of Gregory Bateson. Bateson hoped to learn from Wiener and von Neumann how to think more clearly about important human issues. See Steve Heims, "Gregory Bateson and the Mathematicians: From Interdisciplinary Interaction to Societal Functions," *Journal of the History of the Behavioral Sciences* 13 (1977): 141–159.

40
"There are many highly respectable motives which may lead men to prosecute research, but three which are much more important than the rest. The first (without which the rest must come to nothing) is intellectual curiosity, desire to know the truth. Then, professional pride, anxiety to be satisfied with one's performance, the shame that overcomes any self-respecting craftsman when his work is unworthy of his talent. Finally, ambition, desire for reputation, and the position, even the power or the money, which it brings." G. H. Hardy, *A Mathematician's Apology* (Cambridge: Cambridge University Press, 1969), pp. 78–79.

41
See von Neumann, "The Mathematician" and Wiener, IAM, p. 61.

42
Hardy, *A Mathematician's Apology*, shows mathematical "beauty" and "seriousness" to the lay reader by some carefully chosen examples.

43
J. R. Ravetz, *Scientific Knowledge and Social Problems* (Oxford: Oxford University Press, 1971), is a particularly thoughtful treatment of this topic.

44
J. D. Bernal, *History in Science* (Cambridge, Mass.: MIT Press, 1971), pp. 164–167.

8 The Normal Life of Mathematics Professors

1
IAM, p. 31; interviews by author of Julius Stratton, June 3, 1971, and Karl Wildes, August 18, 1971.

2
Norbert Wiener to Niels Bohr, February 14, 1926 (MIT archives).

3
Stanley Coben, "The Scientific Establishment and the Transmission of Quantum Mechanics to the United States, 1919–32," *American Historical Review* 76 (1971): 442; Charles Weiner, "A New Site for the Seminar," *Perspectives in American History* 2 (1968): 190.

4
Oswald Veblen to John von Neumann, October 15, 1929 (Library of Congress archives). The stipend offered for the one semester was $3,000 plus $1,000 traveling expenses. Wiener's full-year salary for 1929–30 was $4,000.

5
Interview of Eugene Wigner by Charles Weiner, April 1967, and by Thomas Kuhn in November and December 1963 (Quantum archives); interview of Wigner by the author, August 10, 1971. See also Weiner, "A New Site for the Seminar," for the origins of the deliberate policy of inviting two compatible theoreticians together.

6
November 13, 1929; November 19, 1929 (Library of Congress archives).

7
Wigner interviews; interview of Mariette Kövesi by the author, February 25, 1971.

8
Von Neumann to O. Veblen, June 19, 1933 (Library of Congress archives).

9
Herman Goldstine's *The Computer from Pascal to von Neumann* (Princeton: Princeton University Press, 1972) gives some of the early history of the Institute.

10
Abraham Flexner, *I Remember* (New York: Simon & Schuster, 1940), p. 397.

11
At the corner of Library Place and Stockton Street.

12
Von Neumann to Veblen, May 26, 1934 (Library of Congress archives).

13
EP, 140–141.

14
IAM, p. 126.

15
Ibid., p. 128.

16
The American Mathematical Society awarded the Bocher Prize, its highest honor (given only every five years) to Wiener (corecipient, 1932) and von Neumann (1937).

17
P. Halmos in *Bulletin of the American Mathematical Society* 64, no. 3, pt. 2 (1958): 90.

18
"Continuous Geometry," Institute for Advanced Study Lecture Notes (Edwards Bros., 1937), mimeo.

19
Goldstine, *The Computer*, p. 176.

20
Ibid.

21
Ulam, "John von Neumann 1903–1957," p. 4.

22
Proceedings of the National Academy of Sciences 18 (1932): 225–263.

23
Garrett Birkhoff, "Von Neumann and Lattice Theory," *Bulletin of the American Mathematical Society* 64, no. 3, pt. 2

(1958): 51–52; the proofs referred to appear in "Continuous Geometry."

24
Norbert Wiener, *Collected Works* (Cambridge, Mass.: MIT Press, 1975–).

25
See IAM; also Levinson, "Wiener's Life," *Bulletin of the American Mathematical Society* 72, no. 1, pt. 2 (1966): 1–32.

26
Levinson, "Wiener's Life," p. 25.

27
Ibid.

28
Goldstine, *The Computer,* chap. 10; also IAM, chap. 6. Lest Goldstine's comments (p. 91) be misunderstood, it should be noted that in 1940 Wiener was already a champion of the digital rather than the analog point of view (see introduction to *Cybernetics*), although it was not implemented at MIT for several years.

29
US Patent No. 2,024,900, December 17, 1935, later sold to AT&T for $5,000.

30
Lee to Wiener, December 4, 1934 (MIT archives).

31
Wiener to Bush, September 1, 1935. I am indebted to Julius Stratton for this letter.

32
IAM, p. 218. For details of Wiener's activity to aid China see his correspondence on the subject, especially in 1938 (MIT archives).

33
Ibid., p. 220.

34
Veblen to Wiener, May 2 and 7, 1934; Wiener to Veblen, May 4, 1931 (MIT archives).

36
Wiener to F. B. Jewett, president of the National Academy of Sciences, September 22, 1941.

37
It was this encouraging observation about the relation of science to the scientific establishment that he passed on to independent younger colleagues. Edmond Dewan, Armand Siegel, and Asim Yildiz have told me of the encouragement they received from Wiener to disregard the judgment of the scientific establishment.

38
Von Neumann to Wiener, March 28, 1937 (MIT archives).

39
Von Neumann to Wiener, April 9, 1937 (MIT archives).

40
The main objective of ergodic theory is to show that in a large class of instances different kinds of mathematical averages are equal. The theory permits a high degree of abstraction from its application to physics of relating time averages to averages over phase space. Ergodic theory employs the Lebesgue measure (so important in Wiener's Brownian motion theory described in chapter 3), the theory of groups, and the theory of almost periodic functions—each an area of mathematics in which both Wiener and von Neumann had had active interest. According to a 1954 questionnaire, von Neumann regarded his own work in ergodic theory as among his three most important contributions to mathematics; Wiener also regarded ergodic theory as of the highest importance (see, e.g., his *Cybernetics*).

41
Wiener "The Homogeneous Chaos," *American Journal of Mathematics* 60 (1938): 897, n. 11.

42
May 4, 1937. Wiener also wrote more formal letters concerning von Neumann to the president of Tsing Hua University and to Professor Hiong of the mathematics department (MIT archives).

43
I am indebted to Barbara (Wiener) Raisbeck for recalling the "Gentleman Johnny" nickname.

44
EP, p. 11.

45
IAM, p. 213.

46
Edward Dempsey to the author, December 17, 1970.

47
Von Neumann in Europe to Veblen at Princeton during summer and autumn of 1938 (Library of Congress archives).

1
Conference on The Problem of Stellar Energy, organized by G. Gamow and E. Teller and held in Washington, D.C., March 1938. According to L. Fermi, the studies "fostered by the 1938 meeting . . . later inspired physicists to attempt thermonuclear reactions on earth. . . ." L. Fermi, *Illustrious Immigrants* (Chicago: University of Chicago Press, 1971), p. 182.

2
John von Neumann, *The Computer and the Brain* (New Haven, Connecticut: Yale University Press, 1958), p. vii; von Neumann was also a consultant for the Navy Bureau of Ordnance in Washington, D.C.

3
Goldstine, *The Computer*, pp. 177, 179–182.

4
Ibid., p. 182.

5
Ibid., chaps. 7 and 8.

6
Ibid., p. 192.

7
Von Neumann, "First Draft of a Report on the EDVAC," June 30, 1945, mimeo.

8
W. S. McCulloch and W. Pitts, "A Logical Calculus of the Ideas Immanent in Nervous Activity," *Bulletin of Mathematical Biophysics* 5 (1943): 115.

9
See, e.g., James Penick et al., eds., *The Politics of American Science, 1939 to the Present* (Cambridge, Mass.: MIT Press, 1972); D. Greenberg, *The Politics of Pure Science* (New York: New American Library, 1967).

10
J. P. Baxter, *Scientists Against Time* (Boston: Little, Brown, 1946), p. 212.

11
Norbert Wiener, *Extrapolation, Interpolation and Smoothing of Stationary Time Series* (Cambridge, Mass.: MIT Press, 1949).

12
They were presented by Rosenblueth at a conference on Cerebral Inhibition, May 14–15, 1942, in New York City; they first appeared in print as "Behavior, Purpose and Teleology," *Philosophy of Science* 10 (1943): 18–24.

13
Conference on Cerebral Inhibition. It led to McCulloch's suggesting a conference series organized around the interests of Wiener and Rosenblueth and those of McCulloch and Pitts. Frank Fremont-Smith was receptive to McCulloch's suggestion, but it was to be realized only after the war. (Communication to the author from F. Fremont-Smith, October 7, 1968.)

14
See Wiener's correspondence with Pitts, Rosenblueth, and McCulloch during the period 1943–45 (MIT archives).

15
Wiener, Aiken, and von Neumann to S. S. Wilks, W. H. Pitts, E. H. Vestine, W. E. Deming, W. S. McCulloch, R. Lorente de Nó, and Leland E. Cunningham, December 4, 1944 (signed by Wiener).

16
Wiener to Rosenblueth, January 24, 1945. The group, according to von Neumann's report of January 12, had in the spirit of a team effort come to a "decision on a research program" and assigned memoranda on particular subjects: Wiener and Pitts on filtering and production; von Neumann, Goldstine, and others on application to differential equations; and so forth.

17
Ibid.

18
Ibid.

19
Wiener to Rosenblueth, August 11, 1945; Rosenblueth to Wiener, September 3, 1945 (MIT archives).

20
B. Feld, private communication; A. W. Burks, introduction to John von Neumann, *Theory of Self-Reproducing Automata* (Urbana: University of Illinois Press, 1966).

21
Von Neumann to Wiener, April 21, 1945 (MIT archives).

22
Letter, Wiener to Rosenblueth, July 1, 1945 (MIT archives).

23
Wiener to Rosenblueth, August 11, 1945 (MIT archives).

24
Ibid.; Rosenblueth to Wiener, September 3, 1945 (MIT archives).

25
O. Nathan and H. Norden, eds., *Einstein on Peace* (New York: Schocken, 1968), p. 342.

26
Wiener to Karl T. Compton (MIT archives).

27
Von Neumann to Frank Aydelotte, August 25, 1945 (Library of Congress archives).

28
Goldstine, *The Computer*; Freeman Dyson, "The Future of Physics," *Physics Today,* September 1970, p. 24.

29
Goldstine, *The Computer*, p. 220.

30
Ibid., p. 243.

31
The conflict flared up into an acrimonious dispute concerning priorities and patent rights some months later. Goldstine reports on it from his point of view, ibid., pp. 222–224. H. D. Huskey, in a review of Goldstine's book, takes issue with him; see *Science* 180 (1973): 588. My own impression of von Neumann is that he fought energetically for recognition in this instance as in the priority controversies over ergodic theory

and game theory. However, I believe what is at work here is the tendency of historians to oversimplify by exaggerating the contribution of one person to an invention, neglecting the operation of social forces and the multitude of people who played a role in the all but inevitable invention. This tendency in the history of technology has been described and documented by S. C. Gilfillan, *The Sociology of Invention* (Cambridge, Mass.: MIT Press, 1970).

32
See Wiener's fond and colorful description of Bigelow and his job interview in Princeton, IAM, pp. 242–243; elsewhere he had recommended Bigelow as "a first class engineer with an excellent mathematical intelligence" (Wiener to J. R. Kline, April 10, 1941).

33
Norbert Wiener, *Cybernetics* (New York: Wiley 1948), 1st ed., pp. 30–31.

34
In Wiener's autobiography no mention is made of Pitts or McCulloch, presumably because at the time he was writing IAM (1955), his break with them was a painful and angry subject for him. He mentions the Princeton meeting of January 6–7, 1945, but fails to mention von Neumann in that connection (IAM, pp. 269–270), nor does he refer to the plans for bringing von Neumann to MIT. Nor does Goldstine (in *The Computer*) mention von Neumann's consideration of an offer from MIT, perhaps omitting it as unimportant for the reason that "nothing came of it." It is important here because it was a period of close collaboration between Wiener and von Neumann, and it was very important to both of them, at least during 1944–1946; it also shows something of how each engaged in the politics of science.

35
Von Neumann to Wiener, February 1, 1945.

36
IAM, p. 249.

37
Herman Goldstine, in an interview with the author, February 26, 1971.

38
Fermi, *Illustrious Immigrants*, p. 212.

39
Ibid., p. 9.

40
E. Segrè, *Enrico Fermi* (Chicago: University of Chicago Press, 1970), p. 140.

41
In 1953 Julius and Ethel Rosenberg were executed for allegedly giving the highly secret idea of the implosion method to Soviet agents during World War II.

42
S. Groueff, *Manhattan Project*, chapter 48 (Boston: Little, Brown, 1967); Richard Hewlett and Oscar Anderson, *A History of the United States Atomic Energy Commission,* vol. 1, *1939–1946: The New World* (Springfield, Va.: US Dept. of Commerce, National Technical Information Service, 1972). The first Hiroshima bomb was uranium, but shortage of materials made it necessary for the second bomb to be plutonium. The uranium bomb had never been tested for military purposes.

43
Compare Ulam, *Adventures of a Mathematician*, p. 232. The prominence of this ideal among US government intellectuals in the 1960s has been portrayed by David Halberstam, *The Best and the Brightest* (New York: Random House, 1969).

44
Leslie R. Groves, *Now It Can Be Told* (New York: Harper, 1962).

45
See *In the Matter of J. Robert Oppenheimer: Transcript of Hearing before Personnel Security Board and Text of Principal Documents and Letters* (Cambridge, Mass.: MIT Press, 1971), pp. 650–651.

46
Wiener to Bush, director of US Office of Scientific Research and Development, January 6, 1942 (MIT archives).

47
R. Jungk, *Brighter than a Thousand Suns,* appendix A (New York: Grove Press, 1958); J. R. Oppenheimer, *New York Review of Books,* December 17, 1964; O. W. Wilson, *The Great Weapons Heresy* (Boston: Houghton Mifflin, 1970); Alice Kimball Smith, *A Peril and a Hope* (Cambridge, Mass.: MIT Press, 1971).

48
Oppenheimer hearings, p. 654.

49
Goldstine, *The Computer,* p. 168.

50
John Lukacs, *1945: Year Zero* (New York: Doubleday, 1978), is an interesting and revealing description of the shift from German to Russian rule in Hungary.

51
See Smith, *A Peril and a Hope;* also *Leo Szilard: His Version of the Facts,* ed. S. R. Weart and G. Weiss Szilard (Cambridge, Mass.: MIT Press, 1978).

52
See Gar Alperovitz, *Cold War Essays* (New York: Doubleday, 1970), and *Atomic Diplomacy* (New York: Vintage, 1965); Gabriel Kolko, *The Politics of War* (New York: Random House, 1969). It is of interest that immediately upon hearing that the first test of the atomic bomb was successful (July 16, 1945), President Truman's attitude toward the USSR changed from a relatively conciliatory one to a tougher one. See also Charles L. Mee, *Meeting at Potsdam* (New York: M. Evans, 1975), chap. 6.

53
These adjectives are from the toned-down later official report of Brigadier General Farrell to the secretary of war. See also Groueff, *Manhattan Project,* and Jungk, *Brighter than a Thousand Suns,* for some physicists' reactions.

54
Lewis Strauss, *Men and Decisions* (New York: Doubleday, 1962), p. 350. (My emphasis.)

55
Compare Robert J. Lifton, *Boundaries* (New York: Vintage, 1970), especially pp. 26–34; Lifton, *Prophetic Survivors:*

Hiroshima and Beyond (mimeo, 1971); Harvey Cox, *The Seduction of the Spirit* (New York: Simon and Schuster, 1973); and Mircea Eliade, *The Forge and the Crucible* (New York: Harper, 1971).

56
One may compare E. Teller, J. R. Oppenheimer, and L. Szilard, as Lifton does, for a spectrum of responses to the bomb.

57
See chapter 2; originally quoted by Ulam, "John von Neumann, 1903–1957," p. 1.

58
He reportedly gave this as a reason for refusing, after some hesitation, invitations to ally himself with the Association of Los Alamos Scientists (ALAS). ALAS was a political organization of scientists formed during the autumn of 1945 "to promote the attainment and use of scientific and technological advances in the best interest of humanity." The group, which favored steps toward international control of nuclear weapons, differed with General Groves concerning weapons policy. ALAS is described by Smith in *A Peril and a Hope*. The report concerning the reason for von Neumann's lack of interest comes from one of the Los Alamos physicists but is not for attribution. From the time of the end of the war, von Neumann refused all open political commitments and refused to sign any peace-oriented statements. He refused to join the board of advisers of the *Bulletin of Atomic Scientists* and explained to Hyman Goldsmith in a letter (September 23, 1948) that "as a matter of principle . . . I have throughout the last years avoided all participation in public activities which are not of a purely technical nature" (Library of Congress archives). He was behaving like an insider's technical adviser. The philosopher-mathematician Jacob Bronowski has commented on von Neumann in his BBC television series; the edited printed version reads, "There is an age-old conflict between intellectual leadership and civil authority. . . . It is not the business of science to inherit the earth, but to inherit the moral imagination; because without that man and beliefs and science will perish together. . . . The man who personifies these issues for me is John von Neumann . . . he wasted the last years of his life . . . he gave up asking himself how other *people* see things. He became more and more engaged in work for private firms, for industry, for government. They were

enterprises which brought him to the centre of power, but which did not advance either his knowledge or his intimacy with people. . . . Johnny von Neumann was in love with the aristocracy of intellect. And that is a belief which can only destroy the civilization that we know. If we are anything, we must be a democracy of the intellect. We must not perish by the distance between people and government, between people and power, by which Babylon and Egypt and Rome failed" [Bronowski, *The Ascent of Man* (Boston: Little, Brown, 1973), pp. 432–435].

59
Lifton, *Boundaries*, p. 26.

10   A Mutual Interest, But . . .

1
McCulloch to Pitts, February 2, 1946 (MIT archives).

2
Wiener to McCulloch, February 1946. De Santillana was apparently not invited. (McCulloch papers, American Philosophical Society, Philadelphia.)

3
Von Neumann to McCulloch, February 1946 (McCulloch papers). Gödel was sent a formal invitation, but Oskar Morgenstern was apparently not invited.

4
Wiener to McCulloch, February 5, 1946 (McCulloch papers).

5
An unpublished summary of the meeting was prepared by its chairman, Warren McCulloch (McCulloch papers). The transactions of the last five of the series of ten meetings have been published by the Josiah Macy, Jr., Foundation, New York, as *Conference on Cybernetics*, ed. Heinz von Foerster, vols. 6–10 (1950–1955).

6
Von Neumann to Wiener, November 25 and 29, 1946 (MIT archives).

7
Wiener to von Neumann, November 26, 1946 (MIT archives).

8
E. Schrödinger had called the attention of nonbiologists to Delbrück's work through his book *What Is Life?* (Cambridge: Cambridge University Press, 1944).

9
Max Delbrück, in a letter to the author, September 18, 1973, characterizes the discussion bluntly as "vacuous in the extreme and positively inane."

10
Gregory Bateson, interview with the author, August 13, 1968.

11
Lawrence Kubie, interview with the author, March 24, 1969.

12
Armand Siegel, interview with the author, Boston, June 8, 1977.

13
Alex Bavelas, interview with the author, March 17, 1968.

14
The introduction to Wiener's *Cybernetics* (Cambridge, Mass.: MIT Press, 1948) contains a brief description of the Macy conferences, as does Warren McCulloch, "The Origin of Cybernetics," American Society for Cybernetics Forum 6, No. 2 (summer 1974): 5. Further information is found in the correspondence of participants and the recollections of the meetings solicited by the author. See also S. Heims, "Encounter of Behavioral Sciences with New Machine-Organism Analogies in the 1940s," *Journal of the History of the Behavioral Sciences* 11 (1975): 368; "Gregory Bateson and the Mathematician," *Journal of the History of the Behavioral Sciences* 13 (1977): 141.

15
One of these was held in Princeton, December 17–19, 1946, on the general topic of problems of mathematics. Von Neumann chaired the session on new fields, at which Wiener was one of the invited speakers.

16
The history of mathematics and science is full of priority controversies. Robert Merton's work in the sociology of science describes the role of competition. See also F. Reif, "The Competitive World of the Pure Scientist," *Science* 134 (1961): 1957–1962; or J. D. Watson, *The Double Helix* (New York: Atheneum, 1968). J. Ravetz, *Scientific Knowledge and Social Problems* (Oxford: Oxford University Press, 1971), chap. 8, views research such as a new proof of a theorem as the intellectual property of its originator, property that may need protecting and defending.

17
See *Cybernetics*, 1st ed., p. 76; J. von Neumann, review of *Cybernetics, Physics Today*, May 1949, p. 33; also see Herman Goldstine, interview with the author, February 26, 1971.

18
Atlantic Monthly 179 (1947): 46; Norbert Wiener to G. E. Forsythe, December 2, 1946 (MIT archives).

19
March 24, 1949. This brief interchange appeared in the records of the sixth conference on cybernetics but was deleted before publication.

20
John von Neumann, "The General and Logical Theory of Automata," in Cerebral Mechanisms in Behavior, ed. L. Jeffress (New York: Wiley, 1951).

21
Ibid. See also John von Neumann, Computer and the Brain (New Haven: Yale University Press, 1958).

22
See, for example, John G. Kemeny, "Man Viewed as a Machine," Scientific American 192 (1955): 58–67, which was based on von Neumann's Princeton lectures, March 2–5, 1953.

23
John von Neumann, "Probabilistic Logics and the Synthesis of Reliable Organisms from Unreliable Components," Collected Works, ed. A. H. Taub (New York: Macmillan, 1963) 5: 329–378.

24
John von Neumann, Theory of Self-Reproducing Automata, ed. A. W. Burks (Urbana: Illinois University Press, 1966).

25
Wiener to von Neumann, August 10, 1949 (MIT archives).

26
Von Neumann to Wiener, September 4, 1949 (MIT archives).

27
See the works of Merton on the sociology of science, or Jerome Ravetz, Scientific Knowledge and Its Social Problems (New York: Oxford University Press, 1973).

28
Wiener's correspondence with Rosenblueth, as well as with Warren McCulloch, conveys something of the quality of these various personal relations (MIT archives).

29
N. Wiener and A. Rosenblueth, "The Mathematical Formulation of the Problem of Conduction of Impulses in a Network of

Connected Excitable Elements, Specifically in Cardiac Muscle," *Archivos del Instituto de Cardiologia de Mexico* 16 (1946): 205–265.

30
A. Rosenblueth, N. Wiener, W. Pitts, and J. Garcia Ramos, "An Account of the Spike Potential of Axons," *Journal of Cellular and Comparative Physiology* 32 (1948): 275–317.

31
Unpublished; but see *Cybernetics*, 2d ed., pp. 19–21.

32
Cybernetics, 2d ed., p. 143; see also pp. 25–26.

33
Wiener, "Sound Communication with the Deaf," *Philosophy of Science* 16 (1949): 260–262; "Some Problems in Sensory Prosthesis" (with L. Levine), *Science* 110 (1949): 512; "Problems of Sensory Prosthesis," *Bulletin of the American Mathematical Society* 56 (1951): 27–35.

34
See Robert W. Mann, "Commentary on the Papers of Norbert Wiener," in Wiener, *Collected Works* (in press).

35
Dirk Struik, "Norbert Wiener, Colleague and Friend," *American Dialog*, March–April 1966, pp. 34–37. Struik says, "I have seldom seen Wiener so happy, as when he told how he turned the mishap of his fall into a victory for the handicapped."

36
Epilogue to *Progress in Brain Research* 2 (1963): 264–268; also Robert W. Mann, "Prosthesis Control and Feedback via Noninvasive Skin and Invasive Peripheral Nerve Technique," in *Neural Organization and Its Relevance to Prosthetics* (Miami: Symposia Specialists, 1973).

37
A. Rosenblueth, N. Wiener, and J. Bigelow, "Behavior, Purpose and Teleology," *Philosophy of Science* 10 (1943): 18–24; William James's concept of "will" is akin to Wiener's "purpose." See, for example, William Woodward, "William James' Psychology of Will: The Moral Foundation of the New Psychology" (mimeo, 1978), "William James' Theory of Volition" (mimeo, 1979), "William James' Psychology of Will: Its

Impact on American Psychology" in *Explorations in the History of American Psychology*, ed. Josef Brozek (Lewisburg, Pa.: Bucknell University Press, in press).

38
L. R. MacColl, *Fundamental Theory of Servomechanisms* (New York: Van Nostrand, 1945).

39
A large class of servomechanisms is described by a differential equation of the form

$$L_1\left[\frac{dy}{dt}\right] = L_2\left[x(t) - ky\right],$$

where L_1 and L_2 are linear operators, k is a constant, and $x(t)$ any given input. The simplest example is the equation

$$\frac{dy}{dt} = C\left[x(t) - y\right].$$

40
During the period from 1946 to 1950 Wiener's writing on cybernetics came increasingly to deal with human and animal societies. His skeptical view concerning quantitative methods is expressed in *Cybernetics*.

41
Cybernetics, 2d ed., p. 39.

42
HUHB, 2d ed. (New York: Doubleday, 1954), pp. 32, 102–104.

43
Cybernetics, 2d ed., p. 12.

44
Physics Today, May 1949, pp. 33–34.

11 Von Neumann and the Arms Race: Technical Adviser in the Corridors of Power

1
Lewis Strauss, *Men and Decisions* (New York: Doubleday, 1962), p. 188.

2
Charles Mee, *Meeting at Potsdam* (New York: Dell, 1976), chap. 6.

3
Richard J. Barnet, *Roots of War* (New York: Penguin, 1973), p. 17. For discussion of the advisory committee, see Richard Hewlett and Oscar Anderson, *A History of the United States Atomic Energy Commission*, vol. 1, *1939–1946: The New World* (Springfield, Va.: US Dept. of Commerce, National Technical Information Service, 1972).

4
Peter Burchett, *Daily Express*, September 5, 1945; quoted in Philip Noel-Baker, *The Arms Race* (London: Atlantic, 1958), chap. 10. Noel-Baker also gives the Japanese authorities statistical estimates of both the immediate and long-term casualties of the Hiroshima bomb.

5
International Symposium on the Damages and After-Effects of the Atomic Bombing of Hiroshima and Nagasaki, July 21 to Aug. 9, 1977, Tokyo, Hiroshima, and Nagasaki. Report submitted to the 8th Special Session of UN General Assembly Devoted to Disarmament, New York, May–June 1978.

6
US Atomic Energy Commission, *In the Matter of J. Robert Oppenheimer* (Cambridge, Mass.: MIT Press, 1971), p. 649.

7
Ulam, "John von Neumann 1903–1957," *Bulletin of the American Mathematical Society* 64, no. 3, pt. 2 (1958): 6.

8
Alice K. Smith, *A Peril and a Hope* (Cambridge, Mass.: MIT Press, 1970), pp. 113–122.

9
Ibid., pp. 236–237.

10
See, for example, Robert Gilpin, *American Scientists and Nuclear Weapons Policy* (Princeton: Princeton University Press, 1962), p. 52.

11
See, for example, George H. Quester, *Nuclear Diplomacy*, (New York: Dunnellen, 1970), pp. 18–23; also Gilpin, *American Scientists*, and Smith, *A Peril and a Hope.*

12
For details of the disarmament efforts within the UN, see Noel-Baker, *Arms Race*; Gilpin, *American Scientists*; Quester, *Nuclear Diplomacy*; Richard Barnet, *Who Wants Disarmament* (Boston: Beacon, 1960); B. Bechhoefer, *Post-War Negotiations for Arms Control* (Washington, D.C.: Brookings, 1961).

13
Private communication from the physicist who collected the $5.00 membership fee.

14
One World or None, ed. Dexter Masters and Katherine Way (New York: McGraw-Hill, 1946).

15
John von Neumann to Norman Cousins, May 22, 1946 (Library of Congress archives).

16
John von Neumann to Hyman Goldsmith, October 5, 1948 (Library of Congress archives).

17
John von Neumann to John Simpson, November 18, 1948 (Library of Congress archives).

18
AEC, *J. Robert Oppenheimer*, p. 246 (quoted by Oppenheimer).

19
Albert Einstein to Niels Bohr, March 2, 1955; quoted in B. Hoffmann, *Albert Einstein—Creator and Rebel* (New York: New American Library, 1972), p. 259.

20
February 13, 1950; see Albert Einstein, *Ideas and Opinions* (New York: Crown, 1954), pp. 159–160.

21
Ulam, *Adventures of a Mathematician*, p. 232.

22
The most pertinent treatment of the power elite in the US government between 1944 and 1960 is Gabriel Kolko's *The Roots of American Foreign Policy* (Boston: Beacon, 1969), pp. 13–26.

23
Strauss, *Men and Decisions*, p. 82.

24
S. Blumberg, *Energy and Conflict: The Life and Times of Edward Teller* (New York: Putnam, 1976), p. 198.

25
Strauss, *Men and Decisions*, p. 232; also von Neumann to Strauss, October 24, 1945 (Library of Congress archives).

26
Von Neumann to Strauss, May 4, 1946 (Library of Congress archives).

27
In 1950 Detlev Bronk was elected president, under anomalous circumstances, of the National Academy of Sciences; similar political issues were probably at work as at Princeton in 1946. James B. Conant, who like Oppenheimer had worked for limitations in nuclear armament and international control of atomic energy, was the sole candidate for the position, but Wendell Latimer, who shared von Neumann's enthusiasm for the arms race, led an unprecedented floor rebellion that led to Bronk's election. See Daniel S. Greenberg, "The National Academy of Sciences: Profile of an Institution," a three-part article in *Science*, April 14, 21, and 28, 1967.

28
See also Smith, *A Peril and a Hope*; Gilpin, *American Scientists*.

29
AEC, *J. Robert Oppenheimer*, p. 646.

30
Strauss to James Forrestal, August 16, 1945 (quoted in Strauss, *Men and Decisions*, pp. 208–209).

31
Lee DuBridge, "What About the Bikini Tests?" *Bulletin of Atomic Scientists* (1, no. 11), May 15, 1946; Smith, *A Peril and a Hope*, p. 304.

32
The agreement of June 19, 1947, states that von Neumann would need to be available at most thirty days a year in exchange for the fee (Library of Congress archives).

33
Edgar Bottome, *The Balance of Terror* (Boston: Beacon, 1971).

34
The containment policy was clearly articulated by George Kennan, "The Sources of Soviet Conduct," *Foreign Affairs* 25, (July 1947): 566–582.

35
Quoted in H. S. Parmet, *Eisenhower and the American Crusade* (New York: Macmillan, 1972), pp. 570–572.

36
See Seymor Melman, *Pentagon Capitalism: The Management of the New Imperialism* (New York: McGraw-Hill, 1970).

37
Langdon Winner, *Autonomous Technology* (Cambridge, Mass.: MIT Press, 1977), discusses some of the basic issues pertaining to technology and politics. See also my discussion of Wiener's ideas in chapter 12.

38
Los Alamos Scientific Laboratory Report LA-575, June 12, 1946 (unclassified, edited version released May 17, 1971); Herbert York, *The Advisors: Oppenheimer, Teller and the Superbomb* (San Francisco: Freeman, 1976), chap. 2.

39
Quester, *Nuclear Diplomacy*, p. 44; James Forrestal, *Diaries*, ed. W. Millis (New York: Viking, 1951), pp. 157–158. In my discussion of nuclear weapons policies I am relying heavily on Quester, *Nuclear Diplomacy*, and York, *The Advisors*. Gilpin's discussion of the H-bomb decision (*American Scientists*) is out of date, because he lacked the documents available to York.

40

In particular, see the report NSC-68 prepared for President Truman in March 1950; for details see Noel-Baker, *Arms Race*; Bechhoefer, *Arms Control*; and W. Schilling, P. Hammond, and G. Snyder, *Strategy, Politics and Defense Budgets* (New York: Columbia University Press, 1962).

41

Quester, *Nuclear Diplomacy*, p. 68. In 1950 Truman announced that nuclear weapons might be used in Korea.

42

Quoted from C. Blair, "Passing of a Great Mind," *Life*, February 25, 1957. A number of von Neumann's friends and associates have independently told me of von Neumann's strong advocacy of a preventive war at that time, confirming the substance of the statement in *Life*.

43

Ulam, *Adventures of a Mathematician*, p. 209.

44

Gilpin, *American Scientists*, p. 84.

45

For a full discussion and appraisal of the complete report emerging from the October 29 GAC meeting, see York, *The Advisors*. The GAC members were Oppenheimer, Fermi, Conant, DuBridge, Rabi, Hartley Rowe, Seaborg, Cyril Smith, and Oliver Buckley.

46

Strauss to Truman, November 25, 1949; Strauss, *Men and Decisions*, pp. 219–222.

47

Ulam, *Adventures of a Mathematician*, p. 217.

48

York, *The Advisors*, pp. 46–47. Compare C. P. Snow, *Science and Government* (Cambridge, Mass.: Harvard University Press, 1961), p. 1: "One of the most bizarre features of any advanced industrial society in our time is that the cardinal choices have to be made by a handful of men in secret: and, at least in legal form, by men who cannot have first-hand knowledge of what those choices depend upon or what their results may be. . . . And when I say the 'cardinal choices' I mean those which determine in the crudest sense whether we live or die."

49
Von Neumann to H. P. Robertson, May 18, 1950, urging him to become deputy director of the group: "You can perform an enormous service." (Robertson took on the job.) Von Neumann also arranged for his friend Oskar Morgenstern to work with the group; see von Neumann to Vannevar Bush, May 3, 1950 (Library of Congress archives).

50
Von Neumann to H. P. Robertson, May 18, 1950 (Library of Congress archives).

51
Von Neumann to General Vandenberg, June 21, 1951 (Library of Congress archives).

52
AEC, *J. Robert Oppenheimer*, p. 305.

53
John D., Nelson A., Laurence S., Winthrop, and David Rockefeller. See news release of June 26, 1950. Strauss to von Neumann, June 26, 1950 (Library of Congress archives).

54
Strauss, *Men and Decisions*, p. 332.

55
This refers particularly to the still-classified Ulam-Teller-DeHoffman schemes. See York, *The Advisors*, and Ulam, *Adventures of a Mathematician*, for more detail.

56
AEC, *J. Robert Oppenheimer*, p. 655.

57
The "Mike Shot"; see York, *The Advisors*, pp. 82–83.

58
Ulam, "John von Neumann 1903–1957."

59
Author's interview of Cyril Smith, May 8, 1975.

60
Strauss to von Neumann, February 28, 1952 (Library of Congress archives).

61
York, *The Advisors*, pp. 82–83; Robert Jungk, *Brighter than a Thousand Suns* (New York: Harcourt, Brace, 1958), pp. 301–305.

62
Von Neumann to Strauss, March 9, 1953 (Library of Congress archives).

63
Strauss's appointment was approved by the Senate on June 27, and he was sworn in on July 2, 1953.

64
Dwight D. Eisenhower, *The White House Years—Waging Peace* (New York: Doubleday, 1965), pp. 392–396.

65
Ibid., p. 618.

66
John von Neumann, "Defense in Atomic War," paper delivered on December 7, 1955; reprinted in von Neumann, *Collected Works*, vol. 6 (New York: Macmillan, 1963) pp. 523–525.

67
Jungk, *Brighter than a Thousand Suns*, p. 307.

68
Quester, *Nuclear Diplomacy*, p. 90.

69
Only certain unclassified portions of the minutes have been available to the writer. I am indebted to William Woodward of the AEC staff for making these records available to me.

70
See also von Neumann, "Defense in Atomic War."

71
AEC, *J. Robert Oppenheimer*, pp. 643–656; the volume contains the full transcript of the hearings. For the background,

see Philip M. Stern, *The Oppenheimer Case: Security on Trial* (New York: Harper & Row, 1969).

72
Cf. AEC, *J. Robert Oppenheimer*, pp. v, vi.

73
Hearings before Subcommittee on Military Operations, Committee on Government Operations, 83rd Congress, 2d sess., *Organization and Administration of the Military Research and Development Programs*, June 15, 1954; pp. 379–381.

74
Von Neumann's opposition to the harassment of scientists for political reasons is supported by other evidence. For example, he wrote that "requiring loyalty clearance for all scientists receiving financial aid in the same way that it is required from government employees would, I think, be a grave mistake" (Von Neumann to Lyman Spitzer, chairman of the Scientists' Committee on Loyalty Problems, June 9, 1949). In 1950, when Wiener wanted to find a position for a young mathematician, Felix Browder, whose father, Earl Browder, was a leading figure in the American Communist party, he wrote to von Neumann to try to get him a temporary position at the Princeton Institute. Wiener described Felix Browder as completely nonpolitical but seriously handicapped by name in finding work. In fact, von Neumann did not recommend Browder for a job at Princeton [Wiener to von Neumann, September 25, 1950; von Neumann to Wiener, September 29, 1950 (MIT Archives)], so that this incident is evidence only of Wiener's opinion of von Neumann's attitudes. Another bit of evidence is that when questioned about James Alexander, his friend and colleague at the Princeton Institute, and his student I. Halperin, von Neumann defended them in private correspondence with Strauss (1954) against political accusations.

75
David Inglis, *Nuclear Energy: Its Physics and Its Social Challenge* (Reading, Mass.: Addison-Wesley, 1973), p. 214. I have made use of Inglis's description of the Bravo shot. See also Ralph Lapp, *The Voyage of the Lucky Dragon* (New York: Harper, 1958); N. O. Hines, *Proving Ground* (St. Louis, Mo.: Washington University Press, 1962); York, *The Advisors*, pp. 81–85, contains a quantitative summary of the Castle test series. In 1978 the radiation on the island was still so strong that the surviving population had to be moved again. They had

been allowed to return to the island only in 1968 (Reuter dispatch, *Boston Globe*, May 23, 1978).

76
Strauss defends his actions in chapter 19 of *Men and Decisions*.

77
Extensive references to the articles by scientists pointing to the harmfulness of fallout are given in Gilpin, *American Scientists,* pp. 137–143. See also Linus Pauling, *No More War* (New York: Dodd, Mead, 1958). Physician-theologian Albert Schweitzer also made public statements and gave lectures against the continuation of nuclear tests. See Albert Schweitzer, "Declaration of Conscience," *Saturday Review* 40 (May 18, 1957): 58; "Obligation to To-morrow," *Saturday Review* 41 (May 24, 1958): 21.

78
Memorandum to Strauss, 1955; see Strauss, *Men and Decisions*, p. 441.

79
Department of State Bulletin 30 (January 25, 1954): 107–112; John R. Beal, *John Foster Dulles: A Biography* (New York: Harper, 1957).

80
Noel-Baker, *The Arms Race*.

81
New York Times Weekly Review, June 23, 1957 (European edition).

82
White House press conference, March 31, 1954; compare Gilpin, *American Scientists*, pp. 138–143; Lapp, *Lucky Dragon*, pp. 133–135.

83
Quester, *Nuclear Diplomacy*, pp. 97–100; Gilpin, *American Scientists*, pp. 125–130; Barnet, *Disarmament*, pp. 27–32; Parmet, *Eisenhower*, chap. 31; Strauss, *Men and Decisions*, chap. 17; Robert Donovan, *Eisenhower, The Inside Story* (New York: Harper, 1956), chap. 13.

84
Minutes of the thirty-eighth GAC meeting, Jan. 6–8, 1954 (unclassified portions of the minutes obtained from the General Advisory Committee to the US Energy Research and Development Admin., Washington, D.C.).

85
Compare Quester, *Nuclear Diplomacy*, p. 99.

86
Verbatim records of the meetings of the Subcommittee of the United Nations Disarmament Commission, April 19–May 18, 1955, UN no. 3 (1956), Cmd. 9650. Also quoted by Noel-Baker, *The Arms Race*.

87
The discussion of disarmament negotiations relies heavily on Bechhoefer, *Negotiations*, Noel-Baker, *Arms Race*, and Barnet, *Disarmament*. Barnet suggests (pp. 30–35) reasons for the United States having endorsed the British-French proposal in the first place.

88
Dwight D. Eisenhower, Dulles Oral History, Princeton (cf. Parmet, *Eisenhower*, p. 406). See Parmet, chapter 3, for more detail.

89
Parmet, *Eisenhower*, p. 299.

90
See, however, Eisenhower, *Waging Peace*, pp. 106–114.

91
Peter Lyon, *Neutralism* (Atlantic Highlands, N.J.: Humanities Press, Leicester University Press, 1963), p. 107.

92
Barnet, *Disarmament*, p. 2.

93
See, for example, Stern, *Oppenheimer Case*.

94
Samuel Huntington, *The Common Defense* (New York: Columbia University Press, 1961), provides a well-documented study—including the national politics and distribution of

funds—of the armament and arms-limitation programs during the years 1945–1960.

95
Herbert York, *Race to Oblivion: A Participant's View of the Arms Race* (New York: Simon & Schuster, 1971); see also Michael H. Armacost, *The Politics of Weapons Innovation* (New York: Columbia University Press, 1969).

96
York, *Race to Oblivion*, pp. 84–85.

97
Also known as the teapot committee, and more formally as the Strategic Missiles Evaluation Group.

98
Forty-first GAC meeting, July 12–14, 1954.

99
Armacost, *Politics of Weapons*, p. 57; York, *Race to Oblivion*, p. 86. The report containing the recommendations was submitted to the secretary of the Air Force on February 10, 1954.

100
York, *Race to Oblivion*, p. 86.

101
Ernest Schwiebert, *History of the U.S. Air Force Ballistic Missiles* (New York: Praeger, 1965).

102
Armacost, *Politics of Weapons*, p. 58.

103
Eisenhower, *Waging Peace*, p. 208.

104
Eisenhower (ibid.) cites $1.3 billion for the 1957 fiscal year budget for missile development. See also Armacost, *Politics of Weapons*, and York, *Race to Oblivion*.

105
Harvard scientist George Kistiakowsky, concerned with propulsion and reentry problems, became in 1957 the ICBM expert member of the President's Science Advisory Committee; later he became Eisenhower's special assistant for science and technology; see George Kistiakowsky, *A Scientist at the White*

House (Cambridge, Mass.: Harvard University Press, 1976).
MIT engineer Jerome Wiesner, concerned with guidance systems, later became special assistant for science and technology to President Kennedy and subsequently president of MIT. Herbert York became the leading technical research adviser to the Department of Defense, but around 1960 he became disenchanted with the arms race and became an outspoken critic.

106
Von Neumann, "Defense in Atomic War."

107
Einstein, *Ideas*, p. 160.

108
Von Neumann formally accepted the chairmanship in a letter to Theodore von Karman, May 7, 1953 (Library of Congress archives). See also the article by T. F. Walkowicz of the USAF Scientific Advisory Board, *Air Force Magazine*, June 1955, p. 50.

109
Quarles to von Neumann, February 15, 1954; von Neumann to Quarles, February 24, 1954, contains his acceptance of the appointment: "It will be great pleasure and satisfaction to me if I can be of help in this manner" (Library of Congress archives).

110
His formal consultantship with the CIA was terminated on September 6, 1955, because of his position as Atomic Energy commissioner, although he says that he "will be glad to be helpful, when you judge it useful and convenient, but without compensation." Von Neumann to Allen Dulles, April 15, 1955, and von Neumann to H. C. Reynolds, August 29, 1955 (Library of Congress archives).

111
Mario Laserna to von Neumann, November 29, 1954 (Library of Congress archives).

112
Strauss, *Men and Decisions*, p. 235.

113
Ulam, *Adventures of a Mathematician*, pp. 237–238.

114
Von Neumann to Strauss, November 22, 1954.

115
Parmet, *Eisenhower*, p. 466.

116
Ibid., p. 466; for discussion of the Dixon-Yates controversy, see ibid., pp. 465–469; Donovan, *Eisenhower*, pp. 335–341; Strauss, *Men and Decisions*, chap. 14. A Republican Congress in 1954 voted in favor of the Dixon-Yates contract; after the 1954 election, the Democratic Congress cancelled it.

117
The confirmation hearings were held in March 1955 by the Joint Committee on Atomic Energy, consisting of Senators Anderson (chair), Pastore, Gore, Jackson, Hickenlooper, Millikan, Knowland, and Bricker, and Representatives Durham, Price, and Cole.

118
Thomas Murray, *Nuclear Policy for War and Peace* (Cleveland: World, 1960); *Life*, May 6, 1957, p. 181; *Current Biography*, 1950, s.v. "Murray, Thomas (New York: H. W. Wilson Co., 1950).

119
Strauss, *Men and Decisions*, p. 338n.

120
Current Biography, 1949, s.v. "Vance, Harold S."

121
Newsweek, September 27, 1954, pp. 29–30; see also Gilpin, *American Scientists*, p. 166; *Current Biography*, 1954, s.v. "Libby, Willard F."

122
Richard G. Hewlett and Francis Duncan, *A History of the Atomic Energy Commission*, vol. 2, 1947–1952: *Atomic Shield* (US Atomic Energy Commission, 1972); H. Peter Metzger, *The Atomic Establishment* (New York: Simon & Schuster, 1972).

123
This quote from the AEC's semiannual report to Congress, February 1957, is typical.

124
Von Neumann to Mrs. Crocker, September 1, 1955, declining an invitation to speak at a World Affairs Council program, "Diplomats off the Record" (Library of Congress archives).

125
Richard Hewlett, oral communication to the author, October 1976.

126
Metzger, *Atomic Establishment*, p. 81.

127
Harold P. Green and Alan Rosenthal, *Government of the Atom* (Englewood, N.J.: Prentice-Hall, 1963), p. 15.

128
Memorandum to Admiral Foster, July 21, 1955 (US Department of Energy archives).

129
Memorandum for record, August 10, 1955 (Dept. of Energy archives).

130
At the AEC meeting of July 21, 1955, decisions were made on a Detroit Edison 100 MW breeder reactor south of Detroit, a 100 MW "Yankee" pressurized light water reactor in Rowe, Massachusetts, a 75 MW sodium graphite reactor in Nebraska (Consumers Company), and a 180 MW boiling water reactor southwest of Chicago (Nuclear Power Company). The diversity of types of reactors reflects the experimental nature of the program. See also Green and Rosenthal, *Government of the Atom*, pp. 12–16, for background information, as well as the minutes of the July 21 meeting (Dept. of Energy archives).

131
Minutes of the July 21, 1955, meeting of the AEC (Dept. of Energy archives).

132
Public Law 85-256. I quote a brief summary from Inglis, *Nuclear Energy*, p. 122: "With the estimate at hand, then, an accident might possibly cause something like $7 billion in property damage . . . it was found that a consortium of private insurance companies would insure nuclear industries against damage claims for only about 1% of that. The Price-Anderson

Act requires industry to take out that much insurance privately, then provides that the government will carry insurance for $500 million. Beyond that limit there shall be no liability."

133
I thank Silvan S. Schweber for bringing this plausible explanation to my attention.

134
Von Neumann to Paul Klopsteg, November 18, 1955, contains a summary of von Neumann's remarks to the panel of the Advisory Committee on Weather Control.

135
Metzger, *Atomic Establishment.*

136
Memorandum, February 20, 1956 (Dept. of Energy archives).

137
Minutes of June 8, 1955, AEC meeting.

138
See M. Eisenbud and J. Harley, "Radioactive Fallout in the United States," *Science* 121 (1955): 677–680.

139
John von Neumann memorandum to Willard F. Libby, May 25, 1955 (Dept. of Energy archives).

140
Barry Commoner, *Science and Survival* (New York: Viking, 1966), chap. 2; Pauling, *No More War*, chap. 6. Ralph Lapp wrote numerous books, as well as articles appearing in magazines like *The Reporter*, *Life*, and *The Saturday Evening Post*, emphasizing the dangers of fallout.

141
Associated Press dispatch, October 24, 1956; quoted by Commoner, *Science and Survival*, p. 14.

142
Memorandum on AEC Chairman Strauss's proposed letter to Congressman Richards, August 19, 1955 (Dept. of Energy archives). Compare memo to Strauss, September 1955, quoted by Strauss, *Men and Ideas*, p. 441.

143
Von Neumann used that adjective to describe himself in connection with arguments opposing the crash program for development of the superbomb so as to permit renewed efforts at international control and disarmament (AEC), *J. Robert Oppenheimer*, p. 647.

144
Von Neumann to Strauss, November 21, 1951.

145
Bechhoefer, *Post-War Negotiations*, p. 265.

146
Ibid., pp. 258–265; Barnet, *Who Wants Disarmament*, pp. 27–45; Noel-Baker, *Arms Race*; Parmet, *Eisenhower*, chap. 20.

147
Parmet, *Eisenhower*, chap. 34.

148
U.S., Senate, *The Executive Branch and Disarmament Policy*, Senate Foreign Relations Special Study no. 1, February 20, 1956.

149
Barnet, *Who Wants Disarmament*, p. 36.

150
Strauss to von Neumann, August 26, 1955.

151
US Senate, *Executive Branch and Disarmament Policy*.

152
Noel-Baker, *Arms Race*, p. 23.

153
United Nations, Subcommittee on Disarmament meeting, March 19, 1956; Noel-Baker, *Arms Race*, p. 233.

154
UN, Subcommittee on Disarmament meeting, May 4, 1956; Noel-Baker, *Arms Race*, p. 233.

155
Eisenhower, *Waging Peace*, p. 686.

12 *Political Power, Human Nature, and Society*

1

Philip M. Stern, *The Oppenheimer Case* (New York: Harper & Row, 1969).

2

Morgenstern, interview with the author, Princeton, N.J., May 1, 1970.

3

John von Neumann and Oskar Morgenstern, *Theory of Games and Economic Behavior* (Princeton: Princeton University Press, 1944). Von Neumann reiterated this belief in 1955, in "The Impact of Recent Developments in Science on the Economy and on Economics," *Collected Works* (New York: Macmillan, 1963), ed. A. H. Taub, 6:100–101.

4

The reviews of the von Neumann–Morgenstern book indicated the high expectations it raised: "Posterity may regard this book as one of the major scientific achievements of the first half of the twentieth century" [A. H. Copeland, *American Mathematical Society Bulletin* 51 (1945): 498]; "[the reader] will come away from the volume with a wealth of ideas for application and for development of the theory into a fundamental tool of analysis for the social sciences" [H. A. Simon, *American Journal of Sociology* 50 (1945): 558].

5

R. D. Luce and H. Raiffa state in *Games and Decisions* (New York: Wiley, 1957), p. 10, "We have the historical fact that many social scientists have become disillusioned with game theory. Initially there was a naive band-wagon feeling that game theory solved innumerable problems of sociology and economics, or that, at the least it made their solution a practical matter of a few years' work. This has not turned out to be the case."

6

Von Neumann and Morgenstern, *Theory of Games*, 2d ed. (New York: John Wiley, 1946), p. 8.

7

Ibid., p. 44.

8
Anatol Rapaport, *N-Person Game Theory* (Ann Arbor: University of Michigan Press, 1970), chap. 20.

9
Von Neumann and Morgenstern, *Theory of Games*, 2d ed., p. 2.

10
César Graña, *Modernity and Its Discontents* (New York: Harper, 1967), explores the value conflict between the artist and capitalist society.

11
Gunnar Myrdal, *Against the Stream* (New York: Pantheon, 1972), chap. 7. Myrdal's comments apply to a large class of economic theories and are not specifically applied to game theory.

12
Rapaport, *N-Person Game Theory*, p. 301, also concludes that game theory in its present form does not even begin to enable us to deal with the question of how to act in situations of conflict.

13
H. H. Gerth and C. Wright Mills, eds., *From Max Weber* (Oxford: Oxford University Press, 1946), chap. 8, introduce the reader to Weber's thought on bureaucratic institutions. I thank Joy Gordon for calling Weber's writings in this connection to my attention.

14
Joseph Needham and Ho Ping-Yu, "Elixir Poisoning in Medieval China," in Needham, *Clerks and Craftsmen in China and the West* (Cambridge: Cambridge University Press, 1970).

15
"Ends and Means in Political Economy" (originally published in German in 1933), in *Value in Social Theory* (New York: Harper & Row, 1958).

16
Walter Weisskopf, *Alienation and Economics* (New York: Dell, 1973), pp. 23–24.

17

Leonard J. Savage, *The Foundations of Statistics* (New York: Wiley, 1954); the bibliographical supplement to the second edition (New York: Dover, 1972) contains references to Savage's later philosophical reflections.

18

Recorded interview with the author, Yale University, New Haven, Connecticut, June 2, 1970.

19

Von Neumann, *Collected Works*, 6: 100.

20

See Karl Deutsch, *The Nerves of Government* (Glencoe, Ill.: Free Press, 1966), and its bibliography; Gregory Bateson, *Steps to an Ecology of Mind* (New York: Ballantine, 1972) and *Naven* (Stanford: Stanford University Press, 1958), the epilogue to the second edition; Bateson and J. Ruesch, *Communication: The Social Matrix of Psychiatry* (New York: Norton, 1951). Deutsch and Bateson both appear to be faithful to Wiener's point of view. Some other books are S. Beer, *Cybernetics and Management* (New York: Wiley, 1959); Gordon Pask, *An Approach to Cybernetics* (London: Hutchinson, 1961); Richard Meier, *A Communication Theory of Urban Growth* (Cambridge, Mass.: MIT Press, 1962); A. Tustin, *The Mechanism of Economic Systems* (Cambridge, Mass.: Harvard University Press, 1953).

21

Deutsch, *Nerves of Government*, p. 77, quotes Wiener on the relevance of cybernetics to social science: "The existence of social science is based on the ability to treat a social group as an organization and not as an agglomeration. Communication is the cement that makes *organizations*. Communication alone enables a group to think together, to see together, and to act together. All sociology requires the understanding of communications.

"What is true for the unity of a group of people, is equally true for the individual integrity of each person. The various elements which make up each personality are in continual communication with each other and affect each other through control mechanisms which themselves have the nature of communication.

"Certain aspects of the theory of communication have been considered by the engineer. While human and social communication are extremely complicated in comparison to the existing patterns of machine communication, they are subject to the same grammar; and this grammar has received its highest technical development when applied to the simpler content of the machine."

22
HUHB, 1st ed., pp. 112–113.

23
Wiener and Bateson first met on March 8, 1946. For a fuller description of their dialogue, see "Gregory Bateson and the Mathematicians: From Interdisciplinary Interaction to Societal Functions," *Journal of the History of the Behavioral Sciences* 13 (1977): 151.

24
Wiener, *Cybernetics*, chap. 7.

25
Bateson and Ruesch, *Communication*.

26
See Bateson, *Steps to an Ecology of Mind*, pp. 177–339.

27
Norbert Wiener to Lawrence K. Frank, October 26, 1951 (MIT archives).

28
Wiener, *Cybernetics*, 1st ed., p. 191. In connection with statistical runs Wiener had the applicability of his own prediction theory in mind.

29
Von Neumann, *Collected Works*, 6: 100.

30
HUHB, 2d ed., p. 181.

31
See, for example, "Some Moral and Technical Consequences of Automation," *Science* 131 (1960): 1355.

32
Wiener, *Cybernetics*, 1st ed., pp. 185–186.

33
HUHB, 1st ed., pp. 206–210.

34
Gregory Bateson to Norbert Wiener, September 22, 1952 (MIT archives).

35
"Too Damn Close," *Atlantic* 186 (July 1950): 50.

36
HUHB, 2d ed., p. 182.

37
Ibid., p. 181.

38
Wiener, *Cybernetics*, 2d ed., p. 162.

39
See, for example, IAM, pp. 220–221; *God and Golem Inc.: A Comment on Certain Points where Cybernetics Impinges on Religion* (Cambridge, Mass.: MIT Press, 1964), p. 53 (hereafter referred to as G & G). HUHB, 2d ed., pp. 192–193.

40
Wiener, *Cybernetics*, 2d ed., p. 161.

41
Dirk Struik, "Norbert Wiener—Colleague and Friend," *American Dialog*, March–April 1966, p. 35.

42
Wiener, *Cybernetics*, 2d ed., p. 160.

43
Ibid.

44
Oskar Morgenstern, *The Question of National Defense* (New York: Random House, 1959), p. 267.

45
The Rand Corporation—The First Fifteen Years (Santa Monica, Calif.: Rand, 1963).

46
John D. Williams, *The Compleat Strategist: Being a Primer on the Theory of Games of Strategy* (New York: McGraw, 1954);

J. C. C. McKinsey, *Introduction to the Theory of Games* (New York: McGraw, 1952); M. Dresher, *Games of Strategy—Theory and Applications* (Englewood, N.J.: Prentice-Hall, 1961). The books by Williams and Dresher draw their examples primarily from tactical military problems.

47
Col. Oliver G. Haywood, Jr., "ASAF Military Doctrine of Decision and the von Neumann Theory of Games," Rand RM-528, February 2, 1951.

48
George B. Dantzig, *Linear Programming and Extensions* (Princeton: Princeton University Press, 1963), pp. 14–15.

49
Ibid., p. 24; George B. Dantzig, "Memoirs about Linear Programming," November 25, 1973, prepared for Robert Dorfman of Harvard University; John von Neumann, "Discussion of a Maximum," *Collected Works*, 6: 89.

50
Ulam, "John von Neumann 1903–1957," *Bulletin of the American Mathematical Society*, p. 6.

51
Valentine Bargman, interview with the author, Princeton, N.J., December 4, 1972.

52
John von Neumann, "Defense in Atomic War," *Collected Works*, 6: 523.

53
In particular Dennis Gabor (see his *Innovations*, Oxford University Press, 1970) became in his later years concerned with reconsidering the "innovate or die" attitude. It of course recalls von Neumann's boyhood conviction that he had to "produce the unusual or face extinction." It is notable that Edward Teller, Leo Szilard, and Eugene Wigner all took active, innovative political-technological roles in the decade following World War II.

54
Von Neumann, "Defense in Atomic War."

55
Victor F. Weisskopf, "A Race to Death," *New York Times*, May 14, 1978, section E, p. 19.

56
Compare Raymond Aron, "The Evolution of Modern Strategic Thought," in *Problems of Modern Strategy*, Adelphi Papers no. 54 (London: Institute of Strategic Studies, February 1969).

57
Thomas C. Schelling, *Arms and Influence* (New Haven, Connecticut: Yale University Press, 1966).

58
This is described in the articles in *Problems of Modern Strategy*; see also Adelphi Papers no. 55. An exception is Bernard Brodie of Rand, who emphasized and defined a historical perspective.

59
A statistic confirming the continued role of computers in the military is that "when 1972 began, the Federal Government was utilizing 5,400 computers, and 88% of them were accounted for by the Department of Defense." From L. S. Wittner, "IBM and the Pentagon," *The Progressive*, February 1972, pp. 33–34, quoted by H. I. Schiller, *The Mind Managers* (Boston: Beacon, 1973).

60
For a discussion of war games as a means for policy decisions, see, for example, Andrew Wilson, *The Bomb and the Computer* (London: Barrie and Rockliff, 1968), and monographs and reports issued by the Rand Corporation.

61
Gene M. Lyons and Louis Morton, *Schools for Strategy* (New York: Praeger, 1965), p. 30.

62
Ibid., p. 42.

63
For example, Thomas Schelling, *The Strategy of Conflict* (Cambridge, Mass.: Harvard University Press, 1960). For a good introduction to game theory, see A. Rapoport, *Two-Person Game Theory: The Essential Ideas* (Ann Arbor: University of Michigan Press, 1969) and *N-Person Game Theory*; a good but more advanced text is R. D. Luce and H. Raiffa, *Games and Decisions*.

64
See in particular A. Rapoport, *Strategy and Conscience* (New York: Harper & Row, 1964).

65
Philip Green, *Deadly Logic: The Theory of Nuclear Deterrence* (Columbus: Ohio State University Press, 1966), chap. 3.

66
See, for example, Gabriel Kolko, *The Roots of American Foreign Policy* (Boston: Beacon, 1969); David Horwitz, ed., *Containment and Revolution* (Boston: Beacon, 1967).

67
Aron, "Evolution of Modern Strategic Thought," p. 9.

68
P. M. S. Blackett, "Critique of Some Contemporary Defense Thinking," *Encounter*, April 1961, reprinted in P. M. S. Blackett, *Studies of War* (New York: Hill and Wang, 1962).

69
Schelling, *The Strategy of Conflict* (Cambridge, Mass.: Harvard University Press, 1960), preface.

70
Schelling, "Game Theory and the Study of Ethics," *Journal of Conflict Resolution* 12 (March 1968): 34–44.

71
Gene Sharp, *The Politics of Nonviolent Action* (Boston: Porter Sargent, 1973).

72
Daniel Ellsberg, "Theories of Rational Choice and Uncertainty—The Contribution of von Neumann and Morgenstern," honors thesis, Harvard University, 1952; Ellsberg's Ph.D. dissertation also dealt with mathematical decision theory.

73
See, for example, Daniel Ellsberg, "Crude Analysis of Strategic Choices," *American Economic Review* 51 (May 1961): 472, and the study "The Art of Coercion: A Study of Threats in Economic Conflict and War" (March 1959), described in Snyder, *Deterrence and Defense* (Princeton: Princeton University Press, 1961), pp. 16–27.

74

Ellsberg's unpublished report, reproduced by RAND as a courtesy to him as a staff member, "The Day Lôc Tiên Was Pacified," was written in December 1966. The report deals with the villagers' experience of "pacification" in the South Vietnamese hamlet of Lôc Tiên II. Ellsberg alludes to the hurry in which the report was written, so as to justify his direct and personal language, which violates the facelessness expected of a good bureaucrat.

75

Daniel Ellsberg, *Papers on the War* (New York: Simon & Schuster, 1972), p. 18.

76

Ibid., p. 28.

77

Compare Kurt Wolff, *Surrender and Catch* (Hingham, Mass.: Reidel, 1976), wherein such processes are considered.

In the 1950s and 1960s, psychologists seeking to contribute to the prevention of nuclear war developed theories of human conflict and conflict resolution that focused on cognitive processes and emphasized systematic distortions of perception. This approach tended to buttress the views of those opposed to the doctrine of massive retaliation by exposing the assumptions of the policy analysts. In Wiener's social theory the receipt of feedback served as one vehicle for correcting systematic distortions, and the high value placed on channels for the open flow of information served as another. However, whereas the psychologists emphasized subjective elements, those Wiener emphasized were material and political. See an unpublished review article by Lorenz Finison of the department of psychology, Wellesley College: "Organized Psychology, Peace Action, and Theories of Conflict: 1936–62." Most of the psychological work between 1952 and 1957 was published in the *Bulletin of the Research Exchange on the Prevention of War*, and from 1957 on in the *Journal of Conflict Resolution*. John D. Steinbruner, *The Cybernetic Theory of Decision* (Princeton: Princeton University Press, 1974), compares the "cognitive paradigm" to the "cybernetic paradigm" and the "analytic paradigm"; see p. 14, and especially chapter 4. His "cybernetic paradigm," however, is very narrow, taking no more than the servomechanism model from Wiener's suggestions.

78
Compare Aron, "Modern Strategic Thought," p. 2.

79
Blackett, "Contemporary Defence Thinking."

80
"A Call from the Hibakusha of Hiroshima and Nagasaki, 1978" (copy in the Boston University Mugar Memorial Library).

81
It is of interest to compare Steinbruner, *Cybernetic Theory*, who considers the "analytic paradigm" for decision processes as the ideal. Yet his detailed empirical study shows it "is not the most natural or empirically dominant mechanism of decision under complexity" (p. 334).

82
Michael Neumann, interview with the author, Chicago, January 23, 1971.

83
HUHB, 2d ed., chap. 5; Wiener to von Neumann, August 10, 1949.

84
HUHB, 2d ed., p. 95.

85
Norbert Wiener, *The Tempter* (New York: Random House, 1959).

86
Ulam, "John von Neumann 1903–1957," p. 5.

87
Wigner, *Symmetries and Reflections*, p. 261.

88
Warren McCulloch to F. S. C. Northrop, April 15, 1947 (Philosophical Library Archives, Philadelphia).

89
Hans J. Morgenthau, *Politics in the Twentieth Century*, abridged ed. (Chicago: University of Chicago Press, 1962), pp. 190, 195.

90
Robert J. Lifton, *Boundaries* (New York: Vintage, 1970), *Prophetic Survivors: Hiroshima and Beyond* (mimeo, 1971).

91
Ulam, *Adventures of a Mathematician*; Hans Lukas Teuber, interview with the author, Arlington, Mass., October 30, 1968; and others who prefer to be anonymous.

92
Thucydides, *History of the Peloponnesian War* (London: Penguin, 1956), bk. 5, chap. 7, p. 360.

93
Giorgio de Santillana, *The Origins of Scientific Thought* (New York: New American Library, 1961), p. 181.

1
Lewis Coser, *Men of Ideas* (Glencoe: Free Press, 1965), p. viii.

2
Ibid., p. 311.

3
From Jefferson's education bill of 1779, as quoted by Theodore Roszak, ed., *The Dissenting Academy* (New York: Vintage, 1968), p. 5.

4
Cf. Coser, *Men of Ideas*, chap. 26.

5
HUHB, 2d ed., p. 135.

6
N. Levinson, "Wiener's Life," *Bulletin of the American Mathematical Society* 72, no. 1, pt. 2 (1966): 32.

7
Norbert Wiener to Frank B. Jewett, September 22, 1941.

8
Julius Stratton to Rear Admiral D. S. Fahrney, January 27, 1964.

9
Dirk Struik, "Norbert Wiener, Colleague and Friend," *American Dialog,* March–April 1966, 35.

10
Wiener, "The Duty of the Intellectual," *Technology Review* 62 (February 1960): 26–27.

11
Norbert Wiener to J. B. S. Haldane, May 5, 1949.

12
Wiener to James R. Killian, September 13, 1951; in a draft of the May 5 letter to Haldane, Wiener attributes the relative liberalism of MIT to Karl Compton, James Killian, and Julius Stratton.

13

Giorgio de Santillana, "Paralipomena of the Future" in *Reflections on Men and Ideas* (Cambridge, Mass.: MIT Press, 1968), pp. 202–205; however, the style of this piece seems particularly characteristic of Walter Pitts.

In an earlier but undated letter from J. B. S. Haldane to Norbert Wiener, Haldane distinguishes the "Vindobonensis" (inhabitant of Vienna) as a unit of information, based on Wiener's information concept, from another unit based on Fisher's concept of information.

14

Interview of Neal Hartley by the author, Cambridge, Mass., November 15, 1971.

15

Karl Wildes, interview with the author, Cambridge, Mass., August 18, 1971.

16

The most important American philosopher of technology writing in the years prior to World War II was Lewis Mumford. In his *Technics and Civilization* (New York: Harcourt, Brace, 1934) he treated the relation of technology to society, life patterns, values, and history more systematically than Wiener did. However his viewpoint, unlike Wiener's, was primarily that of a member of the general public, not that of a scientist.

17

IAM, pp. 299–307; *Einstein on Peace*, p. 342.

18

Norbert Wiener, "A Scientist Rebels," *Atlantic Monthly* 170 (1947): 46.

19

IAM, pp. 297–298; Wiener to Howard Aiken, February 11, 1947. The issue also appears in an interchange of letters between Wiener and Oskar Morgenstern, who asked Wiener for a copy of his *Analysis of Time Series* (known around MIT as "The Yellow Peril"), which was not generally available at the time (January 3, 1947); it contained work Wiener had done under military auspices during World War II. Wiener answered by inquiring of Morgenstern if he intended to use it for military work (January 21, 1947). Morgenstern answered that he planned to analyze some economic time series together with von Neumann (February 12, 1947). Wiener replied that he would send a copy as soon as he got some (April 30, 1947).

20
Albert Einstein, statement of January 20, 1947, in reply to a question from J. Landau of the Overseas News Agency, originally in German. Translated in *Einstein on Peace*, p. 401

21
Wiener, IAM, p. 329.

22
See, for example, *Newsweek*, November 15, 1948, p. 89; *Time*, December 27, 1948, p. 45, January 23, 1950, pp. 54ff., October 21, 1957, p. 52, and January 11, 1960; *U.S. News and World Report*, February 24, 1964, pp. 84–86.

23
Wiener, *The Tempter* (New York: Random House, 1959); for an early version of the plot, see Wiener to Orson Welles, June 28, 1941.

24
G & G.

25
HUHB, 1st ed., pp. 162–163.

26
Norbert Wiener, "Some Moral and Technical Consequences of Automation," *Science* 131 (1960): 1358.

27
See HUHB, 1st ed., chap. 2. (This material is omitted in the second edition.)

28
Neither Wiener nor the nuclear physicists anticipated at that time the difficulties and hazards that the problems of nuclear waste disposal would create. In 1950 Wiener envisaged the whole countryside as increasingly "man-made" and how in some future age of centralized nuclear power civil liberties might go by the board. See "Too Big for Private Enterprise," *The Nation*, May 20, 1950, pp. 496–497.

29
See, for example, Dennis L. Meadows et al., *The Limits of Growth* (New York: Universe, 1972); M. Mesarovic and E. Pestel, *Mankind at the Turning Point* (New York: Dutton, 1974).

30
Wiener, *Cybernetics,* 1st ed., pp. 36–38. The wave of unemployment in the United States and elsewhere in the late 1970s appears in part to derive from the process of displacement of people in the service sector of the economy by automation.

31
Wiener to Walter Reuther, August 13, 1949 (MIT archives).

32
HUHB, 2d ed., pp. 161–162.

33
G&G, p. 73; see also pp. 77–81. Wiener tended to discourage high expectations from mechanical methods of translation. For example, see Wiener to John Griffith, February 16, 1949; Warren Weaver, *Scene of Change* (New York: Scribner's, 1970), p. 106; Wiener to Warren Weaver, April 30, 1947; Y. Bar-Hillel, *Language and Information* (Reading, Mass.: Addison-Wesley, 1964), pp. 1–15.

34
G&G, p. 52.

35
Alice Kimball Smith, *A Peril and a Hope* (Cambridge, Mass.: MIT Press, 1965), chap. 9.

36
G&G, pp. 58–59; the story is repeated in *Cybernetics*, HUHB, and in "Some Moral and Technical Consequences of Automation." Wiener also repeatedly told the gentler stories of the "Fisherman and the Jinni" from the *Arabian Nights*, which nevertheless make the same point, as well as Goethe's "Sorcerer's Apprentice."

37
Wiener, "Moral and Technical Consequences"; here again, as elsewhere, Wiener casts issues in a form reminiscent of the complementarity principle in physics.

38
Ibid.

39
Wiener, "A Rebellious Scientist after Two Years," *Bulletin of the Atomic Scientists* 4 (November 1948): 338.

40
Wiener, *Cybernetics*, 1st ed., pp. 36–38.

41
See *Time*, December 27, 1948; *Business Week*, February 19, 1949; and *Saturday Review*, April 23, 1949, for reviews of Wiener's *Cybernetics*.

42
Wiener to Reuther, August 13, 1949; Reuther to Wiener (telegram), August 17, 1949 (MIT archives).

43
Wiener-Reuther correspondence, 1949–1950. I am indebted also to Nat Weinberg of the UAW, who attended the Wiener-Reuther breakfast, for his recollections of the event (October 1969).

44
See Reuther's report, "Automation," to the UAW-CIO Economic and Collective Bargaining Conference, November 12–13, 1954; see also the congressional hearings on technology and the American Economy conducted by the Subcommittee on Economic Stabilization in October 1955 in response to union pressure and the public concern Wiener's writings had helped to stimulate.

45
HUHB, 2d ed., p. 162.

46
Gerald Holton, "Scientific Optimism and Societal Concerns: A Note on the Psychology of Scientists," *Annals of the New York Academy of Sciences*, January 23, 1976, p. 92, speaks of "the risk of ostracism or isolation" that a scientist preoccupied "with social action invites, at least until one is amply fortified with professional honors."

47
Russell referred to HUHB as "a book of enormous importance" in his favorable review in *Everybody's* (London), Sep-

tember 15, 1951; Lewis Mumford expressed boundless enthusiasm for HUHB in a letter to Wiener on November 4, 1950:
"I feel as if I should thank you, not just personally, but on be-
half of the human race!"

48
Victor Paschkis, *SSRS, A Review and Outlook* (Bala-Cynwyd,
Penn.: Society for Social Responsibility in Science, 1970).

49
Paschkis to Wiener, February 17, 1950; Wiener to Paschkis,
February 27, 1950, states in part, "I have made the decision
that from now on any political influence I care to assert will be
made in books and articles rather than in meetings," even
though he expresses sympathy with the SSRS. In a letter to the
Faculty-Graduate Committee for Peace, headquartered in
Chicago, Wiener writes in the same vein (March 5, 1951): "I
reserve my political action to books and articles in my name:
That way I take individual rather than group responsibility."
He incidentally suggests that it is a mistake to regard Russia
and China as a "communist bloc," because in all likelihood
China will go its own way in the future.

50
From *Discourse on Method*, quoted by J. Ravetz, in *Scientific
Knowledge and Its Social Problems* (New York: Oxford, 1971),
p. 63.

51
See J. D. Bernal, *The Social Function of Science* (Cambridge,
Mass.: MIT Press, 1967), chap. 7. Leonardo had done military
work but drew the line somewhere. Compare Lewis Mumford,
Technics and Civilization (New York: Harcourt, Brace, 1934),
p. 85.

52
Robert Merton, *Social Theory and Social Structure* (Glencoe:
Free Press, 1957 and 1968); Ravetz, *Scientific Knowledge*.

53
Smith, *A Peril and a Hope*, chap. 12.

54
Goodman, *New Reformation* (New York: Vintage, 1971), p. 7.

55
Joseph Weizenbaum, *Computer Power and Human Reason*
(San Francisco: Freeman, 1976).

56
Langdon Winner, *Autonomous Technology* (Cambridge, Mass.: MIT Press, 1977).

57
Goodman, *New Reformation*, p. 3.

58
Among them Dale Bridenbaugh, Richard Hubbard, and Gregory Minor. The circumstances of their resignation from the General Electric Company are described in H. Caldicott, "Nuclear Madness," *New Age*, June 1978. A similar action in the weapons field by Robert C. Aldridge is also noteworthy. His concerns are expressed in "The Ultimate First Strike Weapon," *The Nation*, February 4, 1978, pp. 324ff. In cases such as these resignation is usually coupled with an effort to publicize the noncooperator's point of view.

59
See S. R. Carpenter, "Developments in the Philosophy of Technology," and E. Byrne, "Society for the Study of the Philosophy of Technology," *Technology and Culture* 19 (January 1978): 93–103.

14 *Von Neumann: Only Human, in Spite of Himself*

1
Eugene Wigner, interview with Thomas Kuhn, November 21, 1963 (Quantum History Archives).

2
I thank poet Beatrice Hawley for her observations on the notion of "genius."

3
Several interviewees commented on Von Neumann's apparent feeling in this regard, among them his medical friend Dr. Shields Warren in an interview with the author, Boston, December 21, 1972.

4
From Wigner's description given to P. Abelson; see P. Abelson, "Relation of Group Activity to Creativity in Science," *Daedalus*, Summer 1965, p. 607.

5
Eugene Wigner and John von Neumann, "Significance of Loewner's Theorem in the Quantum Theory of Collisions," *Annals of Mathematics* 59 (1954): 418–433.

6
Interview of Wigner by Thomas Kuhn, November 21, 1963 (Quantum archives). A similar interchange involving the mathematician Kakutani is quoted by Paul Samuelson, "Maximum Principles in Analytical Economics," *Science*, September 10, 1971, p. 996.

7
See Clay Blair, Jr., "Passing of a Great Mind," *Life*, February 25, 1957, especially the photograph on p. 96; see also the film *John von Neumann* produced by A. Novak under the auspices of the American Mathematical Society.

8
Samuel Grafton's reported interview with Klara von Neumann, in "Married to a Man Who Believes the Mind Can Move the World," *Good Housekeeping*, September 1956, p. 80.

9
Cf. Ulam, *Adventures of a Mathematician* (New York: Scribner's, 1976), pp. 78–79.

10
Others, including Herman Goldstine (interview with the author, Yorktown Heights, N.Y., February 26, 1971) have independently seconded Wigner's views in this regard.

11
Author's interview with Catherine Pedroni, January 12, 1971, New York City.

12
Paul Halmos, interview with the author, January 20, 1971, Bloomington, Indiana.

13
Ulam, *Adventures of a Mathematician*, p. 78: "Some people, especially women, found him lacking in curiosity about subjective or personal feelings and perhaps deficient in emotional development." In my own interviews I found this view of von Neumann not uncommon. Ulam suggests that "a certain shyness" may have prevented him "from having more explicit discussions along these lines." (*Adventures of a Mathematician*, an autobiographical work that was begun as a biography of von Neumann, conveys through many anecdotes the human qualities of von Neumann's friendship and colleagueship.)

14
Lawrence S. Kubie, interview with the author, March 24, 1969, Sparks, Md.

15
Leonard J. Savage, interview with the author, November 10, 1968, New Haven, Ct.

16
Valentine Bargman, interview with the author, December 4, 1972, Princeton, N.J.

17
Natasha (Artim) Brunswick, interview with the author, December 8, 1972, Princeton, N.J.

18
Eugene Wigner, interview with the author, August 10, 1971; cf. Ulam, *Adventures of a Mathematician*, p. 71.

19
Savage interview, November 10, 1968.

20
Compare Ulam, *Adventures of a Mathematician*, p. 79; oral communications to the author from several of von Neumann's colleagues (1971, 1978).

21
Cf. Grafton, "Married to a Man."

22
Pedroni interview, January 12, 1971.

23
Quoted in Blair, "Passing of a Great Mind." See also Ulam's comment on von Neumann's historical interest in *Adventures of a Mathematician*, pp. 80, 102.

24
See Grafton, "Married to a Man"; Blair, "Passing of a Great Mind."

25
Bargman interview, December 4, 1972.

26
Silvan Schweber, oral communication to the author, 1978.

27
Ulam, *Adventures of a Mathematician*, pp. 78–79.

28
Grafton, "Married to a Man."

29
Grafton, "Married to a Man."

30
Goldstine interview, February 26, 1971. "Nebech" is an untranslatable Yiddish word meaning something like "loser."

31
Brunswick interview, December 8, 1972.

32
Oskar Morgenstern, speaking in the film, *John von Neumann*, in which von Neumann is shown wearing such hats.

33
Blair, "Passing of a Great Mind"; cf. Goldstine, *The Computer from Pascal to von Neumann* (Princeton: Princeton University Press, 1972), pp. 181–182.

34
Goldstine interview, February 26, 1971.

35
Morgenstern in the film *John von Neumann.*

36
Jules Charney, interview with the author, Cambridge, Mass., September 30, 1971.

37
Giorgio de Santillana, *Reflections on Men and Ideas* (Cambridge, Mass.: MIT Press, 1968), p. 125.

38
Charney interview, September 30, 1971.

39
Bargman interview, December 4, 1972.

40
Richard Courant, interview with the author, New York, N.Y., March 1, 1971.

41
Goldstine, *Computer*, p. 176.

42
Ulam, *Adventures of a Mathematician*, p. 79.

43
National Academy of Sciences, *Biographical Memoirs* 32 (1958): 439.

44
Savage, Halmos, and Goldstine interviews.

45
Quoted by Blair, "Passing of a Great Mind."

46
Laura Fermi, *Illustrious Immigrants* (Chicago: Chicago University Press, 1961), p. 212.

47
John Stroud to the author, October 8, 1973.

48
Ralph Gerard, interview with the author, Boston, Mass., October 28, 1973. His self-description is from an autobiographical lecture given at MIT, October 29, 1973.

49
J. G. Crowther, *The Social Relations of Science* (London: Macmillan, 1941); Robert K. Merton, *Social Theory and Social Structure* (Glencoe: Free Press, 1957), especially chap. 16; Jerome R. Ravetz, *Scientific Knowledge and Social Problems* (Oxford: Oxford University Press, 1971).

50
Ravetz, *Scientific Knowledge*, p. 63, notes that Francis Bacon and Descartes had been aware of the moral issues implicit in scientific knowledge but that this awareness seems to have been lost when institutionalization occurred.

51
For example, an examination of the German physics community during the 1920s and early 1930s shows that notwithstanding the internationalistic rhetoric required by the scientific ethos, in fact the majority of German physicists were committed to using their science to achieve strongly nationalistic objectives. See Paul Foreman, "Scientific Internationalism and the Weimar Physicists," *Isis* 64 (1973): 151.

52
Alfred N. Whitehead, *The Function of Reason* (Princeton: Princeton University Press, 1929); Paul Feyerabend, *Against Method* (London: Humanities Press, 1975).

53
Merton, *Social Theory*.

54
John B. Bury, *The Idea of Progress* (London: Macmillan, 1921).

55
Thomas Kuhn, *The Structure of Scientific Revolutions* (Chicago: University of Chicago Press, 1962), has examined critically the view that scientific knowledge has a cumulative character.

56
Von Neumann, "The Role of Mathematics in the Sciences and in Society," *Collected Works*, 6: 477.

57
the American Mathematical Society, 64, no. 3, pt. 2 (1958): p. 5.

58
"Method in the Physical Sciences," *Collected Works*, 6: 491.

59
Von Neumann, "The Role of Mathematics."

60
See Daniel Bell, *The Coming of Post-industrial Society* (New York: Basic Books, 1973).

61
For details concerning von Neumann's role in computer development, see Goldstine, *Computer*, and A. W. Burks's introduction to von Neumann, *Theory of Self-Reproducing Automata* (Urbana: University of Illinois Press, 1966). See also von Neumann's "First Draft of a Report on the EDVAC" (Philadelphia: The Moore School, mimeo, 1945), and the report by Burks, Goldstine, and von Neumann, "Preliminary Discussion of the Logical Design of an Electronic Computing Instrument" (Princeton, 1946), describing the proposed Princeton computer, which was intended "conveniently [to] handle problems several orders of magnitude more complex than are now handled by existing machines, electronic or electro-mechanical."

62
Von Neumann, "The NORC and Problems in High Speed Computing" (December 2, 1954), in *Collected Works*, 5: 238. Von Neumann was a consultant to IBM.

63
Goldstine, *Computer*, pp. 331–332.

64
S. C. Gilfillan, *The Sociology of Invention* (Cambridge, Mass.: MIT Press), 1970.

65
See Goldstine, *Computer*, pp. 223–224; see also the review of Goldstine's book by H. D. Huskey in *Science* 180 (1973): 588.

66
Nicholas Metropolis, February 13, 1958, at hearings before the Subcommittee on Research and Development of the Joint

Committee on Atomic Energy, 85th Congress second session, p. 634.

67

John von Neumann, "Can We Survive Technology?" *Fortune*, June 1955.

68

Memorandum to Lewis Strauss, dated Sept. 1955, in Strauss, *Men and Decisions* (Garden City, N.Y.: Doubleday, 1962), p. 441.

69

Warren McCulloch to von Neumann, December 15, 1955 (American Philosophical Society Archives).

70

He also dealt with some of Wiener's predictions, and made his own predictions concerning the impact of the new communications and control technologies on the economy, in a talk to the National Planning Association on December 12, 1955 (*Collected Works*, 6: 100). He begins, without mentioning Wiener directly, by saying that "there has been a great deal of talk . . . that something like a second industrial revolution is impending." In particular von Neumann speaks favorably of the experience of using automata in military decision making and anticipates their corresponding usefulness in economics.

71

Jacques Ellul, *The Technological Society* (New York: Vintage, 1967). (Originally published in French in 1954.) See in particular Robert Merton's introduction to the American edition, which explicates the meaning of "technique" and "the Technical Man." Von Neumann, judging from Merton's description, comes close to being a prototypical technical man.

72

Mircea Eliade, *The Forge and the Crucible* (New York: Harper & Row, 1971), pp. 172–173. (Originally published in French 1956.)

73

Morgenstern interview, May 11, 1970.

74

Teller speaking in the film *John von Neumann*.

75
Eugene Wigner, *Symmetries and Reflections* (Bloomington: Indiana University Press, 1967), p. 261 (written in 1957).

76
Reverend Anselm Strittmatter, interview with the author, Washington, D.C., December 1, 1972.

77
Warren Shields, interview with the author, Boston, Mass., December 21, 1972. I am indebted to Dr. Shields for a medical description of von Neumann's last years.

78
Remarks of Lewis Strauss at von Neumann memorial dinner meeting, Catholic University, Washington, D.C., May 22, 1971. Cf. *Men and Decisions*, p. 236.

15 *Wiener's Later Years: Again the Golem*

1

Margaret Mead, *Virginia Quarterly Review*, Summer 1953, p. 438. Wiener responded to Mead by writing her, "Your review is by far the best and most understanding that my book has received" (Wiener to Mead, June 1, 1953; MIT archives). A review by E. T. Bell, "Prodigy's Escape," the *Nation*, March 28, 1953, pp. 270–271, evoked from Wiener a defense of his father: "You don't appreciate the positive side of my father's education" (Wiener to Bell, March 31, 1953; MIT archives).

2

Page 4 of letter in Wiener archives, MIT. It seems to be part of a letter to Dr. Janet Rioch from Wiener, June 22, 1950. For more on Sidis see EP, pp. 131–135; Morris L. Ernst, *So Far So Good* (New York: Harper, 1948), pp. 51–54.

3

Edmond Dewan, oral communication to the author, December 2, 1970.

4

Wiener to Janet Rioch, June 22, 1950 (MIT archives).

5

Derek, J. de Solla Price, "Automata and the Origins of Mechanism and Mechanistic Philosophy," *Technology and Culture* 5 (1964): 9; "Gods in Black Boxes," in *Computers in Humanistic Research*, ed. Edmund Bowles (Englewood, N.J.: Prentice-Hall, 1967), chap. 1.

6

Price, "Gods in Black Boxes."

7

Norman Faramelli, "Some Implications of Cybernetics for Our Understanding of Man: An Appreciation of the Work of Dr. Norbert Wiener and Implications for Religious Thought," Ph.D. thesis, Temple University, 1968.

8

G&G, p. 95.

9

As Wiener explains in G&G, "in his own image" is from the point of view of function, no matter how different in physical appearance. See pp. 30–31.

10
The one exception is the Rosenberg case. Wiener did not participate in efforts to save the Rosenbergs from execution.

11
EP, p. 34.

12
"The Lonely Nationalism of Rudyard Kipling," *Yale Review* 52 (June 1963): 499.

13
R. E. Park, quoted in Lewis Coser, *Masters of Sociological Thought* (New York: Harcourt Brace Jovanovich, 1971), pp. 365–366; Coser notes that Simmel and Veblen shared this view of marginality.

14
Jerome Ravetz, *Scientific Knowledge and its Social Problems* (New York: Oxford University Press, 1973), p. 5.

15
Eva Ritter, interview with the author, Cambridge, Mass., November 1, 1968.

16
Ibid.

17
Julius Stratton, interview with the author, Cambridge, Mass., June 3, 1971.

18
Stratton recalled that "in the mathematics texts we used then, practically every problem was geared to direct engineering applications—'suppose you had a dam, the amount of water flowing over it . . . etc.' "

19
Stratton was the director of RLE during the years 1946–1949.

20
This picture of the relation between Stratton and Wiener is corroborated by Wiener to Stratton, May 25, 1960.

21
Neal Hartley, interview with the author, Cambridge, Mass., November 15, 1971.

22
Stratton to Wiener, April 15, 1959 (MIT archives).

23
Mr. and Mrs. Alfred Richardson, interview with the author, Ayer, Mass., December 11, 1971.

24
Morris Halle, interview with the author, Cambridge, Mass., May 22, 1978.

25
Albert Hill, interview with the author, Cambridge, Mass., May 26, 1978.

26
See Loren R. Graham, *Science and Philosophy in the Soviet Union* (New York: Vintage, 1974), chapter 9. The shift in attitude toward cybernetics occurred during the years 1954–1958. Since cybernetics impinged on traditional disciplines, its increasing acceptance in the Soviet Union served its scientists as a "Trojan horse" in the "reopening of debates about independent research in such moribund fields of Soviet science as economics, sociology, law, biology, physiology and medicine," enabling scientists to break away from single official schools of science (such as Sechenov-Pavlov in physiology and Michurin-Lysenko in biology) to an increasing pluralism. R. David Gillespie, "Politics of Cybernetics in the Soviet Union," in A. Teich, ed., *Scientists and Public Affairs* (Cambridge, Mass.: MIT Press, 1974).

27
Wiener, "Science and Society," *Voprosy Filosofi*, no. 7 (1961), reprinted in *Technology Review*, July 1961, pp. 49–52.

28
Research Laboratory of Electronics, *R.L.E.: 1946 + 20*, Cambridge, Mass.: Research Laboratory of Electronics 1966.

29
Robert Fano, interview with the author, Cambridge, Mass., April 19, 1978.

30
Edmond Dewan, interview with the author, Bedford, Mass., December 3, 1970.

31
Henry Zimmerman, interview with the author, Cambridge, Mass., May 18, 1978.

32
Armand Siegel, interview with the author, Boston, June 8, 1971.

33
Levinson, "Wiener's Life," *Bulletin of the American Mathematical Society* 72, no. 1, pt. 2 (1966): 25.

34
Siegel interview, June 8, 1971.

35
Dewan interview, December 3, 1970.

36
Siegel interview, June 8, 1971; Dewan interview, December 3, 1970; and others.

37
Asim Yildiz, oral communication to the author, April 1972.

38
Lawrence K. Frank, interview with the author, Belmont, Mass., February 23, 1968. Also Frank to Edmond Dewan (private communication to author from E. Dewan, December 3, 1970).

39
Dirk Struik, interview with the author, Belmont, Mass., December 16, 1970.

40
Wiener to Orson Welles, June 28, 1941; the villains in the plot are AT&T and Michael Pupin of Columbia University.

41
Wiener became a member of the Speckled Band of Boston, Scion Society of the Baker Street Irregulars, in 1948 (MIT archives).

42
Ritter interview, November 1, 1968.

43
Wiener, "The Brain," *Tech Engineering News,* MIT, April 1964, pp 26–32.

44
Wiener, "The Miracle of the Broom Closet," *Tech Engineering News*, MIT, April 1964, pp. 34–36.

45

The neurobiology group—Pitts, Lettvin, and McCulloch—had also in the late 1940s derived encouragement from Wiener for their nonconventional pursuits. Dewan is another. Struik was a nonconformist in a different way. Wiener, the quasi-outsider, had particularly good rapport with students from non-Western countries. Shikao Ikehara from Japan and Yuk Wing Lee from China became his close personal friends. In the 1930s some of the students closest to Wiener were Jews (cf. IAM, p. 180).

46

Siegel interview, June 8, 1971.

47

Struik interview, December 16, 1970.

48

Levinson, "Wiener's Life," p. 24.

49

Walter Rosenblith and Jerome Wiesner, "From Philosophy to Mathematics to Biology," *Bulletin of the American Mathematical Society* 72, no. 1, pt. 2 (1966): 34.

50

W. T. Martin, Foreword to *Selected Papers of Norbert Wiener* (Cambridge, Mass.: MIT Press, 1964).

51

Research Laboratory of Electronics, *R.L.E.: 1946 + 20.*

52

Siegel interview, June 8, 1971. The reminiscence most probably refers to the 1950s.

53

Dewan interview, December 3, 1970. The incident occurred in the early 1960s.

54

Cf. G&G, pp. 66–68.

55

Siegel interview, June 8, 1971.

56

Wiener lecture of May 14, 1963, Bedford, Mass., on "Medical Applications of Cybernetics." (Tape recorded by Edmond Dewan.)

Epilogue

1
Langdon Winner, *Autonomous Technology* (Cambridge, Mass.: MIT Press, 1977), p. 335.

2
See Martin Klein, "Einstein and Some Civilized Discontents," *Physics Today*, January 1965, pp. 38–44.

3
Nicholas Wade, "New Alchemy Institute: Search for an Alternative Agriculture," *Science*, February 25, 1975, pp. 727–729.

4
Warren Hagstrom, *The Scientific Community* (New York: Basic Books, 1965).

5
This view of scientific knowledge has been especially developed by Michael Polanyi in his *Personal Knowledge* (London: Routledge & Kegan Paul, 1958).

6
In the case of mathematics one can illuminate how the subject is to be regarded by means of the formal proofs and theorems of metamathematics, but it is essential to go outside of mathematics itself. Gödel's incompleteness theorem (1930), which showed that if arithmetic is consistent then it must be incomplete, uses metamathematical reasoning to help identify a proper context for mathematics. However, the formal methods of metamathematics are themselves severely limited as tools for clarifying the contexts within which mathematics is to be regarded. Personal and cultural attitudes come into play.

7
G. E. Hutchinson, *The Enchanted Voyage and Other Studies* (New Haven: Yale University Press, 1962), pp. 48–49.

8
Alfred North Whitehead, *The Function of Reason* (Princeton: Princeton University Press, 1929); Paul Feyerabend, *Against Method* (Atlantic Highlands, N.J.: Humanities Press, 1975). One can point to the arbitrariness of the convention that knowing requires the separation of subject and object. How is reality distorted by this convention? Or we can ask how, if our immediate knowledge consists entirely of sensations, could

scientific knowledge be secure as long as its connection to sensations is a continued subject of controversy (mind-body problem)? Moreover, since surely the human nervous system and sensory organs function to select, organize, and condition our cognitions and mental inventions, scientific knowledge and methods are likely to reflect the structure of the human brain to an unknown extent, rather than knowledge of the outside world. In the framework of Darwinian evolution and the knowledge of the utilitarian character of the sensory biology of simpler organisms (e.g., Jerome Lettvin's "What the Frog's Eye Tells the Frog's Brain," *IRE Proceedings* 47 (1959): 1940), the style of knowledge labeled "positivistic-scientific" is merely one style that has evolved and worked well for several hundred years for a portion of the human species. It is species-specific and culture-specific. Among the many civilizations and cultures which have for various time spans inhabited a portion of the globe, one discovers a great variety of ways of perceiving, understanding, and interpreting the world. Moreover, the scientific paradigms of a particular decade or century reflect the consensus of the scientific establishment of that period—a consensus that does not arise out of an Olympian objectivity but out of the accidents of history, social conditions, and fashion [see Thomas Kuhn, *The Structure of Scientific Revolutions* (Chicago: University of Chicago Press, 1962)], and even the collective self-interest of the scientific profession. This always temporary consensus provides an additional element of arbitrariness, and the selection of the logically consistent from the wealth of sensations and immediate experience is also arbitrary.

9
René Dubos, in *Reason Awake* (New York: Columbia University Press, 1970), p. xvii, suggests a shift in the traditional focus of science: "In the final analysis the greatest social contribution of science may well be to help man shape his own destiny by giving him knowledge of the cosmos and of his own nature."

10
The most interesting writings concerning the nuclear arms race which I have found are the books by Richard Barnet, best read in the order of their publication: *Who Wants Disarmament?* (Boston: Beacon, 1969); *After Twenty Years* (with M. G. Raskin) (New York: Random House, 1965); *Intervention and Revolution* (New York: World, 1968); *The Economy of Death* (New York: Atheneum, 1960); *Roots of War* (New York:

Atheneum, 1972); *Global Reach* (with R. E. Mueller) (New York: Simon and Schuster, 1974); *The Giants* (New York: Simon and Schuster, 1977).

11
The Coalition for a New Foreign and Military Policy (120 Maryland Avenue, N.E., Washington, D.C., 20002) has prepared an annotated list of organizations concerned with disarmament.

12
Paha Sapa Report 1, no. 1 (July 1979), issued by Black Hills Alliance, Rapid City, S.D.

13
Ralph P. Hummel, *The Bureaucratic Experience* (New York: St. Martin's, 1977).

14
Paul Goodman, *People or Personnel* (New York: Random House, 1965); see in particular chapter 5 on comparative costs.

INDEX